CCNP Enterprise: Advanced Routing (ENARSI) Lab Manual

Version 8

Cisco Networking Academy

Cisco Press

221 River St

Hoboken, NJ 07030

CCNP Enterprise: Advanced Routing (ENARSI) Lab Manual
Version 8

Cisco Networking Academy

Copyright© 2021 Cisco Systems, Inc.

Published by:
Cisco Press
221 River St
Hoboken, NJ 07030

ScoutAutomatedPrintCode

Library of Congress Control Number: 2020908350

ISBN-13: 978-0-13-687093-7
ISBN-10: 0-13-687093-7

Editor-in-Chief
Mark Taub

Alliances Manager, Cisco Press
Arezou Gol

Director, ITP Product Management
Brett Bartow

Senior Editor
James Manly

Managing Editor
Sandra Schroeder

Project Editor
Mandie Frank

Editorial Assistant
Cindy Teeters

Designer
Chuti Prasertsith

Composition
Bronkella Publishing, Inc.

Proofreader
Debbie Williams

Warning and Disclaimer

This book is designed to provide information about networking. Every effort has been made to make this book as complete and as accurate as possible, but no warranty or fitness is implied.

The information is provided on an "as is" basis. The authors, Cisco Press, and Cisco Systems, Inc. shall have neither liability nor responsibility to any person or entity with respect to any loss or damages arising from the information contained in this book or from the use of the discs or programs that may accompany it.

The opinions expressed in this book belong to the author and are not necessarily those of Cisco Systems, Inc.

Trademark Acknowledgments

All terms mentioned in this book that are known to be trademarks or service marks have been appropriately capitalized. Cisco Press or Cisco Systems, Inc., cannot attest to the accuracy of this information. Use of a term in this book should not be regarded as affecting the validity of any trademark or service mark.

This book is part of the Cisco Networking Academy series from Cisco Press. The products in this series support and complement the Cisco Networking Academy curriculum. If you are using this book outside the Networking Academy, then you are not preparing with a Cisco trained and authorized Networking Academy provider. For more information on the Cisco Networking Academy or to locate a Networking Academy, please visit www.cisco.com/edu.

Special Sales

For information about buying this title in bulk quantities, or for special sales opportunities (which may include electronic versions; custom cover designs; and content particular to your business, training goals, marketing focus, or branding interests), please contact our corporate sales department at corpsales@pearsoned.com or (800) 382-3419.

For government sales inquiries, please contact governmentsales@pearsoned.com.

For questions about sales outside the U.S., please contact intlcs@pearson.com.

Feedback Information

At Cisco Press, our goal is to create in-depth technical books of the highest quality and value. Each book is crafted with care and precision, undergoing rigorous development that involves the unique expertise of members from the professional technical community.

Readers' feedback is a natural continuation of this process. If you have any comments regarding how we could improve the quality of this book, or otherwise alter it to better suit your needs, you can contact us through email at feedback@ciscopress.com. Please make sure to include the book title and ISBN in your message.

We greatly appreciate your assistance.

ılıılıı
CISCO.

Americas Headquarters	Asia Pacific Headquarters	Europe Headquarters
Cisco Systems, Inc.	Cisco Systems (USA) Pte. Ltd.	Cisco Systems International BV Amsterdam,
San Jose, CA	Singapore	The Netherlands

Cisco has more than 200 offices worldwide. Addresses, phone numbers, and fax numbers are listed on the Cisco Website at www.cisco.com/go/offices.

Cisco and the Cisco logo are trademarks or registered trademarks of Cisco and/or its affiliates in the U.S. and other countries. To view a list of Cisco trademarks, go to this URL: www.cisco.com/go/trademarks. Third party trademarks mentioned are the property of their respective owners. The use of the word partner does not imply a partnership relationship between Cisco and any other company. (1110R)

iv

Contents

About This Lab Manual

This is the only authorized Lab Manual for the Cisco Networking Academy CCNP Enterprise: Advanced Routing (ENARSI) v8 Course.

The two courses in this CCNP Enterprise version 8.0 curriculum provide students with knowledge and skills needed to configure, operate, and troubleshoot large scale enterprise networks. The courses cover a broad range of routing, switching, and wireless topics along with security best practices used in software-driven digital networks. CCNP Enterprise certification requires candidates to pass two 120-minute exams: CCNP and CCIE Enterprise Core ENCOR 350-401 and CCNP Enterprise Advanced Routing ENARSI 300-410.

By the end of the CCNP course series, students gain practical, hands-on lab experience preparing them for the CCNP Enterprise certification exams and career-ready skills for professional-level roles in the Information & Communication Technologies (ICT) industry.

CCNP Enterprise: Advanced Routing

This second of the 2-course CCNP Enterprise series focuses on implementation and troubleshooting of advanced routing and redistribution for OSPF, EIGRP, and BGP along with VPN technologies, infrastructure security, and management tools used in Enterprise networks. Comprehensive labs emphasize hands-on learning and practice to reinforce configuration and troubleshooting skills.

This course directly prepares for the Cisco Enterprise Advanced Routing and Services concentration exam (300-410) to earn the Enterprise Advanced Infrastructure Implementation Specialist certification.

By also passing the core exam (350-401 ENCOR), you will earn the CCNP Enterprise certification.

The 40 comprehensive labs in this manual emphasize hands-on learning and practice to reinforce configuration skills.

IPv4/IPv6 Addressing and Routing Review

1.1.2 Lab - Troubleshoot IPv4 and IPv6 Addressing Issues

Topology

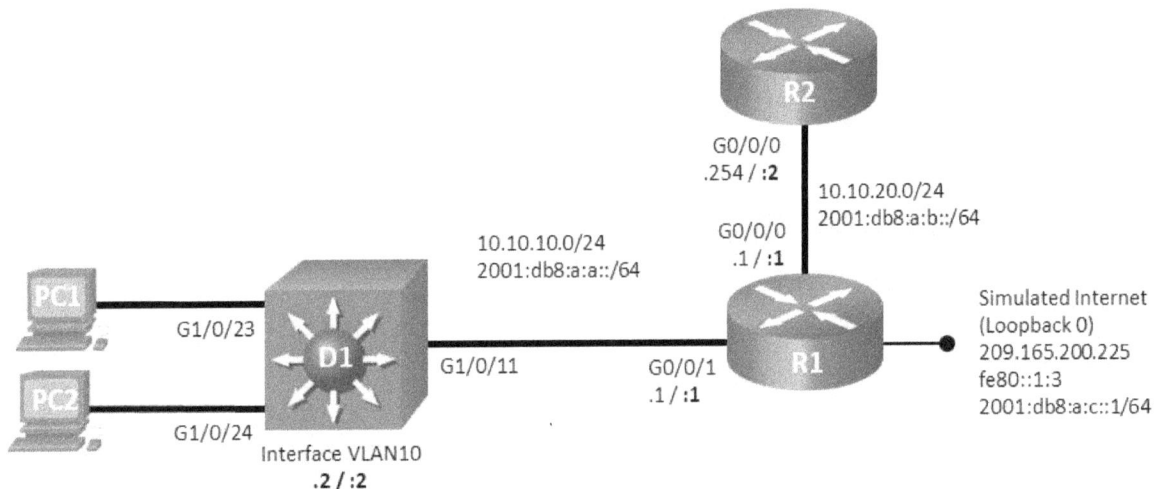

Addressing Table

Device	Interface	IPv4 Address/Mask	IPv6 Address/Prefix	IPv6 Link Local
R1	G0/0/0	10.10.20.1/24	2001:db8:a:b::1/64	fe80::1:1
	G0/0/1	10.10.10.1/24	2001:db8:a:a::1/64	fe80::1:2
	Lo0	209.165.200.225/29	2001:db8:a:c::1/64	fe80::1:3
R2	G0/0/0	10.10.20.254/24	2001:db8:a:b::1/64	fe80::2:1
D1	VLAN 10	10.10.10.2/24	2001:db8:a:a::2/64	fe80::d1:1
PC1	NIC	DHCP	SLAAC	EUI-64
PC2	NIC	DHCP	SLAAC	EUI-64

Objectives

Troubleshoot network issues related to IPv4 and IPv6 addressing.

Background/Scenario

In this topology, router R1 provides connectivity to a simulated internet for VLAN 10. R2 serves as a DHCP server. Switch D1 provides connectivity for VLAN 10. You will be loading configurations with intentional errors onto the network. Your tasks are to FIND the error(s), document your findings and the command(s) or method(s) used to fix them, FIX the issue(s) presented here and then test the network to ensure both of the following conditions are met:

1. the complaint received in the ticket is resolved

2. full reachability is restored

Note: The routers used with CCNP hands-on labs are Cisco 4221 with Cisco IOS XE Release 16.9.4 (universalk9 image). The switches used in the labs are Cisco Catalyst 3650 with Cisco IOS XE Release 16.9.4 (universalk9 image). Other routers, switches, and Cisco IOS versions can be used. Depending on the model and Cisco IOS version, the commands available and the output produced might vary from what is shown in the labs. Refer to the Router Interface Summary Table at the end of the lab for the correct interface identifiers.

Note: Make sure that the switches have been erased and have no startup configurations. If you are unsure, contact your instructor.

Required Resources

- 2 Routers (Cisco 4221 with Cisco IOS XE Release 16.9.4 universal image or comparable)

- 1 Switch (Cisco 3560 with Cisco IOS XE Release 16.9.4 universal image or comparable)

- 2 PCs (Choice of operating system with terminal emulation program installed)

- Console cables to configure the Cisco IOS devices via the console ports

- Ethernet cables as shown in the topology

Instructions

Part 1: Trouble Ticket 1.1.2.1

Scenario:

PC1 is unable to access resources on web server 209.165.200.225.

Use the commands listed below to load the configuration files for this trouble ticket:

Device	Command
R1	`copy flash:/enarsi/1.1.2.1-r1-config.txt run`
R2	`copy flash:/enarsi/1.1.2.1-r2-config.txt run`
D1	`copy flash:/enarsi/1.1.2.1-d1-config.txt run`

- PC1 and PC2 should be configured for and receive an address from an IPv4 DHCP server.

- Passwords on all devices are **cisco12345**. If a username is required, use **admin**.

- When you have fixed the ticket, change the MOTD on EACH DEVICE using the following command:

    banner motd # This is $(hostname) FIXED from ticket <ticket number> #

- Then save the configuration by issuing the **wri** command (on each device).

- Inform your instructor that you are ready for the next ticket.

- After the instructor approves your solution for this ticket, issue the **reset.now** privileged EXEC command. This script will clear your configurations and reload the devices.

Part 2: Trouble Ticket 1.1.2.2

Scenario:

PC1 and PC2 are unable to lease IPv4 addresses from the DHCP server.

Use the commands listed below to load the configuration files for this trouble ticket:

Device	Command
R1	`copy flash:/enarsi/1.1.2.2-r1-config.txt run`
R2	`copy flash:/enarsi/1.1.2.2-r2-config.txt run`
D1	`copy flash:/enarsi/1.1.2.2-d1-config.txt run`

- PC1 and PC2 should be configured for and receive an address from an IPv4 DHCP server.

- Passwords on all devices are **cisco12345**. If a username is required, use **admin**.

- When you have fixed the ticket, change the MOTD on EACH DEVICE using the following command:

   banner motd # This is $(hostname) FIXED from ticket <ticket number> #

- Then save the configuration by issuing the **wri** command (on each device).

- Inform your instructor that you are ready for the next ticket.

- After the instructor approves your solution for this ticket, issue the **reset.now** privileged EXEC command. This script will clear your configurations and reload the devices.

Part 3: Trouble Ticket 1.1.2.3

Scenario:

PC1 and PC2 are unable to resolve IPv6 addresses to hostnames. Upon investigation, it appears that they are not receiving DNS server information from the DHCPv6 server.

Use the commands listed below to load the configuration files for this trouble ticket:

Device	Command
R1	`copy flash:/enarsi/1.1.2.3-r1-config.txt run`
R2	`copy flash:/enarsi/1.1.2.3-r2-config.txt run`
D1	`copy flash:/enarsi/1.1.2.3-d1-config.txt run`

- PC1 and PC2 should be configured to assign an address via SLAAC.

- Passwords on all devices are **cisco12345**. If a username is required, use **admin**.

- When you have fixed the ticket, change the MOTD on EACH DEVICE using the following command:

   banner motd # This is $(hostname) FIXED from ticket <ticket number> #

- Then save the configuration by issuing the **wri** command (on each device).
- Inform your instructor that you are ready for the next ticket.
- After the instructor approves your solution for this ticket, issue the **reset.now** privileged EXEC command. This script will clear your configurations and reload the devices.

Router Interface Summary Table

Router Model	Ethernet Interface #1	Ethernet Interface #2	Serial Interface #1	Serial Interface #2
1800	Fast Ethernet 0/0 (F0/0)	Fast Ethernet 0/1 (F0/1)	Serial 0/0/0 (S0/0/0)	Serial 0/0/1 (S0/0/1)
1900	Gigabit Ethernet 0/0 (G0/0)	Gigabit Ethernet 0/1 (G0/1)	Serial 0/0/0 (S0/0/0)	Serial 0/0/1 (S0/0/1)
2801	Fast Ethernet 0/0 (F0/0)	Fast Ethernet 0/1 (F0/1)	Serial 0/1/0 (S0/1/0)	Serial 0/1/1 (S0/1/1)
2811	Fast Ethernet 0/0 (F0/0)	Fast Ethernet 0/1 (F0/1)	Serial 0/0/0 (S0/0/0)	Serial 0/0/1 (S0/0/1)
2900	Gigabit Ethernet 0/0 (G0/0)	Gigabit Ethernet 0/1 (G0/1)	Serial 0/0/0 (S0/0/0)	Serial 0/0/1 (S0/0/1)
4221	Gigabit Ethernet 0/0/0 (G0/0/0)	Gigabit Ethernet 0/0/1 (G0/0/1)	Serial 0/1/0 (S0/1/0)	Serial 0/1/1 (S0/1/1)
4300	Gigabit Ethernet 0/0/0 (G0/0/0)	Gigabit Ethernet 0/0/1 (G0/0/1)	Serial 0/1/0 (S0/1/0)	Serial 0/1/1 (S0/1/1)

Note: To find out how the router is configured, look at the interfaces to identify the type of router and how many interfaces the router has. There is no way to effectively list all the combinations of configurations for each router class. This table includes identifiers for the possible combinations of Ethernet and Serial interfaces in the device. The table does not include any other type of interface, even though a specific router may contain one. An example of this might be an ISDN BRI interface. The string in parenthesis is the legal abbreviation that can be used in Cisco IOS commands to represent the interface.

Device Config Final – Notes

_____

_____

_____

_____

_____

_____

_____

_____

_____

_____

_____

1.1.3 Lab - Troubleshoot IPv4 and IPv6 Static Routing

Topology

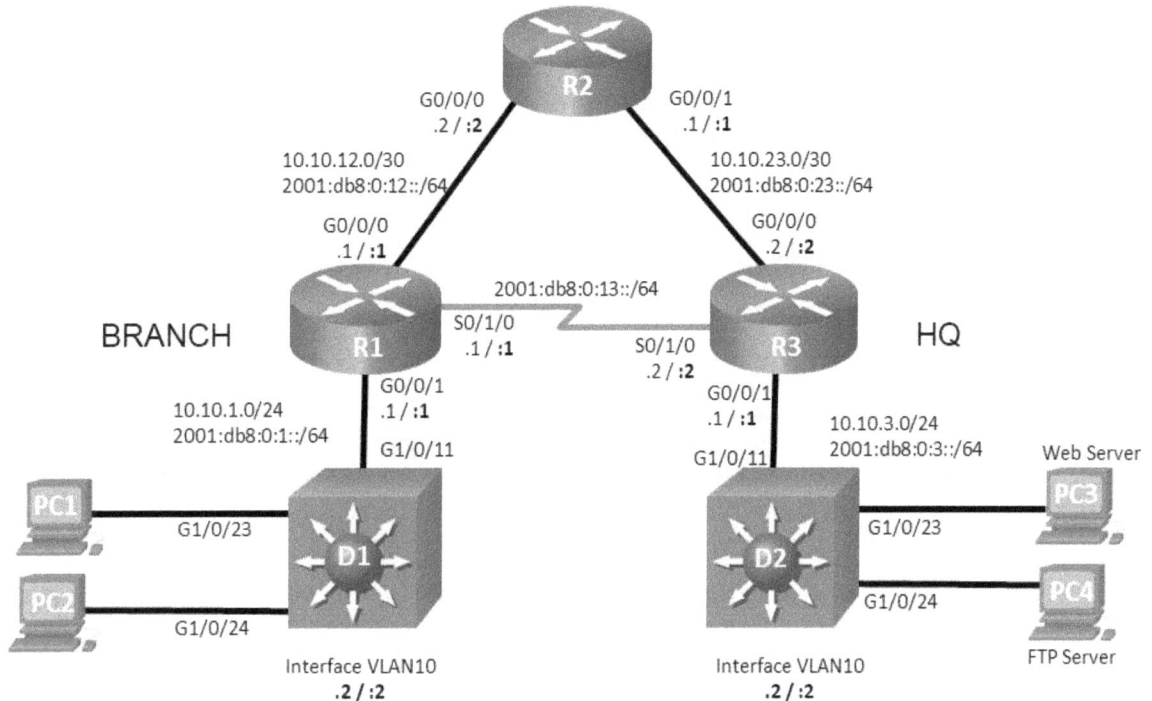

Addressing Table

Device	Interface	IPv4 Address/Mask	IPv6 Address/Prefix	IPv6 Link Local	Default Gateway
R1	G0/0/0	10.10.12.1/24	2001:db8:0:12::1/64	fe80::1:1	N/A
	G0/0/1	10.10.1.1/24	2001:db8:0:1::1/64	fe80::1:2	
	S0/1/0	N/A	2001:db8:0:13::1/64	fe80::1:3	
R2	G0/0/0	10.10.12.2/24	2001:db8:0:12::2/64	fe80::2:1	N/A
	G0/0/1	10.10.23.1/24	2001:db8:0:23::1/64	fe80::2:2	
R3	G0/0/0	10.10.23.2/24	2001:db8:0:23::2/64	fe80::3:1	N/A
	G0/0/1	10.10.3.1/24	2001:db8:0:3::1/64	fe80::3:2	
	S0/1/0		2001:db8:0:13::2/64	fe80::3:3	
D1	VLAN 10	10.10.1.2/24	N/A	N/A	10.10.1.1
D2	VLAN 10	10.10.3.2/24	N/A	N/A	10.10.3.1
PC1	NIC	10.10.1.10/24	2001:db8:0:1::10/64	EUI-64/CGA	10.10.1.1
					2001:db8:0:1::1
PC2	NIC	10.10.1.20/24	2001:db8:0:1::20/64	EUI-64/CGA	10.10.1.1
					2001:db8:0:1::1
Web Server	NIC	10.10.3.5/24	2001:db8:0:3::5/64	EUI-64/CGA	10.10.3.1
					2001:db8:0:3::1
FTP Server	NIC	10.10.3.20/24	2001:db8:0:3::20/64	EUI-64/CGA	10.10.3.1
					2001:db8:0:3::1

Objectives

Troubleshoot network issues related to IPv4 and IPv6 static routing.

Background/Scenario

In this topology, routers R1, R2, and R3 are configured for static routing. Switches D1 and D2 provide LAN connectivity for VLAN 10 for the respective locations. You will be loading configurations with intentional errors onto the network. Your tasks are to FIND the error(s), document your findings and the command(s) or method(s) used to fix them, FIX the issue(s) presented here and then test the network to ensure both of the following conditions are met:

1. the complaint received in the ticket is resolved

2. full reachability is restored

Note: The routers used with CCNA hands-on labs are Cisco 4221 with Cisco IOS XE Release 16.9.4 (universalk9 image). The switches used in the labs are Cisco Catalyst 3560 with Cisco IOS XE Release 16.9.4 (universalk9 image). Other routers, switches, and Cisco IOS versions can be used. Depending on the model and Cisco IOS version, the commands available and the output produced might vary from what is shown in the labs. Refer to the Router Interface Summary Table at the end of the lab for the correct interface identifiers.

Note: Make sure that the switches have been erased and have no startup configurations. If you are unsure, contact your instructor.

Required Resources

- 3 Routers (Cisco 4221 with Cisco IOS XE Release 16.9.4 universal image or comparable)
- 2 Switches (Cisco 3650 with Cisco IOS XE Release 16.9.4 universalk9 image or comparable)
- 4 PCs (Choice of operating system with terminal emulation program installed)
- Console cables to configure the Cisco IOS devices via the console ports
- Ethernet cables as shown in the topology

Instructions

Part 1: Trouble Ticket 1.1.3.1

Scenario:

An FTP Server was recently added to the HQ network. The FTP Server is accessible from all devices in the HQ network. Branch network hosts PC1 and PC2 are able to connect to the Web Server but are unable to connect to the FTP Server using IPv4.

Note: Web or FTP services are not required on the PCs.

Use the commands listed below to load the configuration files for both trouble tickets:

Device	Command
R1	copy flash:/enarsi/1.1.3.1-r1-config.txt run
R2	copy flash:/enarsi/1.1.3.1-r2-config.txt run

Device	Command
R3	`copy flash:/enarsi/1.1.3.1-r3-config.txt run`
D1	`copy flash:/enarsi/1.1.3.1-d1-config.txt run`
D2	`copy flash:/enarsi/1.1.3.1-d2-config.txt run`

- PC 1, PC 2, FTP Server, and Web Server should be configured with the addressing listed in the Addressing Table.

- Passwords on all devices are **cisco12345**. If a username is required, use **admin**.

- After you have fixed the ticket, change the MOTD on EACH DEVICE using the following command:

  banner motd # This is $(hostname) FIXED from ticket <ticket number> #

- Then save the configuration by issuing the **wri** command (on each device).

- Inform your instructor that you are ready for the next ticket.

- After the instructor approves your solution for this ticket, issue the **reset.now** privileged EXEC command. This script will clear your configurations and reload the devices.

Part 2: Trouble Ticket 1.1.3.2

Scenario:

A WAN connection through R2 was recently added to increase the bandwidth that is available between the branch and HQ. It was decided to keep the dedicated T1 connection from R1 to R3 as a backup link for IPv6 traffic. Users at the branch have been complaining that data transfer speeds to PCs at HQ seem to be slow; however, downloads seem to be fine.

Use the commands listed below to load the configuration files for both trouble tickets:

Device	Command
R1	`copy flash:/enarsi/1.1.3.2-r1-config.txt run`
R2	`copy flash:/enarsi/1.1.3.2-r2-config.txt run`
R3	`copy flash:/enarsi/1.1.3.2-r3-config.txt run`
D1	`copy flash:/enarsi/1.1.3.2-d1-config.txt run`
D2	`copy flash:/enarsi/1.1.3.2-d2-config.txt run`

- PC 1, PC 2, FTPServer, and WebServer should be configured with the IPv6 addressing listed in the Addressing Table. It is not necessary to configure the IPv4 addresses.

- Passwords on all devices are **cisco12345**. If a username is required, use **admin**.

- After you have fixed the ticket, change the MOTD on EACH DEVICE using the following command:

  banner motd # This is $(hostname) FIXED from ticket <ticket number> #

- Then save the configuration by issuing the **wri** command (on each device).

- Inform your instructor that you are ready for the next ticket.

- After the instructor approves your solution for this ticket, issue the **reset.now** privileged EXEC command. This script will clear your configurations and reload the devices.

Router Interface Summary Table

Router Model	Ethernet Interface #1	Ethernet Interface #2	Serial Interface #1	Serial Interface #2
1800	Fast Ethernet 0/0 (F0/0)	Fast Ethernet 0/1 (F0/1)	Serial 0/0/0 (S0/0/0)	Serial 0/0/1 (S0/0/1)
1900	Gigabit Ethernet 0/0 (G0/0)	Gigabit Ethernet 0/1 (G0/1)	Serial 0/0/0 (S0/0/0)	Serial 0/0/1 (S0/0/1)
2801	Fast Ethernet 0/0 (F0/0)	Fast Ethernet 0/1 (F0/1)	Serial 0/1/0 (S0/1/0)	Serial 0/1/1 (S0/1/1)
2811	Fast Ethernet 0/0 (F0/0)	Fast Ethernet 0/1 (F0/1)	Serial 0/0/0 (S0/0/0)	Serial 0/0/1 (S0/0/1)
2900	Gigabit Ethernet 0/0 (G0/0)	Gigabit Ethernet 0/1 (G0/1)	Serial 0/0/0 (S0/0/0)	Serial 0/0/1 (S0/0/1)
4221	Gigabit Ethernet 0/0/0 (G0/0/0)	Gigabit Ethernet 0/0/1 (G0/0/1)	Serial 0/1/0 (S0/1/0)	Serial 0/1/1 (S0/1/1)
4300	Gigabit Ethernet 0/0/0 (G0/0/0)	Gigabit Ethernet 0/0/1 (G0/0/1)	Serial 0/1/0 (S0/1/0)	Serial 0/1/1 (S0/1/1)

Note: To find out how the router is configured, look at the interfaces to identify the type of router and how many interfaces the router has. There is no way to effectively list all the combinations of configurations for each router class. This table includes identifiers for the possible combinations of Ethernet and Serial interfaces in the device. The table does not include any other type of interface, even though a specific router may contain one. An example of this might be an ISDN BRI interface. The string in parenthesis is the legal abbreviation that can be used in Cisco IOS commands to represent the interface.

Device Config Final – Notes

_____

_____

_____

_____

_____

_____

_____

_____

_____

_____

_____

_____

_____

2.1.2 Lab - Implement EIGRP for IPv4

Topology

Addressing Table

Device	Interface	IP Address	Subnet Mask
R1	G0/0/0	10.0.12.1	255.255.255.0
	G0/0/1.1	172.16.1.1	255.255.255.0
	G0/0/1.2	192.168.1.1	255.255.255.0
R2	G0/0/0	10.0.12.2	255.255.255.0
	G0/0/1	10.0.23.2	255.255.255.0
R3	G0/0/0	10.0.23.3	255.255.255.0
	G0/0/1	172.16.13.1	255.255.255.0
	Loopback 0	192.168.3.1	255.255.255.0
D2	G1/0/1	172.16.1.2	255.255.255.0
	G1/0/11	172.16.13.2	255.255.255.0
PC1	NIC	DHCP	

Objectives

Part 1: Build the Network and Configure Basic Device Settings

Part 2: Configure and Verify EIGRP for IPv4

Part 3: Tune EIGRP for IPv4

Background/Scenario

EIGRP is an interior gateway routing protocol created by Cisco and published as open source in 2016 in RFC 7868. It is a very efficient distance-vector based protocol. In this lab, you will configure and examine EIGRP in two forms, *Classic* EIGRP and *Named* EIGRP. Named EIGRP is an update to Classic EIGRP that added multiprotocol support and default support for wide metrics.

Note: This lab is an exercise in configuring options available for EIGRP supporting IPv4 and does not necessarily reflect implementation best practices.

Note: The routers used with CCNP hands-on labs are Cisco 4221 with Cisco IOS XE Release 16.9.4 (universalk9 image). The switches used in the labs are Cisco Catalyst 3650 with Cisco IOS XE Release 16.9.4 (universalk9 image). Other routers, switches, and Cisco IOS versions can be used. Depending on the model and Cisco IOS version, the commands available and the output produced might vary from what is shown in the labs. Refer to the Router Interface Summary Table at the end of the lab for the correct interface identifiers.

Note: Make sure that the routers and switches have been erased and have no startup configurations. If you are unsure, contact your instructor.

Required Resources

- 3 Routers (Cisco 4221 with Cisco IOS XE Release 16.9.4 universal image or comparable)
- 2 Switches (Cisco 3650 with Cisco IOS XE Release 16.9.4 universal image or comparable)
- 1 PC (Choice of operating system with a terminal emulation program installed)
- Console cables to configure the Cisco IOS devices via the console ports
- Ethernet cables as shown in the topology

Part 1: Build the Network and Configure Basic Device Settings

In Part 1, you will set up the network topology and configure basic settings on routers.

Step 1. Cable the network as shown in the topology.

Attach the devices as shown in the topology diagram, and cable as necessary.

Step 2. Configure basic settings for each device.

    a. Console into each device, enter global configuration mode, and apply the basic settings. The startup configurations for each device are provided below.

    Router R1

```
hostname R1
no ip domain lookup
banner motd # R1, Implement EIGRP for IPv4 #
```

```
line con 0
 exec-timeout 0 0
 logging synchronous
 exit
line vty 0 4
 privilege level 15
 exec-timeout 0 0
 password cisco123
 login
 exit
interface g0/0/1
 no ip address
 no shutdown
 exit
interface g0/0/1.1
 encapsulation dot1q 1
 ip address 172.16.1.1 255.255.255.0
 no shutdown
 exit
interface g0/0/1.2
 encapsulation dot1q 2
 ip address 192.168.1.1 255.255.255.0
 no shutdown
 exit
interface g0/0/0
 ip address 10.0.12.1 255.255.255.0
 no shutdown
 exit
ip dhcp pool HOSTS
 network 192.168.1.0 255.255.255.0
 default-router 192.168.1.1
 exit
end
```

Router R2

```
hostname R2
no ip domain lookup
banner motd # R2, Implement EIGRP for IPv4 #
line con 0
 exec-timeout 0 0
 logging synchronous
 exit
line vty 0 4
 privilege level 15
 exec-timeout 0 0
 password cisco123
 login
 exit
interface g0/0/0
 ip address 10.0.12.2 255.255.255.0
 no shutdown
```

```
 exit
interface g0/0/1
 ip address 10.0.23.2 255.255.255.0
 no shutdown
 exit
end
```

Router R3

```
hostname R3
no ip domain lookup
banner motd # R3, Implement EIGRP for IPv4 #
line con 0
 exec-timeout 0 0
 logging synchronous
 exit
line vty 0 4
 privilege level 15
 exec-timeout 0 0
 password cisco123
 login
 exit
interface g0/0/0
 ip address 10.0.23.3 255.255.255.0
 no shutdown
 exit
interface g0/0/1
 ip address 172.16.13.1 255.255.255.0
 no shutdown
 exit
interface loopback 0
 ip address 192.168.3.1 255.255.255.0
 no shutdown
 exit
end
```

Switch D1

```
hostname D1
no ip domain lookup
banner motd # D1, Implement EIGRP for IPv4 #
line con 0
 exec-timeout 0 0
 logging synchronous
 exit
line vty 0 4
 privilege level 15
 exec-timeout 0 0
 password cisco123
 login
 exit
vlan 2
 name HOST-VLAN
 exit
```

```
interface range g1/0/1 - 24, g1/1/1 - 4
 shutdown
 exit
interface g1/0/1
 switchport mode access
 spanning-tree portfast
 no shutdown
 exit
interface g1/0/11
 switchport mode trunk
 no shutdown
 exit
interface g1/0/23
 switchport mode access
 switchport access vlan 2
 spanning-tree portfast
 no shutdown
 exit
end
```

Switch D2

```
hostname D2
no ip domain lookup
ip routing
banner motd # D2, Implement EIGRP for IPv4 #
line con 0
 exec-timeout 0 0
 logging synchronous
 exit
line vty 0 4
 privilege level 15
 exec-timeout 0 0
 password cisco123
 login
 exit
interface range g1/0/1 - 24, g1/1/1 - 4
 shutdown
 exit
interface g1/0/1
 no switchport
 ip address 172.16.1.2 255.255.255.0
 no shutdown
 exit
interface g1/0/11
 no switchport
 ip address 172.16.13.2 255.255.255.0
 no shutdown
 exit
end
```

   b.   Set the clock on each device to UTC time.

   c.   Save the running configuration to startup-config.

   d.   Verify that PC1 receives an address via DHCP.

   e.   Verify that PC1 can ping its default gateway.

Part 2: Configure and Verify EIGRP for IPv4

In this part of the lab, you will configure and verify EIGRP in the network. R1 and R3 will used Named EIGRP, while R2 will use Classic EIGRP. After you have established the network, you will examine the differences in how each version of EIGRP deals with metrics.

For the lab, you will use the Autonomous System number 27 on all routers.

Step 1.   Configure Classic EIGRP for IPv4 on R2.

   a.   Start the configuration of Classic EIGRP by issuing the **router eigrp 27** command.

   ```
   R2(config)# router eigrp 27
   ```

   b.   Configure the EIGRP router ID using the **eigrp router-id** command. Use the number 2.2.2.2 for R2.

   ```
   R2(config-router)# eigrp router-id 2.2.2.2
   ```

   c.   Identify the interfaces that should be speaking EIGRP and the networks that should be included in the EIGRP topology table. This is done with the **network** command.

   It is best to be as specific as possible when creating network statements, while balancing efficiency and the number of commands necessary. For our lab example, we will use **network 10.0.0.0 255.255.224.0** to specify the interfaces. This covers less network space than 10.0.0.0/8, while including both interfaces with a single **network** command.

   ```
   R2(config-router)# network 10.0.0.0 255.255.224.0
   R2(config-router)# end
   ```

   d.   Verify the interfaces now involved in EIGRP with the **show ip eigrp interfaces** command.

   ```
   R2# show ip eigrp interfaces
   EIGRP-IPv4 Interfaces for AS(27)
                       Xmit Queue   PeerQ        Mean Pacing Time  Multicast  Pending
   Interface Peers  Un/Reliable  Un/Reliable  SRTT Un/Reliable   Flow Timer Routes
   Gi0/0/0    0         0/0          0/0          0    0/0 0         0
   Gi0/0/1    0         0/0          0/0          0    0/0 0         0
   ```

Step 2.   Configure Named EIGRP for IPv4 on R1 and R3.

   a.   Start the configuration of Named EIGRP by issuing the **router eigrp [name]** command. The name parameter can be a number, but the number does not identify an Autonomous System as it does with Classic EIGRP, it simply identifies the process. For our purposes, name the process BASIC-EIGRP-LAB.

   ```
   R1(config)# router eigrp BASIC-EIGRP-LAB
   ```

   b.   Enter into address-family configuration mode with the **address-family ipv4 unicast autonomous-system 27** command.

   ```
   R1(config-router)# address-family ipv4 unicast autonomous-system 27
   ```

c. Configure the EIGRP router ID using the **eigrp router-id** command. Use the number 1.1.1.1 for R1.

```
R1(config-router-af)# eigrp router-id 1.1.1.1
```

d. Identify the interfaces that should be speaking EIGRP and the networks that should be included in the EIGRP topology table. This is done with the **network** command. In this case, the configuration requires three **network** commands. An example for R1 follows:

```
R1(config-router-af)# network 10.0.12.0 255.255.255.0
R1(config-router-af)# network 172.16.1.0 255.255.255.0
R1(config-router-af)# network 192.168.1.0 255.255.255.0
R1(config-router-af)# end
```

e. Repeat Steps 2a through 2d for R3 and D2. Use 3.3.3.3 for the router ID on R3, and 132.132.132.132 for the router ID on D2. Configure the appropriate network statements on both devices according to the Addressing Table.

Step 3. Verify EIGRP for IPv4.

a. A few seconds after configuring the network statements, you should have seen syslog messages noting that EIGRP adjacencies have been formed.

```
*Feb 18 15:49:34.243: %DUAL-5-NBRCHANGE: EIGRP-IPv4 27: Neighbor 10.0.12.2
(GigabitEthernet0/0/0) is up: new adjacency
```

b. To verify that routing is working, ping from PC1 to interface Loopback 0 on R3 (192.168.3.1). The ping should be successful. You can also randomly ping other addresses in the topology.

c. On R1, examine the EIGRP entries in the IP routing table using the **show ip route eigrp | begin Gateway** command. As you can see, there is one path installed in the routing table for the network, and two paths for the 192.168.3.0/24 network. Take note of the metric values listed.

```
R1# show ip route eigrp | begin Gateway
Gateway of last resort is not set

      10.0.0.0/8 is variably subnetted, 3 subnets, 2 masks
D        10.0.23.0/24 [90/15360] via 10.0.12.2, 00:00:12, GigabitEthernet0/0/0
      172.16.0.0/16 is variably subnetted, 3 subnets, 2 masks
D        172.16.13.0/24
            [90/15360] via 172.16.1.2, 00:00:12, GigabitEthernet0/0/1.1
D     192.168.3.0/24
            [90/16000] via 172.16.1.2, 00:00:12, GigabitEthernet0/0/1.1
            [90/16000] via 10.0.12.2, 00:00:12, GigabitEthernet0/0/0
```

d. Now examine the EIGRP topology table using the **show ip eigrp topology all-links** command. The **all-links** parameter instructs the router to display all available routes, including the ones that are not successors or feasible successors.

Remember that the topology table is EIGRP's database of route information. EIGRP selects the best paths from this database, based on the DUAL algorithm, and offers them to the IP routing table. However, the IP routing table does not have to use those offered paths, because the router may have learned about the same network from a more reliable routing source, which would be a routing source with a lower administrative distance.

```
R1# show ip eigrp topology all-links
EIGRP-IPv4 VR(BASIC-EIGRP-LAB) Topology Table for AS(27)/ID(1.1.1.1)
Codes: P - Passive, A - Active, U - Update, Q - Query, R - Reply,
       r - reply Status, s - sia Status


P 192.168.3.0/24, 2 successors, FD is 2048000, serno 26
        via 10.0.12.2 (2048000/1392640), GigabitEthernet0/0/0
        via 172.16.1.2 (2048000/1392640), GigabitEthernet0/0/1.1
P 172.16.13.0/24, 1 successors, FD is 1966080, serno 16
        via 172.16.1.2 (1966080/1310720), GigabitEthernet0/0/1.1
        via 10.0.12.2 (2621440/1966080), GigabitEthernet0/0/0
P 192.168.1.0/24, 1 successors, FD is 1310720, serno 3
        via Connected, GigabitEthernet0/0/1.2
P 172.16.1.0/24, 1 successors, FD is 1310720, serno 2
        via Connected, GigabitEthernet0/0/1.1
P 10.0.23.0/24, 1 successors, FD is 1966080, serno 27
        via 10.0.12.2 (1966080/1310720), GigabitEthernet0/0/0
        via 172.16.1.2 (2621440/1966080), GigabitEthernet0/0/1.1
P 10.0.12.0/24, 1 successors, FD is 1310720, serno 1
        via Connected, GigabitEthernet0/0/0
```

We will focus on the routes, highlighted in the above output, to 192.168.3.0/24 and 10.0.23.0/24. There are several things to notice:

- The entry for the 192.168.3.0/24.0 network shows two successors, while the entry for 10.0.23.0/24 shows only one successor. Both entries show two paths. The path with the lowest Feasible Distance (FD) is selected as the successor and is offered to the routing table. For 192.168.3.0/24, there are two paths with equal FD (2048000). Therefore, both are successors and both are offered to the routing table. In the case of 10.0.23.0/24, the FD is listed as 19660800. The path via 10.0.12.2 shows that number as the FD (first number in parentheses). The path via 172.16.1.2 shows an FD of 2621440, which is higher than the current feasible distance. So that path, although valid, is a higher cost path and is not offered to the routing table.

- The FD listed in the topology table does not match the metric listed in the routing table. For 192.168.3.0/24, the routing table shows the metric value 16000, while the topology table shows the FD as 2048000. This is due to the routing table having a limit of 4 bytes (32 bits) for metric information while EIGRP on R1 is using EIGRP wide metrics, which are 64 bits. Wide metrics are used by Named EIGRP by default. To work around the 32-bit metric size limitation in the routing table, EIGRP divides the wide-metric value by the EIGRP_RIB_SCALE value, which defaults to 128, as you will see next in the output for the **show ip protocols** command. The value 2048000 divided by 128 is 16000.

Note: A network with mixed EIGRP implementations (Named and Classic in the same routing domain) will have some loss of route clarity which could lead to sub-optimal path selection. The recommended implementation is to use Named EIGRP in all cases.

- There are no feasible successors listed in the topology table for 192.168.3.0/24 or 10.0.23.0/24. The feasibility condition requires that the reported distance (RD) to a destination network be less than the current feasible distance for a next-hop to be considered a feasible successor to the route. In the case of 10.0.23.0/24, the RD of the path via 172.16.1.2 is the second number listed: (2621440/1966080). Because the RD is equal to the current FD, this route is disqualified as a feasible successor. If the path via 172.16.1.2 were to be lost, R1 would have to send queries to find a new way to get to 10.0.23.0/24. Feasible successors only exist in the topology table. Only successors appear in the routing table.

e. To see the Routing Information Base (RIB) Scale and Metric Scale values, as well as other protocol information, issue the **show ip protocols | section eigrp** command.

```
R1# show ip protocols | section eigrp
Routing Protocol is "eigrp 27"
  Outgoing update filter list for all interfaces is not set
  Incoming update filter list for all interfaces is not set
  Default networks flagged in outgoing updates
  Default networks accepted from incoming updates
  EIGRP-IPv4 VR(BASIC-EIGRP-LAB) Address-Family Protocol for AS(27)
    Metric weight K1=1, K2=0, K3=1, K4=0, K5=0 K6=0
    Metric rib-scale 128
    Metric version 64bit
    Soft SIA disabled
    NSF-aware route hold timer is 240
  EIGRP NSF disabled
     NSF signal timer is 20s
     NSF converge timer is 120s
    Router-ID: 1.1.1.1
    Topology : 0 (base)
      Active Timer: 3 min
      Distance: internal 90 external 170
      Maximum path: 4
      Maximum hopcount 100
      Maximum metric variance 1
      Total Prefix Count: 5
      Total Redist Count: 0
```

f. To examine details about a particular path, issue the **show ip eigrp topology [address]** command.

```
R1# show ip eigrp topology 192.168.3.0/24
EIGRP-IPv4 VR(BASIC-EIGRP-LAB) Topology Entry for AS(27)/ID(1.1.1.1) for
192.168.3.0/24
  State is Passive, Query origin flag is 1, 2 Successor(s), FD is 2048000, RIB
is 16000
```

```
Descriptor Blocks:
10.0.12.2 (GigabitEthernet0/0/0), from 10.0.12.2, Send flag is 0x0
    Composite metric is (2048000/1392640), route is Internal
    Vector metric:
        Minimum bandwidth is 1000000 Kbit
        Total delay is 21250000 picoseconds
        Reliability is 255/255
        Load is 1/255
        Minimum MTU is 1500
        Hop count is 2
        Originating router is 3.3.3.3
172.16.1.2 (GigabitEthernet0/0/1.1), from 172.16.1.2, Send flag is 0x0
    Composite metric is (2048000/1392640), route is Internal
    Vector metric:
        Minimum bandwidth is 1000000 Kbit
        Total delay is 21250000 picoseconds
        Reliability is 255/255
        Load is 1/255
        Minimum MTU is 1500
        Hop count is 2
        Originating router is 3.3.3.3
```

Part 3: Tune EIGRP for IPv4

In this part of the lab, you will tune and optimize EIGRP for IPv4 through the use of passive interfaces, authentication, and variance.

Step 1.   Configure specific interfaces as passive.

Passive interfaces are interfaces that only partially participate in a routing protocols operation. The network address for the passive interface is advertised through other interfaces. However, the routing protocol does not actually speak on the passive interface. Use passive interfaces when you have a connected network that you want to advertise, but you do not want protocol neighbors to appear on that interface. For example, interfaces supporting users should always be configured as passive. There are two ways to configure interfaces as passive, either specifically by interface name or by default. Normally a device with many LAN interfaces will use the default option, and then use the **no** form of the command on those specific interfaces that should be sending and receiving EIGRP messages.

a.   On PC1, run Wireshark and set the capture filter to **eigrp**. You should see a hello message roughly every five seconds. If your device were capable of running EIGRP, you might be able to form an adjacency and interact in the routing domain. This is unnecessary traffic on the LAN and a potential security risk. Stop the capture.

b.   On R1, configure af-interface G0/0/1 to be passive.

```
R1(config)# router eigrp BASIC-EIGRP-LAB
R1(config-router)# address-family ipv4 unicast autonomous-system 27
R1(config-router-af)# af-interface g0/0/1.2
R1(config-router-af-interface)# passive-interface
R1(config-router-af-interface)# end
```

c.   On PC1, restart the Wireshark capture with the capture filter still configured for **eigrp**. You should no longer see EIGRP Hello messages.

Step 2.   Configure interfaces to default to passive.

The second option for configuring passive interfaces is to configure them all as passive and then issue the **no passive-interface** command for certain interfaces. This approach is suitable in a security-focused scenario, or when the device has many LAN interfaces. The commands vary depending on whether you are using Classic or Named EIGRP.

   a. In Classic EIGRP configuration, enter the **passive-interface default** command followed by *no passive-interface interface-number* commands for all the interfaces that should be participating in EIGRP. The following provides an example of this on R2. You will temporarily lose EIGRP adjacencies until **passive-interface** is removed from required interfaces.

```
R2(config)# router eigrp 27
R2(config-router)# passive-interface default
R2(config-router)# no passive-interface g0/0/0
R2(config-router)# no passive-interface g0/0/1
R2(config-router)# exit
R2(config)# end
```

   b. In Named EIGRP configuration, you apply the **passive-interface** command to the **af-interface default** configuration, and the **no passive-interface** command to the specific *af-interface interface-number*. The following provides an example of this on R3. You will temporarily lose EIGRP adjacencies until **passive-interface** is removed from required interfaces.

```
R3(config)# router eigrp BASIC-EIGRP-LAB
R3(config-router)# address-family ipv4 unicast autonomous-system 27
R3(config-router-af)# af-interface default
R3(config-router-af-interface)# passive-interface
R3(config-router-af-interface)# exit
R3(config-router-af)# af-interface g0/0/0
R3(config-router-af-interface)# no passive-interface
R3(config-router-af-interface)# exit
R3(config-router-af)# af-interface g0/0/1
R3(config-router-af-interface)# no passive-interface
R3(config-router-af-interface)# end
```

   c. The output of the **show ip protocols | section Passive** command will give you a list of passive interfaces configured for EIGRP, which for R3 will only be the Loopback 0 interface that is simulating a LAN.

```
R3# show ip protocols | section Passive
  Passive Interface(s):
    Loopback0
```

Step 3.   Configure EIGRP authentication.

EIGRP supports authentication on an interface basis. In other words, each interface can be configured to require authentication of the connected peer. This ensures that connected devices that try to form an adjacency are authorized to do so. Classic EIGRP supports key-chain based MD5-hashed keys, while Named EIGRP adds support for SHA256-hashed keys. The two are not compatible.

In this step, you will configure both types of authentication to exercise the range of options available.

a. On R1, R2, R3, and D2, create a key-chain named EIGRP-AUTHEN-KEY with a single key. The key should have the key-string **$3cre7!!**

```
R1(config)# key chain EIGRP-AUTHEN-KEY
R1(config-keychain)# key 1
R1(config-keychain-key)# key-string $3cre7!!
R1(config-keychain-key)# end
```

b. On R2, configure interfaces G0/0/0 and G0/0/1 to encrypt the key chain you just created with MD5. Note that you will lose EIGRP adjacencies until the neighbor interfaces are configured.

```
R2(config)# interface g0/0/0
R2(config-if)# ip authentication key-chain eigrp 27 EIGRP-AUTHEN-KEY
R2(config-if)# ip authentication mode eigrp 27 md5
R2(config-if)# exit
R2(config)# interface g0/0/1
R2(config-if)# ip authentication key-chain eigrp 27 EIGRP-AUTHEN-KEY
R2(config-if)# ip authentication mode eigrp 27 md5
R2(config-if)# end
```

c. Configure interface G0/0/0 on both R1 and R3 to use the key chain with MD5. EIGRP adjacencies with R2 should be restored.

```
R1(config)# router eigrp BASIC-EIGRP-LAB
R1(config-router)# address-family ipv4 unicast autonomous-system 27
R1(config-router-af)# af-interface g0/0/0
R1(config-router-af-interface)# authentication key-chain EIGRP-AUTHEN-KEY
R1(config-router-af-interface)# authentication mode md5
R1(config-router-af-interface)# end
```

d. Use the **show ip eigrp interface detail | section Gi0/0/0** command to verify that authentication is in place and what type of authentication it is.

```
R1# show ip eigrp interface detail | section Gi0/0/0
Gi0/0/0                 1       0/0       0/0       1       0/050
0
  Hello-interval is 5, Hold-time is 15
  Split-horizon is enabled
  Next xmit serial <none>
  Packetized sent/expedited: 14/2
  Hello's sent/expedited: 186/4
  Un/reliable mcasts: 0/11  Un/reliable ucasts: 15/7
  Mcast exceptions: 0  CR packets: 0  ACKs suppressed: 0
  Retransmissions sent: 3  Out-of-sequence rcvd: 0
  Topology-ids on interface - 0
  Authentication mode is md5,  key-chain is "EIGRP-AUTHEN-KEY"
  Topologies advertised on this interface:  base
  Topologies not advertised on this interface:
```

e. On R1 and D2, configure SHA 256 based authentication using the same **$3cre7!!** shared secret.. R1 and D2 are running Named EIGRP, so the configuration is applied in **af-interface** mode. On R1, configure the G0/0/1.1 subinterface. On D2, configure the

G1/0/1 interface. The configuration for R1 is shown. Note that the R1-D2 adjacency will be lost until both ends are configured.

```
R1(config)# router eigrp BASIC-EIGRP-LAB
R1(config-router)# address-family ipv4 unicast autonomous-system 27
R1(config-router-af)# af-interface g0/0/1.1
R1(config-router-af-interface)# authentication mode hmac-sha-256 $3cre7!!
R1(config-router-af-interface)# end
```

_____

_____

_____

_____

_____

f. Use the **show ip eigrp interface detail** command to verify that authentication is in place and what type of authentication it is.

```
R1# show ip eigrp interface detail | section Gi0/0/1.1
Gi0/0/1.1                    1        0/0        0/0        3        0/050
0
   Hello-interval is 5, Hold-time is 15
   Split-horizon is enabled
   Next xmit serial <none>
   Packetized sent/expedited: 11/0
   Hello's sent/expedited: 225/4
   Un/reliable mcasts: 0/11  Un/reliable ucasts: 12/7
   Mcast exceptions: 0  CR packets: 0  ACKs suppressed: 0
   Retransmissions sent: 4  Out-of-sequence rcvd: 0
   Topology-ids on interface - 0
   Authentication mode is HMAC-SHA-256, key-chain is not set
   Topologies advertised on this interface:  base
   Topologies not advertised on this interface:
```

Step 4. Manipulate load balancing with variance

By default, load balancing occurs only over equal-cost paths. EIGRP supports up to four equal cost paths by default but can be configured to support as many as 32 with the **maximum-paths** command.

EIGRP has the added capability to load balance over unequal-cost paths. Load balancing is controlled by the **variance** parameter. Its value is a multiplier that is used to determine how to deal with multiple paths to the same destination.

Variance is set to 1 by default, so any paths up to the configured maximum number of paths that have a feasible distance equal to the best current feasible distance are also offered to the routing table. This provides equal cost load balancing.

The variance parameter can also be set to zero, which dictates that no load balancing takes place.

The variance parameter can be adjusted so that paths that have an FD that is less than or equal to variance times current best FD are also considered as successors and installed into the routing table. There is an extremely important differentiation here -- to be a <u>feasible successor</u>, the

<u>RD</u> of a path must be less than the current best FD. To be considered for <u>unequal load balancing</u>, the FD of the feasible successor is multiplied by the variance value, and if the product of this calculation is less than the current best FD, the feasible successor is promoted to successor.

There are two caveats; first, only feasible successors are considered and second, unequal cost load balancing is unequal; traffic share is proportional to the best metric in the routing table for the given path.

Note: Keep in mind that your routing table may be different than the one created by the examples in this lab. If your results are different, examine them carefully to determine why so that you can thoroughly understand how EIGRP is operating.

---

a. On R3, there are two equal-cost paths to 192.168.1.0/24.

```
R3# show ip route eigrp | section 192.168.1.0
D       192.168.1.0/24
                [90/20480] via 172.16.13.2, 00:08:18, GigabitEthernet0/0/1
                [90/20480] via 10.0.23.2, 00:08:18, GigabitEthernet0/0/0
```

b. To change this and allow for the demonstration of variance, change the interface bandwidth for the R2 interfaces G0/0/0 and G0/0/1 to 800000.

```
R2(config)# interface g0/0/0
R2(config-if)# bandwidth 800000
R2(config-if)# exit
R2(config)# interface g0/0/1
R2(config-if)# bandwidth 800000
R2(config-if)# end
```

c. When you examine the routing table on R3, you see that there is no load balancing occurring. All destinations have a single path.

```
R3# show ip route eigrp | begin Gateway
Gateway of last resort is not set


      10.0.0.0/8 is variably subnetted, 3 subnets, 2 masks
D        10.0.12.0/24 [90/16640] via 10.0.23.2, 00:01:17, GigabitEthernet0/0/0
      172.16.0.0/16 is variably subnetted, 3 subnets, 2 masks
D        172.16.1.0/24
            [90/15360] via 172.16.13.2, 00:01:04, GigabitEthernet0/0/1
D     192.168.1.0/24
            [90/20480] via 172.16.13.2, 00:01:04, GigabitEthernet0/0/1
```

d. However, we know there are multiple paths in the network. The first consideration for manipulating variance is that it only works with feasible successors. Examining the topology table on R3 shows that there is a feasible successor for the 192.168.1.0/24 network. The route via 10.0.23.2 out the G0/0/0 interface has an RD less than the FD for the current successor.

```
R3# show ip eigrp topology | section 192.168.1.0
P 192.168.1.0/24, 1 successors, FD is 2621440
            via 172.16.13.2 (2621440/1966080), GigabitEthernet0/0/1
            via 10.0.23.2 (2785280/2129920), GigabitEthernet0/0/0
```

e. To use the other route for unequal cost load balancing, we can set the variance parameter to 2. This will mean that any path with an RD less than or equal to 5242880 will qualify as a successor (2 x 2621440 = 5242880).

```
R3(config)# router eigrp BASIC-EIGRP-LAB
R3(config-router)# address-family ipv4 unicast autonomous-system 27
R3(config-router-af)# topology base
R3(config-router-af-topology)# variance 2
R3(config-router-af-topology)# end
```

f. The output of the **show ip route eigrp | begin Gateway** command now displays two paths available to the 192.168.1.0/24 network. Notice that the routes have different metrics, but are listed and used just the same. Also, notice adding variance 2 adds a second path to the 10.0.12.0/24 network.

```
R3# show ip route eigrp | begin Gateway
Gateway of last resort is not set

      10.0.0.0/8 is variably subnetted, 3 subnets, 2 masks
D        10.0.12.0/24
            [90/20480] via 172.16.13.2, 00:00:11, GigabitEthernet0/0/1
            [90/16640] via 10.0.23.2, 00:00:11, GigabitEthernet0/0/0
      172.16.0.0/16 is variably subnetted, 3 subnets, 2 masks
D        172.16.1.0/24
            [90/15360] via 172.16.13.2, 00:00:11, GigabitEthernet0/0/1
D     192.168.1.0/24
            [90/20480] via 172.16.13.2, 00:00:11, GigabitEthernet0/0/1
            [90/21760] via 10.0.23.2, 00:00:11, GigabitEthernet0/0/0
```

g. Issue the **show ip route 192.168.1.0** command to see more details about the paths the router has to the 192.168.1.0 network. As a part of this output, you see the traffic share count, which tells you the ratio of traffic that will be sent between these links. In this example, the count is 120 via 172.16.13.2 and 113 via 10.0.23.2. What that means is that 120 packets will be sent via 172.16.13.2 and then 113 packets will be sent via 10.0.23.2.

```
R3# show ip route 192.168.1.0
Routing entry for 192.168.1.0/24
  Known via "eigrp 27", distance 90, metric 20480, type internal
  Redistributing via eigrp 27
  Last update from 10.0.23.2 on GigabitEthernet0/0/0, 00:01:42 ago
  Routing Descriptor Blocks:
  * 172.16.13.2, from 172.16.13.2, 00:01:42 ago, via GigabitEthernet0/0/1
      Route metric is 20480, traffic share count is 120
      Total delay is 30 microseconds, minimum bandwidth is 1000000 Kbit
      Reliability 255/255, minimum MTU 1500 bytes
      Loading 1/255, Hops 2
    10.0.23.2, from 10.0.23.2, 00:01:42 ago, via GigabitEthernet0/0/0
      Route metric is 21760, traffic share count is 113
      Total delay is 30 microseconds, minimum bandwidth is 800000 Kbit
      Reliability 255/255, minimum MTU 1500 bytes
      Loading 1/255, Hops 2
```

Router Interface Summary Table

Router Model	Ethernet Interface #1	Ethernet Interface #2	Serial Interface #1	Serial Interface #2
1800	Fast Ethernet 0/0 (F0/0)	Fast Ethernet 0/1 (F0/1)	Serial 0/0/0 (S0/0/0)	Serial 0/0/1 (S0/0/1)
1900	Gigabit Ethernet 0/0 (G0/0)	Gigabit Ethernet 0/1 (G0/1)	Serial 0/0/0 (S0/0/0)	Serial 0/0/1 (S0/0/1)
2801	Fast Ethernet 0/0 (F0/0)	Fast Ethernet 0/1 (F0/1)	Serial 0/1/0 (S0/1/0)	Serial 0/1/1 (S0/1/1)
2811	Fast Ethernet 0/0 (F0/0)	Fast Ethernet 0/1 (F0/1)	Serial 0/0/0 (S0/0/0)	Serial 0/0/1 (S0/0/1)
2900	Gigabit Ethernet 0/0 (G0/0)	Gigabit Ethernet 0/1 (G0/1)	Serial 0/0/0 (S0/0/0)	Serial 0/0/1 (S0/0/1)
4221	Gigabit Ethernet 0/0/0 (G0/0/0)	Gigabit Ethernet 0/0/1 (G0/0/1)	Serial 0/1/0 (S0/1/0)	Serial 0/1/1 (S0/1/1)
4300	Gigabit Ethernet 0/0/0 (G0/0/0)	Gigabit Ethernet 0/0/1 (G0/0/1)	Serial 0/1/0 (S0/1/0)	Serial 0/1/1 (S0/1/1)

Note: To find out how the router is configured, look at the interfaces to identify the type of router and how many interfaces the router has. There is no way to effectively list all the combinations of configurations for each router class. This table includes identifiers for the possible combinations of Ethernet and Serial interfaces in the device. The table does not include any other type of interface, even though a specific router may contain one. An example of this might be an ISDN BRI interface. The string in parenthesis is the legal abbreviation that can be used in Cisco IOS commands to represent the interface.

Device Config Final – Notes

_____

_____

_____

_____

_____

_____

_____

_____

_____

_____

_____

_____

_____

_____

_____

_____

_____

_____

_____

CHAPTER 3

Advanced EIGRP

3.1.2 Lab - Implement Advanced EIGRP for IPv4 Features

Topology

Addressing Table

Device	Interface	IP Address	Subnet Mask
R1	G0/0/0	10.12.0.1	255.255.255.0
	G0/0/1	192.168.3.1	255.255.255.0
	S0/1/0	10.13.0.1	255.255.255.0
R2	G0/0/0	10.12.0.2	255.255.255.0
	G0/0/1	10.23.0.2	255.255.255.0
R3	G0/0/0	10.23.0.3	255.255.255.0
	S0/1/0	10.13.0.3	255.255.255.0
	Loopback0	172.16.3.1	255.255.255.0
D1	G1/0/11	192.168.3.2	255.255.255.0
	Loopback0	192.168.1.1	255.255.255.0
	Loopback1	192.168.1.1	255.255.255.0
	Loopback2	192.168.0.1	255.255.255.0

Objectives

Part 1: Build the Network and Configure Basic Device Settings

Part 2: Implement EIGRP for IPv4

Part 3: Implement Advanced Features

Background/Scenario

Customizing the operation of EIGRP can yield many benefits, most notably speeding convergence and stabilizing network operations during outages. In this lab you will explore some advanced techniques that can be used to customize and improve EIGRP performance on an enterprise network.

Note: The routers used with CCNP hands-on labs are Cisco 4221s with Cisco IOS XE Release 16.9.4 (universalk9 image). The switches used in the labs are Cisco Catalyst 3650s with Cisco IOS XE Release 16.9.4 (universalk9 image). Other routers, switches, and Cisco IOS versions can be used. Depending on the model and Cisco IOS version, the commands available and the output produced might vary from what is shown in the labs.

Note: Make sure that the routers and switches have been erased and have no startup configurations. If you are unsure, contact your instructor.

Required Resources

- 3 Routers (Cisco 4221 with Cisco IOS XE Release 16.9.4 universal image or comparable)
- 1 Switch (Cisco 3650 with Cisco IOS XE release 16.9.4 universal image or comparable)
- 1 PC (Choice of operating system with a terminal emulation program installed)
- Console cables to configure the Cisco IOS devices via the console ports
- Ethernet cables as shown in the topology

Instructions

Part 1: Build the Network and Configure Basic Device Settings

In Part 1, you will set up the network topology and configure basic settings and interface addressing.

Step 1. Cable the network as shown in the topology.

Attach the devices as shown in the topology diagram, and cable as necessary.

Step 2. Configure basic settings for each device.

   a. Console into each device, enter global configuration mode, and apply the basic settings. The startup configurations for each device are provided below.

   Router R1

```
hostname R1
no ip domain lookup
banner motd # R1, Implement Advanced EIGRP for IPv4 Features #
line con 0
 exec-timeout 0 0
```

```
 logging synchronous
 exit
line vty 0 4
 privilege level 15
 exec-timeout 0 0
 password cisco123
 login
 exit
!
interface g0/0/0
 ip address 10.12.0.1 255.255.255.0
 no shutdown
 exit
interface s0/1/0
 ip address 10.13.0.1 255.255.255.0
 no shutdown
 exit
interface g0/0/1
 ip address 192.168.3.1 255.255.255.0
 no shutdown
 exit
end
```

Router R2

```
hostname R2
no ip domain lookup
banner motd # R2, Implement Advanced EIGRP for IPv4 Features #
line con 0
 exec-timeout 0 0
 logging synchronous
 exit
line vty 0 4
 privilege level 15
 exec-timeout 0 0
 password cisco123
 login
 exit
!
interface g0/0/0
 ip address 10.12.0.2 255.255.255.0
 no shutdown
 exit
interface g0/0/1
 ip address 10.23.0.2 255.255.255.0
 no shutdown
 exit
end
```

Router R3

```
hostname R3
no ip domain lookup
banner motd # R3, Implement Advanced EIGRP for IPv4 Features #
```

```
line con 0
 exec-timeout 0 0
 logging synchronous
 exit
line vty 0 4
 privilege level 15
 exec-timeout 0 0
 password cisco123
 login
 transport input telnet
 exit
interface g0/0/0
 ip address 10.23.0.3 255.255.255.0
 no shutdown
 exit
interface s0/1/0
 ip address 10.13.0.3 255.255.255.0
 no shutdown
 exit
interface loopback 0
 ip address 172.16.3.1 255.255.255.0
 no shutdown
 exit
end
```

Switch D1

```
hostname D1
no ip domain lookup
ip routing
banner motd # D1, Implement Advanced EIGRP for IPv4 Features #
line con 0
 exec-timeout 0 0
 logging synchronous
 exit
line vty 0 4
 privilege level 15
 exec-timeout 0 0
 password cisco123
 login
 exit
interface range g1/0/1-24
 shutdown
 exit
interface g1/0/11
 no switchport
 ip address 192.168.3.2 255.255.255.0
 no shutdown
 exit
interface loopback 0
 ip address 192.168.2.1 255.255.255.0
 no shutdown
```

```
  exit
interface loopback 1
  ip address 192.168.1.1 255.255.255.0
  no shutdown
  exit
interface loopback 2
  ip address 192.168.0.1 255.255.255.0
  no shutdown
  exit
end
```

b.   Set the clock on all devices to UTC time.

c.   Save the running configuration to startup-config on all devices.

Part 2: Implement EIGRP for IPv4

In this part, you will configure classic EIGRP for IPv4 and verify that all routing tables are converged.

Step 1.   Configure classic EIGRP for IPv4.

Configure classic EIGRP for IPv4 on all devices. Use Autonomous System number 98, and advertise only the connected interfaces on each device.

_____

_____

_____

_____

_____

Step 2.   Verify EIGRP for IPv4 routing.

Verify that each device has a complete routing table for all the networks shown in the topology and Addressing Table.

Part 3: Implement Advanced Features

In this part of the lab you will customize several different settings within EIGRP and see the impact of those changes on the network.

Step 1.   Modify timers.

EIGRP uses standard **hello-interval** and **hold-time** timers based on the speed of the interface. If the interface speed is a T1 or less, hellos are sent every 60 seconds and the **hold-time** is set to 180 seconds. If the interface speed is greater than a T1, hellos are sent every 5 seconds and the **hold-time** is set to 15 seconds. These default times might not be appropriate for some network scenarios.

a.   To see what the timers are set to, issue the **show ip eigrp interfaces detail** command.

```
R1# show ip eigrp interfaces detail
EIGRP-IPv4 Interfaces for AS(98)
```

```
                        Xmit Queue   PeerQ       Mean  Pacing Time  Multicast    Pending
        Interface Peers Un/Reliable  Un/Reliable SRTT  Un/Reliable  Flow Timer   Routes
        Gi0/0/0     1       0/0         0/0        1       0/050         0
          Hello-interval is 5, Hold-time is 15
          Split-horizon is enabled
          Next xmit serial <none>
          Packetized sent/expedited: 5/1
          Hello's sent/expedited: 29/2
          Un/reliable mcasts: 0/5  Un/reliable ucasts: 7/3
          Mcast exceptions: 0  CR packets: 0  ACKs suppressed: 0
          Retransmissions sent: 2  Out-of-sequence rcvd: 0
          Topology-ids on interface - 0
          Authentication mode is not set
          Topologies advertised on this interface:  base
          Topologies not advertised on this interface:

        Se0/1/0     1       0/0         0/0        1       0/1250        0
          Hello-interval is 5, Hold-time is 15
          Split-horizon is enabled
          Next xmit serial <none>
          Packetized sent/expedited: 4/0
          Hello's sent/expedited: 32/2
          Un/reliable mcasts: 0/0  Un/reliable ucasts: 6/5
          Mcast exceptions: 0  CR packets: 0  ACKs suppressed: 0
          Retransmissions sent: 0  Out-of-sequence rcvd: 0
          Topology-ids on interface - 0
          Authentication mode is not set
          Topologies advertised on this interface:  base
          Topologies not advertised on this interface:

        Gi0/0/1             1       0/0     0/0        4       0/050              0
          Hello-interval is 5, Hold-time is 15
          Split-horizon is enabled
          Next xmit serial <none>
          Packetized sent/expedited: 2/1
          Hello's sent/expedited: 31/2
          Un/reliable mcasts: 0/2  Un/reliable ucasts: 4/3
          Mcast exceptions: 0  CR packets: 0  ACKs suppressed: 0
          Retransmissions sent: 2  Out-of-sequence rcvd: 0
          Topology-ids on interface - 0
          Authentication mode is not set
          Topologies advertised on this interface:  base
          Topologies not advertised on this interface:
```

  b. For this lab, the timers on R1 interface G0/0/0 and S0/1/0 need to be adjusted to send hellos every 10 seconds and establish a hold time of 30 seconds. EIGRP is unique in that each interface can have a customized **hello-interval** and **hold-time**. The times are not needed to match between the ends of a link. Change the timers using the **ip hello-interval eigrp** *ASN seconds* and the **ip hold-time eigrp** *ASN seconds* interface configuration commands.

```
R1# config t
Enter configuration commands, one per line.  End with CNTL/Z.
R1(config)# interface g0/0/0
R1(config-if)# ip hello-interval eigrp 98 10
R1(config-if)# ip hold-time eigrp 98 30
R1(config-if)# exit
R1(config)# interface s0/1/0
R1(config-if)# ip hello-interval eigrp 98 10
R1(config-if)# ip hold-time eigrp 98 30
R1(config-if)# exit
R1(config)# end
```

c. To verify that the changes were made, check the output of the **show ip eigrp interfaces detail** command.

```
R1# show ip eigrp interfaces detail
EIGRP-IPv4 Interfaces for AS(98)
                    Xmit Queue   PeerQ       Mean  Pacing Time  Multicast   Pending
Interface Peers   Un/Reliable  Un/Reliable  SRTT  Un/Reliable  Flow Timer  Routes
Gi0/0/0    1         0/0          0/0          1      0/050         0
  Hello-interval is 10, Hold-time is 30
  Split-horizon is enabled
  Next xmit serial <none>
  Packetized sent/expedited: 5/1
  Hello's sent/expedited: 84/2
  Un/reliable mcasts: 0/5  Un/reliable ucasts: 7/3
  Mcast exceptions: 0  CR packets: 0  ACKs suppressed: 0
  Retransmissions sent: 2  Out-of-sequence rcvd: 0
  Topology-ids on interface - 0
  Authentication mode is not set
  Topologies advertised on this interface:  base
  Topologies not advertised on this interface:

Se0/1/0             1         0/0          0/0          1      0/1250        0
  Hello-interval is 10, Hold-time is 30
  Split-horizon is enabled
  Next xmit serial <none>
  Packetized sent/expedited: 4/0
  Hello's sent/expedited: 87/2
  Un/reliable mcasts: 0/0  Un/reliable ucasts: 6/5
  Mcast exceptions: 0  CR packets: 0  ACKs suppressed: 0
  Retransmissions sent: 0  Out-of-sequence rcvd: 0
  Topology-ids on interface - 0
  Authentication mode is not set
  Topologies advertised on this interface:  base
  Topologies not advertised on this interface:

Gi0/0/1             1         0/0          0/0          4      0/050         0
  Hello-interval is 5, Hold-time is 15
  Split-horizon is enabled
  Next xmit serial <none>
  Packetized sent/expedited: 2/1
```

```
Hello's sent/expedited: 92/2
Un/reliable mcasts: 0/2  Un/reliable ucasts: 4/3
Mcast exceptions: 0  CR packets: 0  ACKs suppressed: 0
Retransmissions sent: 2  Out-of-sequence rcvd: 0
Topology-ids on interface - 0
Authentication mode is not set
Topologies advertised on this interface:  base
Topologies not advertised on this interface:
```

Step 2.   Create summarized routes in EIGRP.

Large routing tables take more memory and require more CPU time to process. Reducing the size of the routing table is advantageous in all network scenarios. EIGRP supports summarization of routes at any point in the network. There is no boundary router limitation like the limitations imposed in OSPF. However, in order for the summary route to be valid, EIGRP requires that some component of the summary route be in the routing table for the router doing the summarization.

a.   Issue the **show ip route eigrp | begin Gateway** command on R3 and note the group of networks from the 192.168 range of addresses. R1 is advertising this contiguous block of networks individually, instead of sending a summary.

```
R3# show ip route eigrp | begin Gateway
Gateway of last resort is not set


      10.0.0.0/8 is variably subnetted, 5 subnets, 2 masks
D        10.12.0.0/24 [90/3072] via 10.23.0.2, 00:06:35, GigabitEthernet0/0/0
D      192.168.0.0/24 [90/131328] via 10.23.0.2, 00:06:08, GigabitEthernet0/0/0
D      192.168.1.0/24 [90/131328] via 10.23.0.2, 00:06:08, GigabitEthernet0/0/0
D      192.168.2.0/24 [90/131328] via 10.23.0.2, 00:06:08, GigabitEthernet0/0/0
D      192.168.3.0/24 [90/3328] via 10.23.0.2, 00:06:35, GigabitEthernet0/0/0
```

b.   On R1, configure a summary of the networks between R1 and D1, as well as the networks on D1 on the interfaces connecting to R3 and R2.

```
R1# conf t
R1(config)# interface g0/0/0
R1(config-if)# ip summary-address eigrp 98 192.168.0.0 255.255.252.0
*Mar  6 00:07:33.742: %DUAL-5-NBRCHANGE: EIGRP-IPv4 98: Neighbor 10.12.0.2
(GigabitEthernet0/0/0) is resync: summary configured
R1(config-if)# exit
R1(config)# interface s0/1/0
R1(config-if)# ip summary-address eigrp 98 192.168.0.0 255.255.252.0
*Mar  6 00:07:35.220: %DUAL-5-NBRCHANGE: EIGRP-IPv4 98: Neighbor 10.13.0.3
(Serial0/1/0) is resync: summary configured
R1(config-if)# end
```

c.   Now examine the routing table on R3 again using the **show ip route eigrp | begin Gateway** command. In the output, you now see a single route taking the place of what had been four distinct routes.

```
R3# show ip route eigrp | begin Gateway
Gateway of last resort is not set


      10.0.0.0/8 is variably subnetted, 5 subnets, 2 masks
D        10.12.0.0/24 [90/3072] via 10.23.0.2, 00:07:59, GigabitEthernet0/0/0
D      192.168.0.0/22 [90/3328] via 10.23.0.2, 00:00:25, GigabitEthernet0/0/0
```

Step 3.   Control EIGRP query propagation with EIGRP stub routers.

EIGRP uses query messages to find a path to networks in the autonomous system. The query messages always require an acknowledgement. But a router will only send a response if it has a potential route that satisfies the query. If it does not have a route, it sends its own queries to its neighbors. This process can lead to long delays in reconvergence after an outage.

Query scoping refers to using various techniques to control how far across a network queries have to be sent. Summarization is one way of controlling query propagation. Another way to control query propagation is to use EIGRP stub routers where appropriate. When a router is single-homed to the rest of the network, and no other networks exist beyond that router, there is no real point in sending it a query looking for lost networks. The stub router declares itself as a stub to the router connected to the rest of the network, which is considered a hub router. The hub router then forwards no queries to the stub router because it knows there are no other networks, beyond those reported, existing beyond the stub router. In the topology for this lab, switch D1 is a stub router and R1 is its hub router.

a.   To verify that switch D1 is receiving EIGRP queries, issue the **shutdown** command on R2 interface G0/0/1. On switch D1, issue the **show ip eigrp events** command. This command outputs a timestamped list of actions that EIGRP is taking. In the output, you will find an entry that says switch D1 received a query trying to find the 10.23.0.0/24 network. Take note of the time stamp.

```
D1# show ip eigrp events
Event information for AS 98:
1     00:10:46.753 NDB delete: 10.23.0.0/24 1
2     00:10:46.753 RDB delete: 10.23.0.0/24 192.168.3.1
3     00:10:46.737 Metric set: 10.23.0.0/24 metric(Infinity)
4     00:10:46.737 Poison squashed: 10.23.0.0/24 lost if
5     00:10:46.737 Poison squashed: 10.23.0.0/24 rt net gone
6     00:10:46.737 Route installing: 10.23.0.0/24 192.168.3.1
7     00:10:46.737 Send reply: 10.23.0.0/24 192.168.3.1
8     00:10:46.737 Not active net/1=SH: 10.23.0.0/24 1
9     00:10:46.737 FC not sat Dmin/met: metric(Infinity) metric(1782272)
10    00:10:46.737 Find FS: 10.23.0.0/24 metric(1782272)
11    00:10:46.737 Rcv query met/succ met: metric(Infinity) metric(Infinity)
12    00:10:46.737 Rcv query dest/nh: 10.23.0.0/24 192.168.3.1
```

b.   Issue the **no shutdown** command on R2 interface G0/0/1.

c.   Configure D1 as an EIGRP stub router.

```
D1# config t
Enter configuration commands, one per line.  End with CNTL/Z.
D1(config)# router eigrp 98
D1(config-router)# eigrp stub
D1(config)# end
D1#
Mar  6 00:11:40.624: %DUAL-5-NBRCHANGE: EIGRP-IPv4 98: Neighbor 192.168.3.1
(GigabitEthernet1/0/11) is down: peer info changed
Mar  6 00:11:45.174: %DUAL-5-NBRCHANGE: EIGRP-IPv4 98: Neighbor 192.168.3.1
(GigabitEthernet1/0/11) is up: new adjacency
```

d. Verify that R1 sees switch D1 as a stub by examining the output of the **show ip eigrp neighbor detail** command.

```
R1# show ip eigrp neighbor detail
EIGRP-IPv4 Neighbors for AS(98)
H   Address          Interface          Hold Uptime   SRTT   RTO  QSeq
                                        (sec)         (ms)        CntNum
2   192.168.3.2      Gi0/0/1             14 00:00:39    3    100  010
    Version 25.0/2.0, Retrans: 1, Retries: 0, Prefixes: 3
    Topology-ids from peer - 0
    Topologies advertised to peer:    base

    Stub Peer Advertising (CONNECTED SUMMARY ) Routes
    Suppressing queries
1   10.13.0.3        Se0/1/0             11 00:11:44    1    100  028
    Time since Restart 00:04:06
    Version 25.0/2.0, Retrans: 0, Retries: 0, Prefixes: 3
    Topology-ids from peer - 0
    Topologies advertised to peer:    base

0   10.12.0.2        Gi0/0/0             12 00:12:24    1    100  030
    Time since Restart 00:04:07
    Version 25.0/2.0, Retrans: 2, Retries: 0, Prefixes: 2
    Topology-ids from peer - 0
    Topologies advertised to peer:    base


Max Nbrs: 0, Current Nbrs: 0
```

e. Issue the **shutdown** command on R2 interface G0/0/1. Take note of the timestamp on the syslog message reporting that the interface is down.

f. On switch D1, issue the **show ip eigrp events** command. You will see that no query was received looking for the 10.23.0.0/24 network. The query was stopped at R1, speeding up the convergence process.

```
D1# show ip eigrp events
Event information for AS 98:
1    00:12:53.776 NDB delete: 10.23.0.0/24 1
2    00:12:53.776 Poison squashed: 10.23.0.0/24 rt net gone
3    00:12:53.776 RDB delete: 10.23.0.0/24 192.168.3.1
4    00:12:53.776 Not active net/1=SH: 10.23.0.0/24 1
5    00:12:53.776 FC not sat Dmin/met: metric(Infinity) metric(1782272)
6    00:12:53.776 Find FS: 10.23.0.0/24 metric(1782272)
7    00:12:53.776 Rcv update met/succmet: metric(Infinity) metric(Infinity)
8    00:12:53.776 Rcv update dest/nh: 10.23.0.0/24 192.168.3.1
9    00:12:53.776 Ignored route, hopcount: 10.23.0.0 255
10   00:12:50.077 Metric set: 172.16.3.0/24 metric(1910016)
11   00:12:50.077 Route installed: 172.16.3.0/24 192.168.3.1
12   00:12:50.077 Route installing: 172.16.3.0/24 192.168.3.1
13   00:12:50.077 Find FS: 172.16.3.0/24 metric(Infinity)
14   00:12:50.077 Free reply status: 172.16.3.0/24
15   00:12:50.077 Clr handle num/bits: 0 0x0
```

```
16    00:12:50.077 Clr handle dest/cnt: 172.16.3.0/24 0
17    00:12:50.077 Rcv reply met/succ met: metric(1910016) metric(1909760)
18    00:12:50.077 Rcv reply dest/nh: 172.16.3.0/24 192.168.3.1
19    00:12:50.077 Metric set: 10.23.0.0/24 metric(1782272)
20    00:12:50.076 Route installed: 10.23.0.0/24 192.168.3.1
21    00:12:50.076 Route installing: 10.23.0.0/24 192.168.3.1
22    00:12:50.076 Find FS: 10.23.0.0/24 metric(Infinity)
23    00:12:50.076 Free reply status: 10.23.0.0/24
```

    g. Issue the **no shutdown** command on R2 interface G0/0/1.

Step 4. Filter EIGRP routes with a distribute list.

EIGRP supports several different filtering capabilities. The simplest and most direct is to use a distribute list. A distribute list refers to an access list which can be applied to all EIGRP updates being sent by a certain router, or it can be applied to a specific interface to modify updates as they exit. For this exercise, we will filter the 10.12.0.0/24 network from updates being sent out of R2 interface G0/0/1. This will cause a change in R3's routing table.

    a. Examine the routing table on R3 by issuing the **show ip route eigrp | begin Gateway** command. In the output, you can see that R3 has calculated the path via R2 at 10.23.0.2 to be the best path to reach the 10.12.0.0/24 network.

```
R3# show ip route eigrp | begin Gateway
Gateway of last resort is not set


      10.0.0.0/8 is variably subnetted, 5 subnets, 2 masks
D        10.12.0.0/24 [90/3072] via 10.23.0.2, 00:00:17, GigabitEthernet0/0/0
D        192.168.0.0/22 [90/3328] via 10.23.0.2, 00:00:53, GigabitEthernet0/0/0
```

    b. Our intent is to change the configuration at R2 so that R3 only learns about the 10.12.0.0/24 network from R1. Create an access list that denies the 10.12.0.0/24 network and permits all other networks.

```
R2# config t
R2(config)# ip access-list standard EIGRP-FILTER
R2(config-std-nacl)# deny 10.12.0.0 0.0.255.255
R2(config-std-nacl)# permit any
R2(config-std-nacl)# exit
```

    c. Enter EIGRP router configuration mode and configure the distribute list to reference the access list you just created, further specifying that the filter should be effective outbound on interface G0/0/1.

```
R2(config)# router eigrp 98
R2(config-router)# distribute-list EIGRP-FILTER out g0/0/1
R2(config-router)# end
*Mar 6 00:19:56.379: %DUAL-5-NBRCHANGE: EIGRP-IPv4 98: Neighbor 10.23.0.3
(GigabitEthernet0/0/1) is resync: intf route configuration changed
```

    d. On R3, issue the **show ip route eigrp | begin Gateway** command. As you can see, the successor for the 10.12.0.0/24 network has changed to R1 at 10.13.0.1.

```
R3# show ip route eigrp | begin Gateway
Gateway of last resort is not set


      10.0.0.0/8 is variably subnetted, 5 subnets, 2 masks
D        10.12.0.0/24 [90/1792256] via 10.13.0.1, 00:01:30, Serial0/1/0
D        192.168.0.0/22 [90/3328] via 10.23.0.2, 00:03:00, GigabitEthernet0/0/0
```

Router Interface Summary Table

Router Model	Ethernet Interface #1	Ethernet Interface #2	Serial Interface #1	Serial Interface #2
1800	Fast Ethernet 0/0 (F0/0)	Fast Ethernet 0/1 (F0/1)	Serial 0/0/0 (S0/0/0)	Serial 0/0/1 (S0/0/1)
1900	Gigabit Ethernet 0/0 (G0/0)	Gigabit Ethernet 0/1 (G0/1)	Serial 0/0/0 (S0/0/0)	Serial 0/0/1 (S0/0/1)
2801	Fast Ethernet 0/0 (F0/0)	Fast Ethernet 0/1 (F0/1)	Serial 0/1/0 (S0/1/0)	Serial 0/1/1 (S0/1/1)
2811	Fast Ethernet 0/0 (F0/0)	Fast Ethernet 0/1 (F0/1)	Serial 0/0/0 (S0/0/0)	Serial 0/0/1 (S0/0/1)
2900	Gigabit Ethernet 0/0 (G0/0)	Gigabit Ethernet 0/1 (G0/1)	Serial 0/0/0 (S0/0/0)	Serial 0/0/1 (S0/0/1)
4221	Gigabit Ethernet 0/0/0 (G0/0/0)	Gigabit Ethernet 0/0/1 (G0/0/1)	Serial 0/1/0 (S0/1/0)	Serial 0/1/1 (S0/1/1)
4300	Gigabit Ethernet 0/0/0 (G0/0/0)	Gigabit Ethernet 0/0/1 (G0/0/1)	Serial 0/1/0 (S0/1/0)	Serial 0/1/1 (S0/1/1)

Note: To find out how the router is configured, look at the interfaces to identify the type of router and how many interfaces the router has. There is no way to effectively list all the combinations of configurations for each router class. This table includes identifiers for the possible combinations of Ethernet and Serial interfaces in the device. The table does not include any other type of interface, even though a specific router may contain one. An example of this might be an ISDN BRI interface. The string in parenthesis is the legal abbreviation that can be used in Cisco IOS commands to represent the interface.

Device Config Final – Notes

_____

_____

_____

_____

_____

_____

_____

_____

_____

_____

_____

_____

_____

_____

_____

_____

_____

CHAPTER 4

Troubleshooting EIGRP for IPv4

4.1.2 Lab - Troubleshoot EIGRP for IPv4

Topology

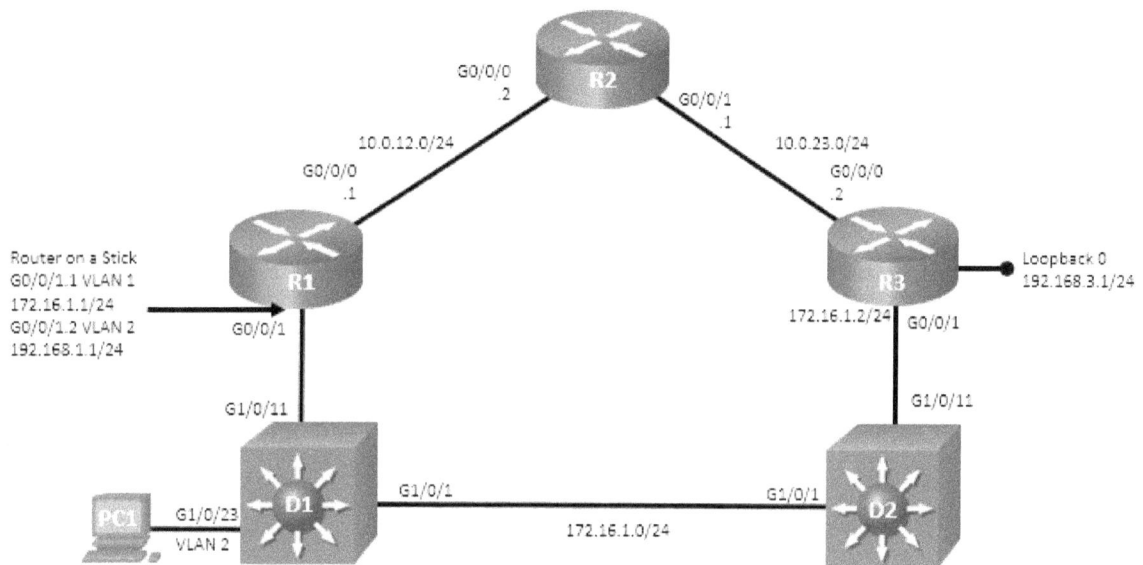

Addressing Table

Device	Interface	IP Address	Subnet Mask
R1	G0/0/0	10.0.12.1	255.255.255.0
	G0/0/1.1	172.16.1.1	255.255.255.0
	G0/0/1.2	192.168.1.1	255.255.255.0
R2	G0/0/0	10.0.12.2	255.255.255.0
	G0/0/1	10.0.23.1	255.255.255.0
R3	G0/0/0	10.0.23.2	255.255.255.0
	G0/0/1	172.16.1.2	255.255.255.0
	Loopback 0	192.168.3.1	255.255.255.0
PC1	NIC	DHCP	

Objectives

Troubleshoot network issues related to the configuration and operation of EIGRP for IPv4.

Background/Scenario

In this topology, routers R1, R2, and R3 are EIGRP neighbors. Switches D1 and D2 provide connectivity between R1 and R3. Router R1 is configured for inter-VLAN routing and DHCP to provide support for PC 1. You will be loading configurations with intentional errors onto the network. Your tasks are to FIND the error(s), document your findings and the command(s) or method(s) used to fix them. FIX the issue(s) presented here and then test the network to ensure both of the following conditions are met:

1. the complaint received in the ticket is resolved

2. full reachability is restored

Note: The routers used with CCNP hands-on labs are Cisco 4221 with Cisco IOS XE Release 16.9.4 (universalk9 image). The switches used in the labs are Cisco Catalyst 3650s with Cisco IOS XE Release 16.9.4 (universalk9 image) and Cisco Catalyst 2960s with Cisco IOS Release 15.2(2) (lanbasek9 image). Other routers, switches, and Cisco IOS versions can be used. Depending on the model and Cisco IOS version, the commands available and the output produced might vary from what is shown in the labs. Refer to the Router Interface Summary Table at the end of the lab for the correct interface identifiers.

Note: Make sure that the switches have been erased and have no startup configurations. If you are unsure, contact your instructor.

Required Resources

- 3 Routers (Cisco 4221 with Cisco IOS XE Release 16.9.4 universal image or comparable)
- 2 Switches (Cisco 3560 with Cisco IOS XE Release 16.9.4 universal image or comparable)
- 1 PC (Choice of operating system with terminal emulation program installed)
- Console cables to configure the Cisco IOS devices via the console ports
- Ethernet cables as shown in the topology

Instructions

Part 1: Trouble Ticket 4.1.2.1

Scenario:

R2 was added to the network to increase bandwidth for traffic between the R1 and R3 LANs. However, R2, is not sending routes to R1 or forming adjacencies with R3.

Use the commands listed below to load the configuration files for this trouble ticket:

Device	Command
R1	`copy flash:/enarsi/4.1.2.1-r1-config.txt run`
R2	`copy flash:/enarsi/4.1.2.1-r2-config.txt run`
R3	`copy flash:/enarsi/4.1.2.1-r3-config.txt run`
D1	`copy flash:/enarsi/4.1.2.1-d1-config.txt run`
D2	`copy flash:/enarsi/4.1.2.1-d2-config.txt run`

- PC 1 should be configured for and receive an address from an IPv4 DHCP server.

- Passwords on all devices are **cisco12345**. If a username is required, use **admin**.

- After you have fixed the ticket, change the MOTD on EACH DEVICE using the following command:

  banner motd # This is $(hostname) FIXED from ticket <ticket number> #

- Then save the configuration by issuing the **wri** command (on each device).

- Inform your instructor that you are ready for the next ticket.

- After the instructor approves your solution for this ticket, issue the **reset.now** privileged EXEC command on each device. This script will clear your configurations and reload the devices.

Part 2: Trouble Ticket 4.1.2.2

Scenario:

Your company hired an outside security consultant to work on the network. His task was to ensure that routing was being done securely. After he completed his work, R2 stopped sharing route information about networks at R1.

Use the commands listed below to load the configuration files for this trouble ticket:

Device	Command
R1	`copy flash:/enarsi/4.1.2.2-r1-config.txt run`
R2	`copy flash:/enarsi/4.1.2.2-r2-config.txt run`
R3	`copy flash:/enarsi/4.1.2.2-r3-config.txt run`
D1	`copy flash:/enarsi/4.1.2.2-d1-config.txt run`
D2	`copy flash:/enarsi/4.1.2.2-d2-config.txt run`

- PC1 should be configured for and receive an address from an IPv4 DHCP server.

- Passwords on all devices are **cisco12345**. If a username is required, use **admin**.

- After you have fixed the ticket, change the MOTD on EACH DEVICE using the following command:

  banner motd # This is $(hostname) FIXED from ticket <ticket number> #

- Then save the configuration by issuing the **wri** command (on each device).

- Inform your instructor that you are ready for the next ticket.

- After the instructor approves your solution for this ticket, issue the **reset.now** privileged EXEC command. This script will clear your configurations and reload the devices.

Part 3: Trouble Ticket 4.1.2.3

Topology Update

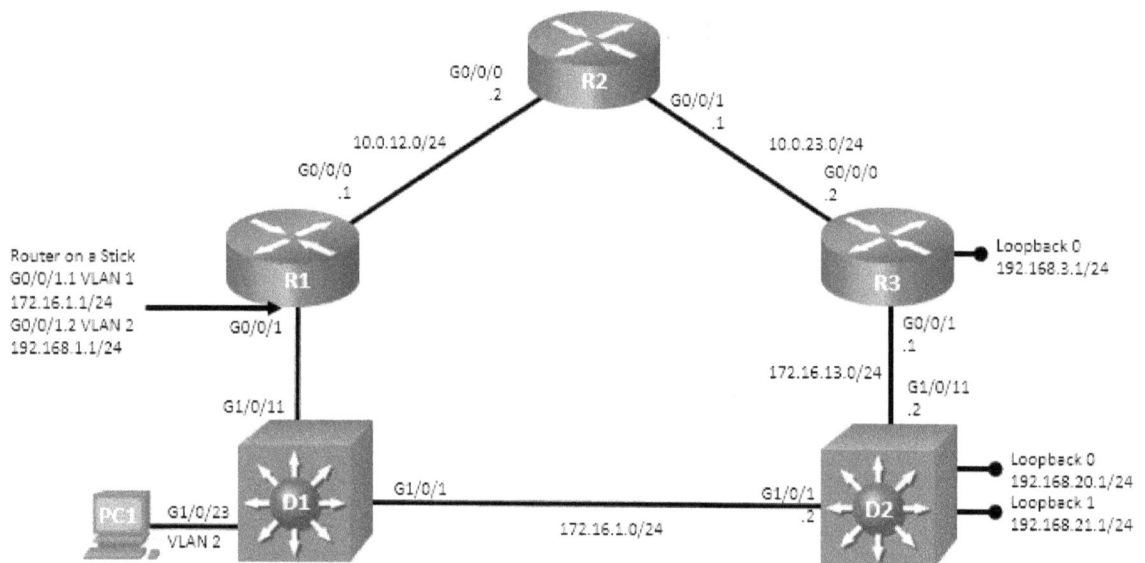

Addressing Table Update

Device	Interface	IP Address	Subnet Mask
R1	G0/0/0	10.0.12.1	255.255.255.0
	G0/0/1.1	172.16.1.1	255.255.255.0
	G0/0/1.2	192.168.1.1	255.255.255.0
R2	G0/0/0	10.0.12.2	255.255.255.0
	G0/0/1	10.0.23.1	255.255.255.0
R3	G0/0/0	10.0.23.2	255.255.255.0
	G0/0/1	172.16.1.2	255.255.255.0
	Loopback 0	192.168.3.1	255.255.255.0
D2	G1/0/1	172.16.1.2	255.255.255.0
	G1/0/11	172.16.13.2	255.255.255.0
	Loopback 0	192.168.20.1	255.255.255.0
	Loopback 1	192.168.21.1	255.255.255.0
PC1	NIC	DHCP	

Scenario:

Switch D2 was converted to support Inter-VLAN routing and is connected to the network via two routed ports. Switch D2 was also configured to join the EIGRP domain, but it is not forming adjacencies.

Use the commands listed below to load the configuration files for this trouble ticket:

Device	Command
R1	`copy flash:/enarsi/4.1.2.3-r1-config.txt run`
R2	`copy flash:/enarsi/4.1.2.3-r2-config.txt run`
R3	`copy flash:/enarsi/4.1.2.3-r3-config.txt run`
D1	`copy flash:/enarsi/4.1.2.3-d1-config.txt run`
D2	`copy flash:/enarsi/4.1.2.3-d2-config.txt run`

- PC1 should be configured for and receive an address from an IPv4 DHCP server.
- Passwords on all devices are **cisco12345**. If a username is required, use **admin**.
- After you have fixed the ticket, change the MOTD on EACH DEVICE using the following command:

  banner motd # This is $(hostname) FIXED from ticket <ticket number> #
- Then save the configuration by issuing the **wri** command (on each device).
- Inform your instructor that you are ready for the next ticket.
- After the instructor approves your solution for this ticket, issue the **reset.now** privileged EXEC command. This script will clear your configurations and reload the devices.

Router Interface Summary Table

Router Model	Ethernet Interface #1	Ethernet Interface #2	Serial Interface #1	Serial Interface #2
1800	Fast Ethernet 0/0 (F0/0)	Fast Ethernet 0/1 (F0/1)	Serial 0/0/0 (S0/0/0)	Serial 0/0/1 (S0/0/1)
1900	Gigabit Ethernet 0/0 (G0/0)	Gigabit Ethernet 0/1 (G0/1)	Serial 0/0/0 (S0/0/0)	Serial 0/0/1 (S0/0/1)
2801	Fast Ethernet 0/0 (F0/0)	Fast Ethernet 0/1 (F0/1)	Serial 0/1/0 (S0/1/0)	Serial 0/1/1 (S0/1/1)
2811	Fast Ethernet 0/0 (F0/0)	Fast Ethernet 0/1 (F0/1)	Serial 0/0/0 (S0/0/0)	Serial 0/0/1 (S0/0/1)
2900	Gigabit Ethernet 0/0 (G0/0)	Gigabit Ethernet 0/1 (G0/1)	Serial 0/0/0 (S0/0/0)	Serial 0/0/1 (S0/0/1)
4221	Gigabit Ethernet 0/0/0 (G0/0/0)	Gigabit Ethernet 0/0/1 (G0/0/1)	Serial 0/1/0 (S0/1/0)	Serial 0/1/1 (S0/1/1)
4300	Gigabit Ethernet 0/0/0 (G0/0/0)	Gigabit Ethernet 0/0/1 (G0/0/1)	Serial 0/1/0 (S0/1/0)	Serial 0/1/1 (S0/1/1)

Note: To find out how the router is configured, look at the interfaces to identify the type of router and how many interfaces the router has. There is no way to effectively list all the combinations of configurations for each router class. This table includes identifiers for the possible combinations of Ethernet and Serial interfaces in the device. The table does not include any other type of interface, even though a specific router may contain one. An example of this might be an ISDN BRI interface. The string in parenthesis is the legal abbreviation that can be used in Cisco IOS commands to represent the interface.

Device Config Final – Notes

5.1.2 Lab - Implement EIGRP for IPv6

Topology

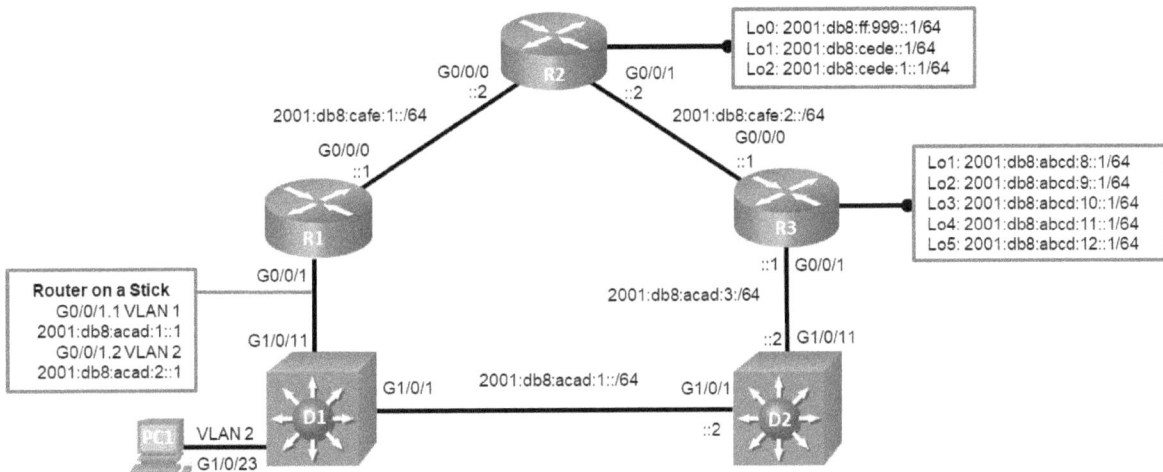

Lo0: 2001:db8:ff:999::1/64
Lo1: 2001:db8:cede::1/64
Lo2: 2001:db8:cede:1::1/64

G0/0/0 ::2   R2   G0/0/1 ::2

2001:db8:cafe:1::/64

G0/0/0 ::1

2001:db8:cafe:2::/64
G0/0/0 ::1

R1

Lo1: 2001:db8:abcd:8::1/64
Lo2: 2001:db8:abcd:9::1/64
Lo3: 2001:db8:abcd:10::1/64
Lo4: 2001:db8:abcd:11::1/64
Lo5: 2001:db8:abcd:12::1/64

R3

Router on a Stick
G0/0/1.1 VLAN 1
2001:db8:acad:1::1
G0/0/1.2 VLAN 2
2001:db8:acad:2::1

G0/0/1

::1 G0/0/1

2001:db8:acad:3::/64

::2 G1/0/11

G1/0/11

G1/0/1   2001:db8:acad:1::/64   G1/0/1

D1   D2   ::2

PC1 VLAN 2
G1/0/23

Addressing Table

Device	Interface	IPv6 Address/Prefix Length	Link Local Address
R1	G0/0/0	2001:db8:cafc:1::1/64	fe80::1:1
	G0/0/1.1	2001:db8:acad:1::1/64	fe80::1:2
	G0/0/1.2	2001:db8:acad:2::1/64	fe80::1:3
R2	G0/0/0	2001:db8:cafe:1::2/64	fe80::2:1
	G0/0/1	2001:db8:cafe:2::2/64	fe80::2:2
	Loopback 0	2001:db8:ff:999:153/64	fe80::2:3
	Loopback 1	2001:db8:cede::1/64	fe80::2:4
	Loopback 2	2001:db8:cede:1::1/64	fe80::2:5
R3	G0/0/0	2001:db8:cafe:2::1/64	fe80::3:1
	G0/0/1	2001:db8:acad:3::1/64	fe80::3:2
	Loopback 1	2001:db8:abcd:8::1/64	fe80::3:3
	Loopback 2	2001:db8:abcd:9::1/64	fe80::3:4
	Loopback 3	2001:db8:abcd:10::1/64	fe80::3:5
	Loopback 4	2001:db8:abcd:11::1/64	fe80::3:6
	Loopback 5	2001:db8:abcd:12::1/64	fe80::3:7
D2	G1/0/1	2001:db8:acad:1::2/64	fe80::d2:1
	G1/0/11	2001:db8:acad:3::2/64	fe80::d2:2
PC1	NIC	SLAAC	EUI-64

Objectives

Part 1: Build the Network and Configure Basic Device Settings

Part 2: Implement EIGRP for IPv6 and Named EIGRP

Part 3: Tune and Optimize EIGRP for IPv6

Background/Scenario

EIGRP for IPv6 has the same overall operation and features as EIGRP for IPv4. However, there are a few major differences between them:

- IPv6 unicast routing must be enabled before the routing process can be configured.
- In the absence of the router having any IPv4 addresses, a 32-bit router ID must be configured for the routing process to start.
- EIGRP for IPv6 is configured directly on the router interfaces.

In this lab, you will configure the network with EIGRP for IPv6. You will also configure passive interfaces, propagate a default route, configure a summary route, implement routing protocol authentication, modify load balancing, and filter routes with a prefix list.

Note: This lab is an exercise in configuring the options available with EIGRP for IPv6. It does not necessarily reflect implementation best practices.

Note: The routers used with CCNP hands-on labs are Cisco 4221 with Cisco IOS XE Release 16.9.4 (universalk9 image). The switches used in the labs are Cisco Catalyst 3650 with Cisco IOS XE Release 16.9.4 (universalk9 image). Other routers, switches, and Cisco IOS versions can be used. Depending on the model and Cisco IOS version, the commands available and the output produced might vary from what is shown in the labs. Refer to the Router Interface Summary Table at the end of the lab for the correct interface identifiers.

Note: Make sure that the routers and switches have been erased and have no startup configurations. If you are unsure, contact your instructor.

Required Resources

- 3 Routers (Cisco 4221 with Cisco IOS XE Release 16.9.4 universal image or comparable)
- 2 Switches (Cisco 3650 with Cisco IOS XE Release 16.9.4 universal image or comparable)
- 1 PC (Choice of operating system with terminal emulation program installed)
- Console cables to configure the Cisco IOS devices via the console ports
- Ethernet cables as shown in the topology

Instructions

Part 1: Build the Network and Configure Basic Device Settings

In Part 1, you will set up the network topology and configure basic settings and interface addressing on routers.

Step 1.   Cable the network as shown in the topology.

Attach the devices as shown in the topology diagram, and cable as necessary.

Step 2.   Configure basic settings for each device.

   a.   Console into each device, enter global configuration mode, and apply the basic settings. The startup configurations for each device are provided below.

Router R1

```
hostname R1
no ip domain lookup
ipv6 unicast-routing
banner motd # R1, Implement EIGRP for IPv6 #
line con 0
 exec-timeout 0 0
 logging synchronous
 exit
line vty 0 4
 privilege level 15
 exec-timeout 0 0
 password cisco123
 login
 exit
interface g0/0/0
 ipv6 address 2001:db8:cafe:1::1/64
 ipv6 address fe80::1:1 link-local
 no shutdown
 exit
interface g0/0/1
 no ip address
 no shutdown
 exit
interface g0/0/1.1
 description VLAN 1 Interface
 encapsulation dot1q 1
 ipv6 address fe80::1:2 link-local
 ipv6 address 2001:db8:acad:1::1/64
 no shutdown
 exit
interface g0/0/1.2
 description VLAN 2 Interface
 encapsulation dot1q 2
 ipv6 address fe80::1:3 link-local
 ipv6 address 2001:db8:acad:2::1/64
 no shutdown
 exit
end
```

Router R2

```
hostname R2
no ip domain lookup
```

```
ipv6 unicast-routing
banner motd # R2, Implement EIGRP for IPv6 #
line con 0
 exec-timeout 0 0
 logging synchronous
 exit
line vty 0 4
 privilege level 15
 exec-timeout 0 0
 password cisco123
 login
 exit
interface g0/0/0
 ipv6 address fe80::2:1 link-local
 ipv6 address 2001:db8:cafe:1::2/64
 no shutdown
 exit
interface g0/0/1
 ipv6 address fe80::2:2 link-local
 ipv6 address 2001:db8:cafe:2::2/64
 no shutdown
 exit
interface loopback 0
 description Internet host
 ipv6 address fe80::2:3 link-local
 ipv6 address 2001:db8:ff:999::153/64
 no shutdown
 exit
interface loopback 1
 ipv6 address fe80::2:4 link-local
 ipv6 address 2001:db8:cede::1/64
 no shutdown
 exit
interface loopback 2
 ipv6 address fe80::2:5 link-local
 ipv6 address 2001:db8:cede:1::1/64
 no shutdown
 exit
end
```

Router R3

```
hostname R3
no ip domain lookup
ipv6 unicast-routing
banner motd # R3, Implement EIGRP for IPv6 #
line con 0
 exec-timeout 0 0
 logging synchronous
 exit
line vty 0 4
 privilege level 15
```

```
 exec-timeout 0 0
 password cisco123
 login
 exit
interface g0/0/0
 ipv6 address fe80::3:1 link-local
 ipv6 address 2001:db8:cafe:2::1/64
 no shutdown
 exit
interface g0/0/1
 ipv6 address fe80::3:2 link-local
 ipv6 address 2001:db8:acad:1::3/64
 no shutdown
 exit
interface loopback 1
 ipv6 address fe80::3:3 link-local
 ipv6 address 2001:db8:abcd:8::1/64
 no shutdown
interface loopback 2
 ipv6 address fe80::3:4 link-local
 ipv6 address 2001:db8:abcd:9::1/64
 no shutdown
interface loopback 3
 ipv6 address fe80::3:5 link-local
 ipv6 address 2001:db8:abcd:10::1/64
 no shutdown
interface loopback 4
 ipv6 address fe80::3:6 link-local
 ipv6 address 2001:db8:abcd:11::1/64
 no shutdown
interface loopback 5
 ipv6 address fe80::3:7 link-local
 ipv6 address 2001:db8:abcd:12::1/64
 no shutdown
end
```

Router D1

```
hostname D1
no ip domain lookup
banner motd # D1, Implement EIGRP for IPv6 #
line con 0
 exec-timeout 0 0
 logging synchronous
line vty 0 4
 privilege level 15
 exec-timeout 0 0
 password cisco123
 login
vlan 2
 name HOST-VLAN
interface range g1/0/1 - 24, g1/1/1 - 2
```

```
  switchport mode access
  shutdown
interface range g1/0/1, g1/0/11
 switchport mode trunk
 no shutdown
 exit
interface g1/0/23
 switchport mode access
 switchport access vlan 2
 spanning-tree portfast
 no shutdown
 exit
end
```

Router D2

```
hostname D2
no ip domain lookup
ipv6 unicast-routing
banner motd # D2, Implement EIGRP for IPv6 #
line con 0
 exec-timeout 0 0
 logging synchronous
 exit
line vty 0 4
 privilege level 15
 exec-timeout 0 0
 password cisco123
 login
 exit
interface range g1/0/1 - 24, g1/1/1 - 2
 shutdown
 exit
interface g1/0/1
 no switchport
 ipv6 address fe80::d1:1 link-local
 ipv6 address 2001:Db8:acad:1::2/64
 no shutdown
 exit
interface g1/0/11
 no switchport
 ipv6 address fe80::d1:2 link-local
 ipv6 address 2001:db8:acad:3::2/64
 no shutdown
 exit
end
```

b. Set the clock on each device to UTC time.

c. Save the running configuration to startup-config.

d. Verify that PC1 generates an IPv6 address.

e. Verify that PC1 can ping its default gateway at fe80::1:3.

Part 2: Implement EIGRP for IPv6 and Named EIGRP

In this part of the lab, you will configure and verify EIGRP in the network. Routers R1 and R3 will use Named EIGRP, while router R2 will use Classic EIGRP. After you have established the network, you will examine the differences in how each version of EIGRP deals with metrics.

For the lab, you will use the Autonomous System number 43 on all routers.

Step 1.   Configure EIGRP for IPv6 on R2

a.   Start the configuration of Classic EIGRP by issuing the **ipv6 router eigrp 43** command.

```
R2(config)# ipv6 router eigrp 43
```

b.   Configure the EIGRP Router ID using the **eigrp router-id** command. Use the number 2.2.2.2 for R2.

```
R2(config-router)# eigrp router-id 2.2.2.2
```

c.   Identify the interfaces that should be configured with EIGRP and the networks that should be included in the EIGRP topology table. This is done on the interfaces with the **ipv6 eigrp** command.

```
R2(config)# interface g0/0/0
R2(config-if)# ipv6 eigrp 43
R2(config-if)# exit
R2(config)# interface g0/0/1
R2(config-if)# ipv6 eigrp 43
R2(config-if)# exit
R2(config)# interface loopback0
R2(config-if)# ipv6 eigrp 43
R2(config-if)# exit
R2(config)# interface loopback1
R2(config-if)# ipv6 eigrp 43
R2(config-if)# exit
R2(config)# interface loopback 2
R2(config-if)# ipv6 eigrp 43
R2(config-if)# end
```

d.   Verify the interfaces now involved in EIGRP with the **show ipv6 eigrp interfaces** command.

```
R2# show ipv6 eigrp interfaces
EIGRP-IPv6 Interfaces for AS(43)
                    Xmit Queue   PeerQ       Mean  Pacing Time  Multicast  Pending
Interface   Peers   Un/Reliable  Un/Reliable SRTT  Un/Reliable  Flow Timer Routes
Gi0/0/0     0       0/0          0/0         0     0/0 0        0
Gi0/0/1     0       0/0          0/0         0     0/0 0        0
Lo0         0       0/0          0/0         0     0/0 0        0
Lo1         0       0/0          0/0         0     0/0 0        0
Lo2         0       0/0          0/0         0     0/0 0        0
```

Step 2.   Configure Named EIGRP for IPv6 on R1 and R3

a.   Start the configuration of Named EIGRP by issuing the **router eigrp [name]** command. The name parameter can be a number, but the number does not identify an

Autonomous System as it does with Classic EIGRP, it simply identifies the process. For our purposes, name the process **EIGRP_IPV6**.

```
R1(config)# router eigrp EIGRP_IPV6
R1(config-router)#
```

b. Enter into address-family configuration mode with the **address-family ipv6 unicast autonomous-system 43** command. It is not necessary to configure EIGRP for IPv6 on the interfaces. In named-mode configuration, EIGRP for IPv6 is automatically enabled on all interfaces that are configured with an IPv6 address.

```
R1(config-router)# address-family ipv6 unicast autonomous-system 43
```

c. Configure the EIGRP Router ID using the **eigrp router-id** command. Use the number 1.1.1.1.

```
R1(config-router-af)# eigrp router-id 1.1.1.1
```

d. Repeat the configuration process on R3 and D2. For the R3 router ID use 3.3.3.3, and for the D2 router ID use 132.132.132.132.

Step 3.   Verify EIGRP for IPv6

a. A few seconds after configuring the network statements on R1, R3 and D2, you should have seen EIGRP adjacencies being formed, as noted at the console by messages similar to the one below.

```
*Mar  9 13:42:47.969: %DUAL-5-NBRCHANGE: EIGRP-IPv6 43: Neighbor FE80::3:2
(GigabitEthernet1/0/11) is up: new adjacency
```

b. To verify that routing is working, ping from PC1 to Interface Loopback 1 on R3 (2001:db8:abcd:8::1). The ping should be successful.

c. On R1, examine the EIGRP entries in the IPv6 routing table using the **show ipv6 route eigrp** command. As you can see, R1 is aware of all of the networks that have been configured in the topology. The remote networks that were learned from EIGRP and that appear in the routing table were learned from routers R2 and R3 as indicated by the link local address that is displayed for these entries. Note that in some cases, EIGRP for IPv6 has two equal cost routes for a network.

```
R1# show ipv6 route eigrp
IPv6 Routing Table - default - 17 entries
<output omitted>
D   2001:DB8:FF:999::/64 [90/2570240]
      via FE80::2:1, GigabitEthernet0/0/0
D   2001:DB8:ABCD:8::/64 [90/16000]
      via FE80::2:1, GigabitEthernet0/0/0
      via FE80::D1:1, GigabitEthernet0/0/1.1
D   2001:DB8:ABCD:9::/64 [90/16000]
      via FE80::2:1, GigabitEthernet0/0/0
      via FE80::D1:1, GigabitEthernet0/0/1.1
D   2001:DB8:ABCD:10::/64 [90/16000]
      via FE80::2:1, GigabitEthernet0/0/0
      via FE80::D1:1, GigabitEthernet0/0/1.1
D   2001:DB8:ABCD:11::/64 [90/16000]
      via FE80::2:1, GigabitEthernet0/0/0
      via FE80::D1:1, GigabitEthernet0/0/1.1
```

```
D   2001:DB8:ABCD:12::/64 [90/16000]
     via FE80::2:1, GigabitEthernet0/0/0
     via FE80::D1:1, GigabitEthernet0/0/1.1
D   2001:DB8:ACAD:3::/64 [90/15360]
     via FE80::D1:1, GigabitEthernet0/0/1.1
D   2001:DB8:CAFE:2::/64 [90/15360]
     via FE80::2:1, GigabitEthernet0/0/0
D   2001:DB8:CEDE::/64 [90/2570240]
     via FE80::2:1, GigabitEthernet0/0/0
D   2001:DB8:CEDE:1::/64 [90/2570240]
     via FE80::2:1, GigabitEthernet0/0/0
```

d. Now examine the EIGRP topology table using the **show ipv6 eigrp topology all-links** command. The all-links parameter instructs the router to display routes that are not successors or feasible successors. We will focus on the routes to 2001:db8:abcd:10::/64 and 2001:db8:cafe:2::/64. There are several things to notice.

Remember that the topology table is EIGRP's database of route information. EIGRP selects the best paths, based on the DUAL algorithm, and offers them to the IP routing table. However, the IP routing table does not have to use those offered paths, because the router may have learned about the same network from a more reliable routing source, which would be a routing source with a lower administrative distance.

```
R1# show ipv6 eigrp topology all-links
EIGRP-IPv6 VR(EIGRP_IPV6) Topology Table for AS(43)/ID(1.1.1.1)
Codes: P - Passive, A - Active, U - Update, Q - Query, R - Reply,
       r - reply Status, s - sia Status


P 2001:DB8:CEDE:1::/64, 1 successors, FD is 328990720, serno 7
        via FE80::2:1 (328990720/327761920), GigabitEthernet0/0/0
P 2001:DB8:CEDE::/64, 1 successors, FD is 328990720, serno 6
        via FE80::2:1 (328990720/327761920), GigabitEthernet0/0/0
P 2001:DB8:ABCD:10::/64, 2 successors, FD is 2048000, serno 16
        via FE80::2:1 (2048000/1392640), GigabitEthernet0/0/0
        via FE80::D1:1 (2048000/1392640), GigabitEthernet0/0/1.1
P 2001:DB8:ACAD:1::/64, 1 successors, FD is 1310720, serno 2
        via Connected, GigabitEthernet0/0/1.1
        via FE80::D1:1 (1966080/1310720), GigabitEthernet0/0/1.1
P 2001:DB8:ABCD:12::/64, 2 successors, FD is 2048000, serno 18
        via FE80::2:1 (2048000/1392640), GigabitEthernet0/0/0
        via FE80::D1:1 (2048000/1392640), GigabitEthernet0/0/1.1
P 2001:DB8:CAFE:2::/64, 1 successors, FD is 1966080, serno 4
        via FE80::2:1 (1966080/1310720), GigabitEthernet0/0/0
        via FE80::D1:1 (2621440/1966080), GigabitEthernet0/0/1.1
P 2001:DB8:ABCD:9::/64, 2 successors, FD is 2048000, serno 15
        via FE80::2:1 (2048000/1392640), GigabitEthernet0/0/0
        via FE80::D1:1 (2048000/1392640), GigabitEthernet0/0/1.1
P 2001:DB8:ABCD:11::/64, 2 successors, FD is 2048000, serno 17
        via FE80::2:1 (2048000/1392640), GigabitEthernet0/0/0
        via FE80::D1:1 (2048000/1392640), GigabitEthernet0/0/1.1
P 2001:DB8:ACAD:2::/64, 1 successors, FD is 1310720, serno 3
        via Connected, GigabitEthernet0/0/1.2
```

```
P 2001:DB8:ABCD:8::/64, 2 successors, FD is 2048000, serno 14
        via FE80::2:1 (2048000/1392640), GigabitEthernet0/0/0
        via FE80::D1:1 (2048000/1392640), GigabitEthernet0/0/1.1
P 2001:DB8:CAFE:1::/64, 1 successors, FD is 1310720, serno 1
        via Connected, GigabitEthernet0/0/0
        via FE80::2:1 (1966080/1310720), GigabitEthernet0/0/0
P 2001:DB8:FF:999::/64, 1 successors, FD is 328990720, serno 5
        via FE80::2:1 (328990720/327761920), GigabitEthernet0/0/0
P 2001:DB8:ACAD:3::/64, 1 successors, FD is 1966080, serno 13
        via FE80::D1:1 (1966080/1310720), GigabitEthernet0/0/1.1
```

We will focus on the routes, highlighted in the above output, to 2001:db8:abcd:10::/64 and 2001:db8:cafe:2::/64. There are several things to notice:

- The entry for the 2001:db8:abcd:10::/64 network shows two successors, while the entry for 2001:db8:cafe:2::/64 shows only one successor. Both entries show two paths. The path with the lowest Feasible Distance (FD) is selected as the successor and is offered to the routing table. For 2001:db8:abcd:10::/64, there are two paths with equal FD, so they are both successors and both are offered to the global routing table. In the case of 2001:db8:cafe:2::/64, the FD is listed as 19660800. The path via fe80::2:1 shows that number as the FD (first number in parentheses). The path via fe80::d1:1 shows an FD of 2621440, which is higher than the current FD. So that path, although valid, is a higher cost path and is not offered to the routing table.

- The FD listed in the topology table does not match the metric listed in the routing table. For 2001:db8:abcd:10::/64, the routing table shows the metric value 16000 while the topology table shows the FD as 2048000. This is due to the routing table having a limit of 4 bytes (32 bits) for metric information, while EIGRP on R1 is using EIGRP wide metrics, which are 64 bits. Wide metrics are used by Named EIGRP by default. To work around the 32-bit metric size limitation in the routing table, EIGRP divides the wide-metric value by the EIGRP_RIB_SCALE value, which defaults to 128. The value 2048000 divided by 128 is 16000.

Note: A network with mixed EIGRP implementations (Named and Classic in the same routing domain) will have some loss of route clarity, which could lead to sub-optimal path selection. The recommended implementation is to use Named EIGRP in all cases.

- There are no feasible successors listed in the topology table for 2001:db8:abcd:10::/64 or 2001:db8:cafe:2::/64. The feasibility condition requires that the reported distance (RD) to a destination network be less than the current FD for a next-hop to be considered a feasible successor to the route. In the case of 2001:db8:cafe:2::/64, the RD of the path via fe80::d1:1 is listed as 1966080, which is equal to the current FD, which disqualifies this path as a feasible successor. If the path via fe80::2:1 were to be lost, R1 would have to send queries to find a new way to get to 2001:db8:cafe:2::/64. Feasible successors appear only in the topology table. Only successors appear in the routing table.

e. To see the RIB Scale and metric version values, as well as other protocol information, issue the **show ipv6 protocols | section EIGRP_IPv6** command.

```
R1# show ipv6 protocols | section EIGRP_IPV6
EIGRP-IPv6 VR(EIGRP_IPV6) Address-Family Protocol for AS(43)
```

```
Metric weight K1=1, K2=0, K3=1, K4=0, K5=0 K6=0
Metric rib-scale 128
Metric version 64bit
Soft SIA disabled
NSF-aware route hold timer is 240
EIGRP NSF disabled
   NSF signal timer is 20s
   NSF converge timer is 120s
Router-ID: 1.1.1.1
Topology : 0 (base)
  Active Timer: 3 min
  Distance: internal 90 external 170
  Maximum path: 16
  Maximum hopcount 100
  Maximum metric variance 1
  Total Prefix Count: 13
  Total Redist Count: 0
```

f. To examine details about a particular path, issue the **show ipv6 eigrp topology [address]** command. Among other things in this output, you can see the values used in calculating the metric.

```
R1# show ipv6 eigrp topology 2001:db8:cafe:2::/64
EIGRP-IPv6 VR(EIGRP_IPV6) Topology Entry for AS(43)/ID(1.1.1.1) for
2001:DB8:CAFE:2::/64
  State is Passive, Query origin flag is 1, 1 Successor(s), FD is 1966080, RIB
is 15360
  Descriptor Blocks:
  FE80::2:1 (GigabitEthernet0/0/0), from FE80::2:1, Send flag is 0x0
      Composite metric is (1966080/1310720), route is Internal
      Vector metric:
        Minimum bandwidth is 1000000 Kbit
        Total delay is 20000000 picoseconds
        Reliability is 255/255
        Load is 1/255
        Minimum MTU is 1500
        Hop count is 1
        Originating router is 2.2.2.2
  FE80::D1:1 (GigabitEthernet0/0/1.1), from FE80::D1:1, Send flag is 0x0
      Composite metric is (2621440/1966080), route is Internal
      Vector metric:
        Minimum bandwidth is 1000000 Kbit
        Total delay is 30000000 picoseconds
        Reliability is 255/255
        Load is 1/255
        Minimum MTU is 1500
        Hop count is 2
        Originating router is 3.3.3.3
```

Part 3: Tune and Optimize EIGRP for IPv6

In this part of the lab, you will tune and optimize EIGRP for IPv6 through the use of passive interfaces, default router redistribution, summary routes, authentication, load balancing, and route filtering.

Step 1.   Configure specific interfaces as passive.

Passive interfaces are interfaces that only partially participate in the operation of a routing protocol. The network that a passive interface is connected to is advertised, while the routing protocol does not actually transmit routing protocol-specific traffic on that interface. Use passive interfaces when you have a connected network that you want to advertise, but you do not want protocol neighbors to appear on that interface. Interfaces supporting users should always be configured as passive. There are two ways to configure interfaces as passive. The first is specifically by interface. The other is to make all interfaces default to passive default. Normally a device with many LAN interfaces will use the default option, and then use the **no** form of the command on the specific interfaces that should be sending and receiving EIGRP messages.

   a.   On PC1, run Wireshark and set the display capture filter to **eigrp**. You should see a hello message roughly every five seconds. If PC 1 is capable of running EIGRP for IPv6, you might be able to form an adjacency and interact in the routing domain. This is not desirable.

   b.   On R1, configure **af-interface g0/0/1** to be passive.

```
R1(config)# router eigrp EIGRP_IPV6
R1(config-router)# address-family ipv6 unicast autonomous-system 43
R1(config-router-af)# af-interface g0/0/1.2
R1(config-router-af-interface)# passive-interface
R1(config-router-af-interface)# end
```

   c.   On PC1, restart the Wireshark capture with the **eigrp** capture filter. You should no longer see EIGRP Hello messages.

Step 2.   Configure interfaces from default to passive.

The second option for configuring passive interfaces is to configure them all as passive and then issue the **no passive-interface** command for certain interfaces. This approach is suitable in a security-focused scenario, or when the device has many LAN interfaces. The commands vary depending on whether you are using Classic or Named EIGRP.

   a.   In Classic EIGRP configuration, issue the **passive-interface default** command followed by the **no passive-interface [interface designation]** command on the interfaces that should be participating in EIGRP. As an example, configure this on R2, and then make interfaces G0/0/0 and G0/0/1 active. Note that you will lose EIGRP adjacencies until the interfaces are active.

```
R2(config)# ipv6 router eigrp 43
R2(config-rtr)# passive-interface default
R2(config-rtr)# no passive-interface g0/0/0
R2(config-rtr)# no passive-interface g0/0/1
R2(config-rtr)# end
```

   b.   In Named EIGRP configuration, you apply the **passive-interface** command to the af-interface default configuration, and then the **no passive-interface** command to the af-interface specific interface. On R3, set the af-interface default as passive and then configure G0/0/0 and S0/1/0 as active. Note that you will lose EIGRP adjacencies until the interfaces are active.

```
R3(config)# router eigrp EIGRP_IPV6
R3(config-router)# address-family ipv6 unicast autonomous-system 43
R3(config-router-af)# af-interface default
R3(config-router-af-interface)# passive-interface
R3(config-router-af-interface)# exit-af-interface
R3(config-router-af)# af-interface g0/0/0
R3(config-router-af-interface)# no passive-interface
R3(config-router-af-interface)# exit-af-interface
R3(config-router-af)# af-interface g0/0/1
R3(config-router-af-interface)# no passive-interface
R3(config-router-af-interface)# end
```

c. The output of **show ip protocols | include (passive)** will give you a list of passive interfaces configured for EIGRP.

```
R3# show ipv6 protocols | include (passive)
    Loopback5 (passive)
    Loopback4 (passive)
    Loopback3 (passive)
    Loopback2 (passive)
Loopback1 (passive)
```

Step 3. Propagate a default route.

EIGRP for IPv6 can be configured to propagate a default route to other EIGRP routers in the AS. This lab will explore two methods of propagating a default route, either by redistributing a default static route or by sharing a summary default route.

In this topology, interface Loopback 0 on R2 has been configured to simulate an internet destination. Therefore, we will configure a default route on R2 and then configure EIGRP for IPv6 to redistribute the route.

a. Configure a static default route on R2 with an exit interface of Loopback0s IPv6 address.

```
R2(config)# ipv6 route ::/0 2001:db8:ff:999::1
```

b. Go into EIGRP configuration and add the **redistribute static** command.

```
R2(config)# ipv6 router eigrp 43
R2(config-rtr)# redistribute static
R2(config-rtr)# end
```

c. At R1, issue the **show ipv6 route eigrp | begin EX ::** command. Notice the default route is present as an EIGRP external route with an AD of 170. Further, notice that individual routes for the 2001:db8:cede::/64 and 2001:db8:cede:1::/64 networks, representing R2 interfaces Lo1 and Lo2, are present in the routing table.

```
R1# show ipv6 route eigrp | begin EX  ::
EX  ::/0 [170/2570240]
    via FE80::2:1, GigabitEthernet0/0/0
D   2001:DB8:FF:999::/64 [90/2570240]
    via FE80::2:1, GigabitEthernet0/0/0
D   2001:DB8:ABCD:8::/64 [90/16000]
    via FE80::2:1, GigabitEthernet0/0/0
    via FE80::D1:1, GigabitEthernet0/0/1.1
```

```
D    2001:DB8:ABCD:9::/64 [90/16000]
        via FE80::2:1, GigabitEthernet0/0/0
        via FE80::D1:1, GigabitEthernet0/0/1.1
D    2001:DB8:ABCD:10::/64 [90/16000]
        via FE80::2:1, GigabitEthernet0/0/0
        via FE80::D1:1, GigabitEthernet0/0/1.1
D    2001:DB8:ABCD:11::/64 [90/16000]
        via FE80::2:1, GigabitEthernet0/0/0
        via FE80::D1:1, GigabitEthernet0/0/1.1
D    2001:DB8:ABCD:12::/64 [90/16000]
        via FE80::2:1, GigabitEthernet0/0/0
        via FE80::D1:1, GigabitEthernet0/0/1.1
D    2001:DB8:ACAD:3::/64 [90/15360]
        via FE80::D1:1, GigabitEthernet0/0/1.1
D    2001:DB8:CAFE:2::/64 [90/15360]
        via FE80::2:1, GigabitEthernet0/0/0
D    2001:DB8:CEDE::/64 [90/2570240]
        via FE80::2:1, GigabitEthernet0/0/0
D    2001:DB8:CEDE:1::/64 [90/2570240]
        via FE80::2:1, GigabitEthernet0/0/0
```

d. On R2, remove the **redistribute static** command from EIGRP and remove the static default route.

e. On R2, configure the **ipv6 summary-address** command on the GigabitEthernet0/0/0 and GigabtEthernet0/0/1 interfaces. Specify the eigrp 43 and the route ::/0

```
R2(config)# interface GigabitEthernet0/0/0
R2(config-if)# ipv6 summary-address eigrp 43 ::/0
R2(config-if)# interface GigabitEthernet0/0/1
R2(config-if)# ipv6 summary-address eigrp 43 ::/0
```

f. Go to router R1 and use the **show ipv6 route eigrp** command to see the default route that has been injected into the routing table. Notice in the output that the route now appears as an internal EIGRP route with an AD of 90. Also notice that individual routes for the 2001:db8:cede::/64 and 2001:db8:cede:1::/64 networks, representing R2 interfaces Lo1 and Lo2, are no longer present in the routing table. The **ipv6 summary-address ::/0** command replaced all individual routes that R2 was advertising.

Note: If you were to add another summary address on R2, similar to what you will do in the next sub-step, that summary would be advertised as well.

```
R1# show ipv6 route eigrp
<output omitted>
D    ::/0 [90/15360]
        via FE80::2:1, GigabitEthernet0/0/0
D    2001:DB8:ABCD:8::/64 [90/16000]
        via FE80::D1:1, GigabitEthernet0/0/1.1
D    2001:DB8:ABCD:9::/64 [90/16000]
        via FE80::D1:1, GigabitEthernet0/0/1.1
D    2001:DB8:ABCD:10::/64 [90/16000]
        via FE80::D1:1, GigabitEthernet0/0/1.1
D    2001:DB8:ABCD:11::/64 [90/16000]
        via FE80::D1:1, GigabitEthernet0/0/1.1
```

```
D   2001:DB8:ABCD:12::/64 [90/16000]
      via FE80::D1:1, GigabitEthernet0/0/1.1
D   2001:DB8:ACAD:3::/64 [90/15360]
      via FE80::D1:1, GigabitEthernet0/0/1.1
D   2001:DB8:CAFE:2::/64 [90/20480]
      via FE80::D1:1, GigabitEthernet0/0/1.1
```

Step 4. Configure an EIGRP for IPv6 Summary Address.

Router R3 is configured with five loopback interfaces that simulate five IPv6 LANs. Those LAN addresses appear in the other EIGRP routers as five individual routes. In order to limit the impact of these five LANs on routing tables and routing protocol traffic, the routes can be configured with a single route summary address that will enable all five networks to be reached without requiring separate information to be shared for each network.

a. To optimize EIGRP for IPv6, on R3 summarize the loopback addresses as a single route and advertise the summary route in R3's EIGRP updates to R1 and R2. Use the same summarization method that is used for IPv4 by finding the bits that all five addresses have in common. The IPv6 loopback addresses could be summarized as 2001:db8:abcd::/61, but common practice is not to split the summary at the nibble level. Therefore, summary masks will normally be 48, 52, 56, and 60 bits. For our exercise, we will specify a 56 bit mask, even though that summary would indicate more networks than R3 is hosting. After configuring the summary route on the interface, notice that the neighbor adjacency between R3 and R2 and R1 is resynchronized (restarted).

```
R3(config)# router eigrp EIGRP_IPV6
R3(config-router)# address-family ipv6 unicast autonomous-system 43
R3(config-router-af)# af-interface g0/0/0
R3(config-router-af-interface)# summary-address 2001:db8:abcd::/56
R3(config-router-af-interface)# exit
R3(config-router-af)# af-interface g0/0/1
R3(config-router-af-interface)# summary-address 2001:db8:abcd::/56
R3(config-router-af-interface)# end
```

b. Examine the routing table of R1 to verify that R1 is receiving only one summary route for the loopback interfaces.

```
R1# show ipv6 route eigrp
<output omitted>
D   ::/0 [90/15360]
      via FE80::2:1, GigabitEthernet0/0/0
D   2001:DB8:ABCD::/56 [90/16000]
      via FE80::D1:1, GigabitEthernet0/0/1.1
D   2001:DB8:ACAD:3::/64 [90/15360]
      via FE80::D1:1, GigabitEthernet0/0/1.1
D   2001:DB8:CAFE:2::/64 [90/20480]
      via FE80::D1:1, GigabitEthernet0/0/1.1
```

Step 5. Configure EIGRP authentication.

EIGRP for IPv6 supports authentication on an interface basis. In other words, each interface can be configured to require authentication of the connected peer. This ensures that connected devices that try to form an adjacency are authorized to do so. Classic EIGRP supports key-chain based MD5-hashed keys, while Named EIGRP adds support for SHA256-hashed keys. The two are not compatible.

In this step, you will configure both types of authentication to exercise the range of options available.

a. On R1, R2, and R3, create a key-chain named EIGRPv6-AUTHEN-KEY with a single key. The key should have the key-string $3cre7!!

```
R1(config)# key-chain EIGRPv6-AUTHEN-KEY
R1(config-keychain)# key 1
R1(config-keychain-key)# key-string $3cre7!!
R1(config-keychain-key)# end
```

b. On R2, configure interfaces G0/0/0 and G0/0/1 to use the key-chain that you just created with MD5. Note that you will lose EIGRP adjacencies until the neighbor interfaces are configured.

```
R2(config)# interface g0/0/0
R2(config-if)# ipv6 authentication key-chain eigrp 43 EIGRPv6-AUTHEN-KEY
R2(config-if)# ipv6 authentication mode eigrp 43 md5
R2(config-if)# exit
R2(config)# interface g0/0/1
R2(config-if)# ipv6 authentication key-chain eigrp 43 EIGRPv6-AUTHEN-KEY
R2(config-if)# ipv6 authentication mode eigrp 43 md5
R2(config-if)# end
```

c. Configure interfaces GigabitEthernet0/0/0 on R1 and R3 to use the key-chain with MD5. EIGRP adjacencies with R2 should be restored.

```
R1(config)# router eigrp EIGRP_IPV6
R1(config-router)# address-family ipv6 unicast autonomous-system 43
R1(config-router-af)# af-interface g0/0/0
R1(config-router-af-interface)# authentication key-chain EIGRPv6-AUTHEN-KEY
R1(config-router-af-interface)# authentication mode md5
R1(config-router-af-interface)# end
```

d. Use the **show ip eigrp interface detail | section Gi0/0/0** command to verify authentication is in place and what type it is.

```
R1# show ipv6 eigrp interface detail | section Gi0/0/0
Gi0/0/0                1        0/0        0/0        1        0/050              0
   Hello-interval is 5, Hold-time is 15
   Split-horizon is enabled
   Next xmit serial <none>
   Packetized sent/expedited: 14/2
   Hello's sent/expedited: 186/4
   Un/reliable mcasts: 0/11  Un/reliable ucasts: 15/7
   Mcast exceptions: 0  CR packets: 0  ACKs suppressed: 0
   Retransmissions sent: 3  Out-of-sequence rcvd: 0
   Topology-ids on interface - 0
   Authentication mode is md5,  key-chain is "EIGRP-AUTHEN-KEY"
   Topologies advertised on this interface:  base
   Topologies not advertised on this interface:
```

e. On R1, R3 and D2, configure HMAC-SHA-256 based authentication using the same shared secret, **$3cre7!!**, on R1 interface G0/0/1.1, R3 interface G0/0/1, and D2 interfaces G1/0/1 and G1/0/11. Note that EIGRP adjacency will be lost until both ends of a link are configured.

```
R1(config)# router eigrp EIGRP_IPV6
R1(config-router)# address-family ipv6 unicast autonomous-system 43
R1(config-router-af)# af-interface g0/0/1.1
R1(config-router-af-interface)# authentication mode hmac-sha-256 $3cre7!!
R1(config-router-af-interface)# end
```

f. Use the **show ipv6 eigrp interface detail** command to verify authentication is in place and what type it is.

```
R1# show ipv6 eigrp interface detail | section Gi0/0/1.1
Gi0/0/1.1              1        0/0       0/0        3        0/050         0
  Hello-interval is 5, Hold-time is 15
  Split-horizon is enabled
  Next xmit serial <none>
  Packetized sent/expedited: 27/1
  Hello's sent/expedited: 582/3
  Un/reliable mcasts: 0/28  Un/reliable ucasts: 35/11
  Mcast exceptions: 0  CR packets: 0  ACKs suppressed: 0
  Retransmissions sent: 2  Out-of-sequence rcvd: 0
  Topology-ids on interface - 0
  Authentication mode is HMAC-SHA-256, key-chain is not set
  Topologies advertised on this interface:  base
  Topologies not advertised on this interface:
```

Step 6. Manipulate load balancing with variance.

By default, load balancing occurs only over equal-cost paths. EIGRP supports up to four equal cost paths by default but can be configured to support as many as 32 with the **maximum-paths** command.

EIGRP has the added capability to load balance over unequal-cost paths. Load balancing is controlled by the **variance** parameter. Its value is a multiplier that is used to determine how to deal with multiple paths to the same destination.

Variance is set to 1 by default, so any paths up to the configured maximum number of paths that have an FD equal to the best current FD are also offered to the routing table. This provides equal cost load balancing.

The variance parameter can also be set to zero, which dictates that no load balancing takes place.

The variance parameter can be adjusted so that paths that have an FD that is less than or equal to variance times current best FD are also considered as successors and installed into the routing table. There is an extremely important differentiation here -- to be a <u>feasible successor</u>, the <u>RD</u> of a path must be less than the current best FD. To be considered for <u>unequal load balancing</u>, the FD of the feasible successor is multiplied by the variance value, and if the product of this calculation is less than the current best FD, the feasible successor is promoted to successor.

There are two caveats; first, only feasible successors are considered and second, with unequal cost load balancing, traffic share is proportional to the best metric in the routing table for the given path.

Note: Keep in mind that your routing table may be different than the one created by the examples in this lab. If your results are different, examine them carefully to determine why so that you can thoroughly understand how EIGRP is operating.

a. Before manipulating variance, R3 needs to see individual routes from R2 instead of a summary. Therefore, remove the summary routes on R2 so that it will again advertise more specific EIGRP routes to R3.

```
R2(config-if)# interface g0/0/0
R2(config-if)# no ipv6 summary-address eigrp 43 ::/0
R2(config-if)# exit
R2(config)# interface g0/0/1
R2(config-if)# no ipv6 summary-address eigrp 43 ::/0
R2(config-if)# end
```

b. On R3, verify that there are again two equal-cost paths to 2001:db8:acad:2::64. In this example, the IPv6 address must be entered in ALL CAPS.

```
R3# show ipv6 route eigrp | section 2001:DB8:ACAD:2::/64
D    2001:DB8:ACAD:2::/64 [90/20480]
       via FE80::D1:2, GigabitEthernet0/0/1
       via FE80::2:2, GigabitEthernet0/0/0
```

c. To change this and allow for the demonstration of variance, change the interface bandwidth for the R2 interfaces G0/0/0 and G0/0/1 to 800000.

```
R2(config)# interface g0/0/0
R2(config-if)# bandwidth 800000
R2(config-if)# exit
R2(config)# interface g0/0/1
R2(config-if)# bandwidth 800000
R2(config-if)# end
```

d. When you examine the routing table on R3, you see that there is no load balancing occurring. All destinations have a single path.

```
R3# show ipv6 route eigrp | section 2001:DB8:ACAD:2::/64
D    2001:DB8:ACAD:2::/64 [90/20480]
       via FE80::D1:2, GigabitEthernet0/0/1
```

e. However, we know there are multiple paths in the network. The first consideration for manipulating variance is that it only works with feasible successors. Examining the topology table on R3 shows that there is a feasible successor for the 2001:db8:acad:2::/64 network. The route via fe80::2:2 out the G0/0/0 interface has an RD less than the FD for the current successor.

```
R3# show ipv6 eigrp topology | section 2001:DB8:ACAD:2::/64
P 2001:DB8:ACAD:2::/64, 1 successors, FD is 2621440
        via FE80::D1:2 (2621440/1966080), GigabitEthernet0/0/1
        via FE80::2:2 (2785280/2129920), GigabitEthernet0/0/0
```

f. To use the other route for unequal cost load balancing, we can set the variance parameter to 2. This will mean that any path with an RD less than or equal to 5242880 will qualify as a successor (2 x 2621440 = 5242880).

```
R3(config)# router eigrp EIGRP_IPV6
R3(config-router)# address-family ipv6 unicast autonomous-system 43
R3(config-router-af)# topology base
R3(config-router-af-topology)# variance 2
```

```
R3(config-router-af-topology)# exit
R3(config-router-af)#  exit
R3(config-router)# exit
R3(config)# end
```

g. The output of the **show ipv6 route eigrp** command now displays two paths available to the 2001:db8:acad:2::/64 network. Notice that the routes have different metrics, but are listed and used just the same. Also, notice adding variance 2 adds a second path to the 2001:db8:cafe:1::/64 network.

```
R3# show ipv6 route eigrp
<output omitted>
D    2001:DB8:FF:999::/64 [90/2570240]
        via FE80::2:2, GigabitEthernet0/0/0
D    2001:DB8:ABCD::/56 [5/1280]
        via Null0, directly connected
D    2001:DB8:ACAD:2::/64 [90/20480]
        via FE80::D1:2, GigabitEthernet0/0/1
        via FE80::2:2, GigabitEthernet0/0/0
D    2001:DB8:ACAD:3::/64 [90/15360]
        via FE80::D1:2, GigabitEthernet0/0/1
D    2001:DB8:CAFE:1::/64 [90/16640]
        via FE80::2:2, GigabitEthernet0/0/0
        via FE80::D1:2, GigabitEthernet0/0/1
D    2001:DB8:CEDE::/64 [90/2570240]
        via FE80::2:2, GigabitEthernet0/0/0
D    2001:DB8:CEDE:1::/64 [90/2570240]
        via FE80::2:2, GigabitEthernet0/0/0
```

Step 7.   Filter EIGRP routes using a prefix list.

In this step, you will configure a filter at R2 to block propagation of the network 2001:db8:cafe:1::/64 to R3.

a. On R3, issue the **show ipv6 route 2001:db8:cafe:1::/64** command. The output should list two successors, one via fe80::2:2 and one via fe90::d2:2. We want to filter route via fe80::2:1.

```
R3# show ipv6 route 2001:db8:cafe:1::/64
Routing entry for 2001:DB8:CAFE:1::/64
  Known via "eigrp 43", distance 90, metric 16640, type internal
  Route count is 2/2, share count 0
  Routing paths:
    FE80::2:2, GigabitEthernet0/0/0
      From FE80::2:2
      Last updated 00:05:00 ago
    FE80::D1:2, GigabitEthernet0/0/1
      From FE80::D1:2
      Last updated 00:04:35 ago
```

b. On R2, create an IPv6 prefix list that matches the 2001:db8:cafe:1::/64 network.

```
R2(config)# ipv6 prefix-list DROP-CAFE-1 seq 10 deny 2001:db8:cafe:1::/64
R2(config)# ipv6 prefix-list DROP-CAFE-1 seq 20 permit ::/0
R2(config)# end
```

c. On R2, apply the prefix list as a distribute list for updates exiting the G0/0/1 interface towards R3.

```
R2(config)# ipv6 router eigrp 43
R2(config-rtr)# distribute-list prefix-list DROP-CAFE-1 out g0/0/1
R2(config-rtr)# exit
R2(config)# end
```

d. On R3, issue the **show ipv6 route 2001:db8:cafe:1::/64** command. The output should now list one successor fe80::d1:2. Verify that R3 no longer has a successor route via fe80::2:2 to the 2001:db8:cafe:1::/64 network.

```
R3# show ipv6 route 2001:db8:cafe:1::/64
Routing entry for 2001:DB8:CAFE:1::/64
  Known via "eigrp 43", distance 90, metric 20480, type internal
  Route count is 1/1, share count 0
  Routing paths:
    FE80::D1:2, GigabitEthernet0/0/1
      From FE80::D1:2
      Last updated 00:15:29 ago
```

Router Interface Summary Table

Router Model	Ethernet Interface #1	Ethernet Interface #2	Serial Interface #1	Serial Interface #2
1800	Fast Ethernet 0/0 (F0/0)	Fast Ethernet 0/1 (F0/1)	Serial 0/0/0 (S0/0/0)	Serial 0/0/1 (S0/0/1)
1900	Gigabit Ethernet 0/0 (G0/0)	Gigabit Ethernet 0/1 (G0/1)	Serial 0/0/0 (S0/0/0)	Serial 0/0/1 (S0/0/1)
2801	Fast Ethernet 0/0 (F0/0)	Fast Ethernet 0/1 (F0/1)	Serial 0/1/0 (S0/1/0)	Serial 0/1/1 (S0/1/1)
2811	Fast Ethernet 0/0 (F0/0)	Fast Ethernet 0/1 (F0/1)	Serial 0/0/0 (S0/0/0)	Serial 0/0/1 (S0/0/1)
2900	Gigabit Ethernet 0/0 (G0/0)	Gigabit Ethernet 0/1 (G0/1)	Serial 0/0/0 (S0/0/0)	Serial 0/0/1 (S0/0/1)
4221	Gigabit Ethernet 0/0/0 (G0/0/0)	Gigabit Ethernet 0/0/1 (G0/0/1)	Serial 0/1/0 (S0/1/0)	Serial 0/1/1 (S0/1/1)
4300	Gigabit Ethernet 0/0/0 (G0/0/0)	Gigabit Ethernet 0/0/1 (G0/0/1)	Serial 0/1/0 (S0/1/0)	Serial 0/1/1 (S0/1/1)

Note: To find out how the router is configured, look at the interfaces to identify the type of router and how many interfaces the router has. There is no way to effectively list all the combinations of configurations for each router class. This table includes identifiers for the possible combinations of Ethernet and Serial interfaces in the device. The table does not include any other type of interface, even though a specific router may contain one. An example of this might be an ISDN BRI interface. The string in parenthesis is the legal abbreviation that can be used in Cisco IOS commands to represent the interface.

Device Config Final – Notes

5.1.3 Lab - Troubleshoot EIGRP for IPv6

Topology

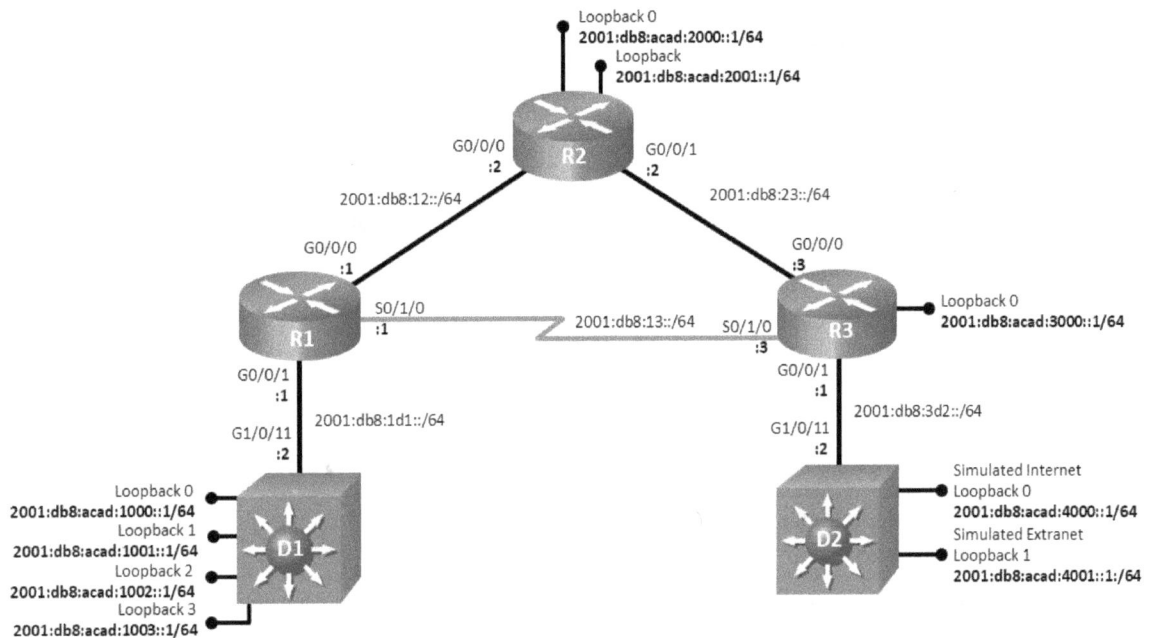

Addressing Table

Device	Interface	IPv6 Address/Prefix Length	Link-Local Address
R1	G0/0/0	2001:db8:12::1/64	fe80::1:1
	G0/0/1	2001:db8:1d1::1/64	fe80::1:2
	S0/1/0	2001:db8:13::1/64	fe80::1:3
R2	G0/0/0	2001:db8:12::2/64	fe80::2:1
	G0/0/1	2001:db8:23::2/64	fe80::2:2
	Loopback 0	2001:db8:acad:2000::1/64	fe80::2:3
	Loopback 1	2001:db8:acad:2001::1/64	fe80::2:4
R3	G0/0/0	2001:db8:23::3/64	fe80::3:1
	G0/0/1	2001:db8:3d2::1/64	fe80::3:2
	S0/1/0	2001:db8:13::3/64	fe80::3:3
	Loopback 0	2001:db8:acad:3000::1/64	fe80::3:4
D1	G1/0/11	2001:db8:1d1::2/64	fe80::d1:1
	Loopback 0	2001:db8:acad:1000::1/64	fe80::d1:2
	Loopback 1	2001:db8:acad:1001::1/64	fe80::d1:3
	Loopback 2	2001:db8:acad:1002::1/64	fe80::d1:4
	Loopback 3	2001:db8:acad:1003::1/64	fe80::d1:5
D2	G1/0/11	2001:db8:3d2::2/64	fe80::d2:1
	Loopback 0	2001:db8:acad:4000::1/64	fe80::d2:2
	Loopback 1	2001:db8:acad:4001::1/64	fe80::d2:3

Objectives

Troubleshoot network issues related to the configuration and operation of EIGRP for IPv6.

Background/Scenario

In this topology, R1, R2, R3, D1, and D2 are EIGRP neighbors. Switch D1 provides connectivity for a large branch of the network, R2 provides connectivity for a small branch, R3 supports a single LAN enclave, and switch D2 provides both an internet connection and an extranet connection. You will be loading configurations with intentional errors onto the network. Your tasks are to FIND the error(s), document your findings and the command(s) or method(s) used to fix them, FIX the issue(s) presented here and then test the network to ensure both of the following conditions are met:

1. the complaint received in the ticket is resolved

2. full reachability is restored

Note: The routers used with CCNP hands-on labs are Cisco 4221 with Cisco IOS XE Release 16.9.4 (universalk9 image). The switches used in the labs are Cisco Catalyst 3650 with Cisco IOS XE Release 16.9.4 (universalk9 image). Other routers, switches, and Cisco IOS versions can be used. Depending on the model and Cisco IOS version, the commands available and the output produced might vary from what is shown in the labs. Refer to the Router Interface Summary Table at the end of the lab for the correct interface identifiers.

Note: Make sure that the switches have been erased and have no startup configurations. If you are unsure, contact your instructor.

Required Resources

- 3 Routers (Cisco 4221 with Cisco IOS XE Release 16.9.4 universal image or comparable)

- 2 Switches (Cisco 3560 with Cisco IOS XE Release 16.9.4 universal image or comparable)

- 1 PC (Choice of operating system with terminal emulation program installed)

- Console cables to configure the Cisco IOS devices via the console ports

- Ethernet and serial cables as shown in the topology

Instructions

Part 1: Trouble Ticket 5.1.3.1

Scenario:

You took a spring break vacation, leaving the junior network administrator with some seemingly simple tasks. The routing table was to be reduced in size while the extranet on D2 and the enclave on R3 would not be visible in the routing table at all. Routing to the internet via D2 should work. Future additions to the large branch at R1 should be accounted for in the routing table so that when those networks are added (the network addresses are 2001:db8:acad:1004::/64, 2001:db8:acad:1005::/64, 2001:db8:acad:1006::/64, and 2001:db8:acad:1007::/64), there is no change to the routing table. The junior network administrator was not very successful. Now it is up to you to find and fix the misconfigurations before the CIO returns.

Use the commands listed below to load the configuration files for this trouble ticket:

Device	Command
R1	`copy flash:/enarsi/5.1.3.1-r1-config.txt run`
R2	`copy flash:/enarsi/5.1.3.1-r2-config.txt run`
R3	`copy flash:/enarsi/5.1.3.1-r3-config.txt run`
D1	`copy flash:/enarsi/5.1.3.1-d1-config.txt run`
D2	`copy flash:/enarsi/5.1.3.1-d2-config.txt run`

- Passwords on all devices are **cisco12345**. If a username is required, use **admin**.
- When you have fixed the ticket, change the MOTD on EACH DEVICE using the following command:

  banner motd # This is $(hostname) FIXED from ticket <ticket number> #
- Then save the configuration by issuing the **wri** command (on each device).
- Inform your instructor that you are ready for the next ticket.
- After the instructor approves your solution for this ticket, issue the **reset.now** privileged EXEC command. This script will clear your configurations and reload the devices.

Router Interface Summary Table

Router Model	Ethernet Interface #1	Ethernet Interface #2	Serial Interface #1	Serial Interface #2
1800	Fast Ethernet 0/0 (F0/0)	Fast Ethernet 0/1 (F0/1)	Serial 0/0/0 (S0/0/0)	Serial 0/0/1 (S0/0/1)
1900	Gigabit Ethernet 0/0 (G0/0)	Gigabit Ethernet 0/1 (G0/1)	Serial 0/0/0 (S0/0/0)	Serial 0/0/1 (S0/0/1)
2801	Fast Ethernet 0/0 (F0/0)	Fast Ethernet 0/1 (F0/1)	Serial 0/1/0 (S0/1/0)	Serial 0/1/1 (S0/1/1)
2811	Fast Ethernet 0/0 (F0/0)	Fast Ethernet 0/1 (F0/1)	Serial 0/0/0 (S0/0/0)	Serial 0/0/1 (S0/0/1)
2900	Gigabit Ethernet 0/0 (G0/0)	Gigabit Ethernet 0/1 (G0/1)	Serial 0/0/0 (S0/0/0)	Serial 0/0/1 (S0/0/1)
4221	Gigabit Ethernet 0/0/0 (G0/0/0)	Gigabit Ethernet 0/0/1 (G0/0/1)	Serial 0/1/0 (S0/1/0)	Serial 0/1/1 (S0/1/1)
4300	Gigabit Ethernet 0/0/0 (G0/0/0)	Gigabit Ethernet 0/0/1 (G0/0/1)	Serial 0/1/0 (S0/1/0)	Serial 0/1/1 (S0/1/1)

Note: To find out how the router is configured, look at the interfaces to identify the type of router and how many interfaces the router has. There is no way to effectively list all the combinations of configurations for each router class. This table includes identifiers for the possible combinations of Ethernet and Serial interfaces in the device. The table does not include any other type of interface, even though a specific router may contain one. An example of this might be an ISDN BRI interface. The string in parenthesis is the legal abbreviation that can be used in Cisco IOS commands to represent the interface.

Device Config Final – Notes

6.1.2 Lab - Implement Single-Area OSPFv2

Topology

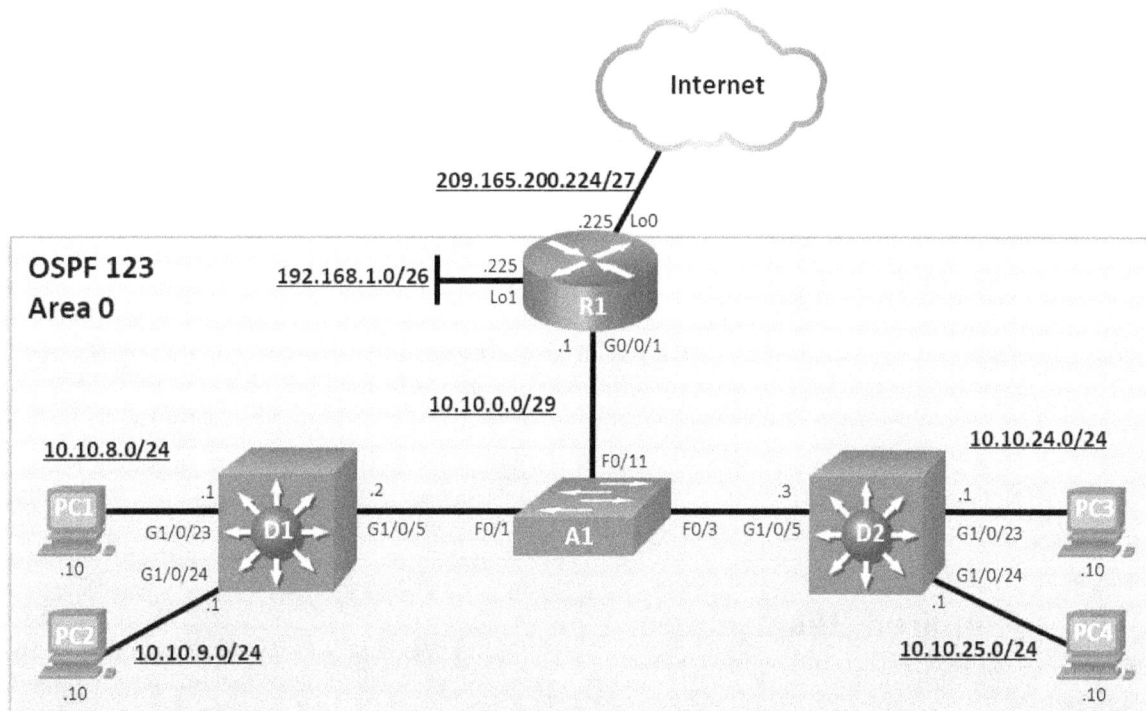

Addressing Table

Device	Interface	IPv4 Address
R1	G0/0/1	10.10.0.1/29
	Loopback0	209.165.200.225/27
	Loopback1	192.168.1.1/26
D1	G1/0/5	10.10.0.2/29
	G1/0/23	10.10.8.1/24
	G1/0/24	10.10.9.1/24
D2	G1/0/5	10.10.0.3/29
	G1/0/23	10.10.24.1/24
	G1/0/24	10.10.25.1/24
PC1	NIC	10.10.8.10/24
PC2	NIC	10.10.9.10/24
PC3	NIC	10.10.24.10/24
PC4	NIC	10.10.25.10/24

Objectives

Part 1: Build the Network and Configure Basic Device Settings and Interface Addressing

Part 2: Configure and Verify Single Area OSPF for IPv4 on R1, D1, and D2

Part 3: Configure Default Route Propagation on R1 and Verify the Propagation

Part 4: Implement OSPF Network Optimizing Features

Part 5: DR and BDR Placement

Background/Scenario

In this lab, you will configure single-area OSPF version 2 for IPv4 on a multiaccess Ethernet LAN. This lab was specifically designed to use two Layer 3 switches instead of three routers to highlight how a Layer 3 switch can also be used to provide routing services.

Note: This lab is an exercise in developing, deploying, and verifying how OSPF operates and does not reflect networking best practices.

Note: The router used with CCNP hands-on labs is a Cisco 4221and the two Layer 3 switches are Catalyst 3560 switches. Other Layer 3 switches and Cisco IOS versions can be used. Depending on the model and Cisco IOS version, the commands available and the output produced might vary from what is shown in the labs.

Note: Make sure that the switches have been erased and have no startup configurations. If you are unsure, contact your instructor.

Required Resources

- 1 Router (Cisco 4221 with Cisco IOS XE Release 16.9.4 universal image or comparable)
- 2 Switches (Cisco 3650 with Cisco IOS XE release 16.9.4 universal image or comparable)
- 1 Switch (Cisco 2960 with Cisco IOS Release 15.2(2) lanbasek9 image or comparable)
- 4 PCs (Windows with terminal emulation program, such as Tera Term)
- Console cables to configure the Cisco IOS devices via the console ports
- Ethernet cables as shown in the topology

Instructions

Part 1: Build the Network and Configure Basic Device Settings and Interface Addressing

In Part 1, you will set up the network topology and configure basic settings and interface addressing on the router and Layer 3 switches.

Note: The Layer 2 switch should only have a default configuration.

Step 1.   Cable the network as shown in the topology.

Attach the devices as shown in the topology diagram, and cable as necessary.

Step 2.   Configure basic settings for the router and the two Layer 3 switches.

a.   Console into each router and Layer 3 switch, enter global configuration mode, and apply the basic settings and interface addressing using the following startup configurations for each device.

Router R1

```
hostname R1
no ip domain lookup
line con 0
 logging sync
 exec-time 0 0
 exit
interface Loopback0
 ip address 209.165.200.225 255.255.255.224
 no shut
 exit
interface Loopback1
 ip address 192.168.1.1 255.255.255.192
 no shut
 exit
interface GigabitEthernet0/0/1
 ip address 10.10.0.1 255.255.255.248
 no shut
 exit
```

Switch D1

```
hostname D1
no ip domain lookup
line con 0
 logging sync
 exec-time 0 0
 exit
interface g1/0/5
 no switchport
 ip address 10.10.0.2 255.255.255.248
 no shut
 exit
interface g1/0/23
 no switchport
 ip address 10.10.8.1 255.255.255.0
 no shut
 exit
interface g1/0/24
 no switchport
 ip address 10.10.9.1 255.255.255.0
 no shut
 exit
```

Switch D2

```
hostname D2
no ip domain lookup
line con 0
 logging sync
 exec-time 0 0
 exit
interface g1/0/5
 no switchport
 ip address 10.10.0.3 255.255.255.248
 no shut
 exit
interface g1/0/23
 no switchport
 ip address 10.10.24.1 255.255.255.0
 no shut
 exit
interface g1/0/24
 no switchport
 ip address 10.10.25.1 255.255.255.0
 no shut
 exit
```

b. Save the running configuration to startup-config.

c. Verify the interface status using the **show ip interface brief** command.

```
R1# show ip interface brief | include manual
GigabitEthernet0/0/1    10.10.0.1       YES manual up        up
Loopback0               209.165.200.225 YES manual up        up
Loopback1               192.168.1.1     YES manual up        up


D1# show ip interface brief | include manual
GigabitEthernet1/0/5    10.10.0.2       YES manual up        up
GigabitEthernet1/0/23,  10.10.8.1       YES manual up        up
GigabitEthernet1/0/24   10.10.9.1       YES manual up        up


D2# show ip interface brief | include manual
GigabitEthernet1/0/5    10.10.0.3       YES manual up        up
GigabitEthernet1/0/23   10.10.24.1      YES manual up        up
GigabitEthernet1/0/24   10.10.25.1      YES manual up        up
```

d. Verify direct connectivity between the highlighted IP addresses of R1, D1, and D2.

```
R1# ping 10.10.0.2
Type escape sequence to abort.
Sending 5, 100-byte ICMP Echos to 10.10.0.2, timeout is 2 seconds:
..!!!
Success rate is 60 percent (3/5), round-trip min/avg/max = 2/2/3 ms


R1# ping 10.10.0.3
Type escape sequence to abort.
```

```
Sending 5, 100-byte ICMP Echos to 10.10.0.3, timeout is 2 seconds:
..!!!
Success rate is 60 percent (3/5), round-trip min/avg/max = 2/2/3 ms
```

All three devices should be able to reach the other directly connected networks (i.e., 10.10.0.0/29). Troubleshoot if necessary.

Part 2: Configure Single-Area OSPFv2

In this part, you will implement single-area OSPF on a multiaccess Ethernet network.

OSPF can be enabled using the traditional **network** router config command and wildcard mask. The wildcard mask enables the configuration to be as specific or vague as necessary. For example:

- **network** *ip-address* **0.0.0.0 area** *area-id* – Configuring the **network** statement with an IP address explicitly enables OSPF on that interface.

- **network** *network wildcard-mask* **area** *area-id* – The wildcard mask can explicitly match a subnet, or it can be less specific to match several subnets as required.

- **network 0.0.0.0 255.255.255.255 area** *area-id* – This is the vaguest method as the 0.0.0.0 network with 255.255.255.255 wildcard mask matches all enabled interfaces.

An alternate method to using the **network** router configuration command is to use the interface specific method. Instead of the **network** statement, an interface is enabled for OSPF using the **ip ospf** *process-id* **area** *area-id* interface configuration command. Although simpler to use, the disadvantage is that the configuration is not centralized and increases in complexity as the number of interfaces on the routers increases.

Note: There is a newer method to configure OSPF using address families. Address families are covered in OSPFv3 and in CCNP Enterprise: Advanced Routing.

Step 1. Implement OSPF on D1 using Explicit IP addresses.

D1 will advertise its OSPF networks using the OSPF **network** *ip-address* **0.0.0.0 area** *area-id* command method and quad-zero wildcard mask. This enables OSPF for Area 0 only on the interfaces that explicitly match the IP addresses configured.

a. Layer 3 switches are not enabled to perform routing by default. Therefore, routing must be enabled using the **ip routing** global configuration command.

```
D1(config)# ip routing
```

b. Next, enter the OSPF router configuration mode using process ID 123.

```
D1(config)# router ospf 123
```

c. When using the quad-zero method, it is not necessary to calculate the actual wildcard mask. You simply advertise the IP address of the interface with a quad-zero wildcard mask and OSPF will advertise using the subnet mask of the interface. Configure OSPF to advertise the network address of the G1/0/5 interface (i.e., 10.10.0.2) with the quad-zero mask.

```
D1(config-router)# network 10.10.0.2 0.0.0.0 area 0
```

d. Next, enable OSPF on the G1/0/23 and G1/0/24 interfaces using a quad-zero mask.

```
D1(config-router)# network 10.10.8.1 0.0.0.0 area 0
D1(config-router)# network 10.10.9.1 0.0.0.0 area 0
```

These networks are now being advertised to other OSPF routers.

e. Verify the OSPF configuration on D1 using the **show ip protocols** command.

```
D1# show ip protocols
*** IP Routing is NSF aware ***

Routing Protocol is "ospf 123"
  Outgoing update filter list for all interfaces is not set
  Incoming update filter list for all interfaces is not set
  Router ID 10.10.9.1
  Number of areas in this router is 1. 1 normal 0 stub 0 nssa
  Maximum path: 4
  Routing for Networks:
    10.10.0.2 0.0.0.0 area 0
    10.10.8.1 0.0.0.0 area 0
    10.10.9.1 0.0.0.0 area 0
  Routing Information Sources:
    Gateway         Distance      Last Update
  Distance: (default is 110)
```

The OSPF router ID chosen was the highest active IPv4 address configured on D1. The Routing for Networks section in the output above confirms that the configured statements are accurately advertising the D1 networks.

Step 2.   Implement OSPF on D2 using Wildcard Masks.

D2 will advertise its OSPF networks using the **network** router configuration command and wildcard masks.

a. Like D1, D2 must be enabled for routing using the **ip routing** global configuration command.

```
D2(config)# ip routing
```

b. Next, enter the OSPF router configuration mode using process ID 123. Note that process IDs are only locally significant. Therefore, the process ID of other OSPF routers do not need to match. However, using the same process ID makes it simpler to remember and reduces potential configuration mistakes.

```
D2(config)# router ospf 123
```

c. Configure D2 to advertise the g1/0/5 /29 interface in OSPF area 0. The wildcard mask can be calculated by deducting the subnet mask (i.e., /29 = 255.255.255.248) from 255.255.255.255, resulting in a wildcard mask of 0.0.0.7.

```
D2(config-router)# network 10.10.0.0 0.0.0.7 area 0
D2(config-router)#
*Mar  1 00:16:46.465: %OSPF-5-ADJCHG: Process 123, Nbr 10.10.9.1 on
GigabitEthernet1/0/5 from LOADING to FULL, Loading Done
```

Notice the informational message confirming that D2 has established a neighbor relationship with D1 (i.e., 10.10.9.1).

d. Next, configure D2 to advertise the two /24 networks in OSPF area 0. This can be accomplished using two **network** statements with specific wildcard mask for each subnet.

```
D2(config-router)# network 10.10.24.0 0.0.0.255 area 0
D2(config-router)# network 10.10.25.0 0.0.0.255 area 0
```

Note: The two networks could also be enabled using the **network 10.10.24.0 0.0.254.255** statement instead.

There are no informational messages this time because these interfaces are not connected to other OSPF-enabled routers. However, these networks are now being advertised to other OSPF routers.

e. Verify the OSPF configuration on D2 using the **show ip protocols** command.

```
D2# show ip protocols
*** IP Routing is NSF aware ***

Routing Protocol is "ospf 123"
  Outgoing update filter list for all interfaces is not set
  Incoming update filter list for all interfaces is not set
  Router ID 10.10.25.1
  Number of areas in this router is 1. 1 normal 0 stub 0 nssa
  Maximum path: 4
  Routing for Networks:
    10.10.0.0 0.0.0.7 area 0
    10.10.24.0 0.0.0.255 area 0
    10.10.25.0 0.0.0.255 area 0
  Routing on Interfaces Configured Explicitly (Area 0):

  Routing Information Sources:
    Gateway         Distance      Last Update
    10.10.9.1            110      00:11:05
  Distance: (default is 110)
```

Again, the OSPF router ID chosen was the highest active IPv4 address configured on D2. The Routing for Networks section confirms that the configured statements are accurately advertising the D2 networks. We now also have another routing information source, 10.10.9.1 (i.e., D1).

Step 3.   Implement OSPF on R1 using the Interface Specific method.

R1 will use the OSPF interface specific method to advertise the Lo1 interface and the G0/0/1 interface. Interface Lo0 will be advertised in Part 3. The interface specific method is simple because there is no need to enter **network** statements or calculate wildcard masks. You simply enter the **ip ospf** *process-id* **area** *area-id* interface configuration command on an interface.

Note: Alternatively, the **network 0.0.0.0 255.255.255.255 area 0** router configuration command would be simpler. However, it would enable OSPF on all interfaces including the Lo0 interface that will be advertised in Part 3.

a. The loopback interface on R1 is only configured to simulate another network for OSPF to advertise. However, the default behavior of OSPF for loopback interfaces is to advertise a 32-bit host route. To ensure that the /26 network is advertised, the **ip ospf network point-to-point** interface command must be configured on the loopback 1 interface. Change the network type on the loopback interfaces so that they are advertised with the correct subnet.

```
R1(config)# interface loopback 1
R1(config-if)# ip ospf network point-to-point
```

b. Next enable the loopback interface for OSPF using the **ip ospf 123 area 0** command as shown.

```
R1(config-if)# ip ospf 123 area 0
R1(config-if)# exit
```

c. Enter interface G0/0/1 and enable it for OSPF.

```
R1(config)# interface g0/0/1
R1(config-if)# ip ospf 123 area 0
R1(config-if)# end
R1#
R1#
*Dec 22 18:32:48.873: %OSPF-6-DFT_OPT: Protocol timers for fast convergence are
Enabled.
R1#
*Dec 22 18:32:49.683: %OSPF-5-ADJCHG: Process 123, Nbr 10.10.9.1 on
GigabitEthernet0/0/1 from LOADING to FULL, Loading Done
*Dec 22 18:32:49.683: %OSPF-5-ADJCHG: Process 123, Nbr 10.10.25.1 on
GigabitEthernet0/0/1 from LOADING to FULL, Loading Done
*Dec 22 18:32:49.755: %SYS-5-CONFIG_I: Configured from console by console
```

Notice how the informational messages are confirming that neighbor adjacencies have been established with D1 (i.e., 10.10.9.1) and D2 (i.e., 10.10.25.1).

d. Verify the OSPF configuration on R1 using the **show ip protocols** command.

```
R1# show ip protocols | section ospf
Routing Protocol is "ospf 123"
  Outgoing update filter list for all interfaces is not set
  Incoming update filter list for all interfaces is not set
  Router ID 209.165.200.225
  Number of areas in this router is 1. 1 normal 0 stub 0 nssa
  Maximum path: 4
  Routing for Networks:
  Routing on Interfaces Configured Explicitly (Area 0):
    Loopback1
    GigabitEthernet0/0/1
  Routing Information Sources:
    Gateway         Distance      Last Update
  10.10.9.1           110         00:03:47
   10.10.25.1         110         00:03:47
      10.10.25.1         110        00:03:47
  Distance: (default is 110)
```

Again, the router ID chosen is the highest active IPv4 loopback address configured on R1. The Routing for Networks section confirms that routing was explicitly configured on the interfaces. It also displays a new routing source; 10.10.25.1.

Step 4. Assign Router IDs on R1, D1, and D2.

The OSPF router ID is dynamically assigned in order of preference:

- Manually configured using the **router-id** *router-id* router configuration command.

- If it is not manually assigned, then the highest enabled loopback IP address is used as the router ID.

- If there are no loopback interfaces configured, then the highest IP address of any active physical interfaces in the up state becomes the RID when the OSPF process initializes.

It is best to assign a static OSPF router ID for troubleshooting purposes.

To force an existing OSPF network to use the new router IP, the OSPF process must be reset using the **clear ip ospf process** privileged EXEC command.

a. Assign R1 the router ID 1.1.1.1 as shown.

```
R1(config)# router ospf 123
R1(config-router)# router-id 1.1.1.1
% OSPF: Reload or use "clear ip ospf process" command, for this to take effect
R1(config-router)# end
```

b. Next clear the OSPF process as shown.

```
R1# clear ip ospf process
Reset ALL OSPF processes? [no]: yes
R1#
*Dec 22 19:10:30.681: %OSPF-5-ADJCHG: Process 123, Nbr 10.10.9.1 on
GigabitEthernet0/0/1 from FULL to DOWN, Neighbor Down: Interface down or
detached
*Dec 22 19:10:30.681: %OSPF-5-ADJCHG: Process 123, Nbr 10.10.25.1 on
GigabitEthernet0/0/1 from FULL to DOWN, Neighbor Down: Interface down or
detached
*Dec 22 19:10:30.692: %OSPF-5-ADJCHG: Process 123, Nbr 10.10.9.1 on
GigabitEthernet0/0/1 from LOADING to FULL, Loading Done
*Dec 22 19:10:30.692: %OSPF-5-ADJCHG: Process 123, Nbr 10.10.25.1 on
GigabitEthernet0/0/1 from LOADING to FULL, Loading Done
```

c. Confirm that R1 is now using the new router ID as shown.

```
R1# show ip protocol | include Router ID
  Router ID 1.1.1.1
```

d. Repeat the process on D1 and D2. Use router ID 2.2.2.2 for D1 and 3.3.3.3 for D2. Also confirm that D1 and D2 are using the new router ID.

_____

_____

_____

_____

_____

Step 5. Verify OSPF settings on R1, D1, and D2.

It is imperative to know how to validate that OSPF is operating as configured. The **show running-config** command only displays the initial OSPF configuration. It does not validate the operation and functionality of OSPF.

Along with the **show ip protocols** command, there are several other useful OSPF-related **show** commands to verify that OSPF is operating as expected.

a. The **show ip route ospf** privileged EXEC command is used to verify the operation of OSPF. The command displays OSPF routes learned with an O, the administrative distance, the assigned metric, the next-hop IP address, and the local exit interface to reach the network.

```
R1# show ip route ospf | begin Gateway
Gateway of last resort is not set


      10.0.0.0/8 is variably subnetted, 6 subnets, 3 masks
O        10.10.8.0/24 [110/2] via 10.10.0.2, 00:14:49, GigabitEthernet0/0/1
O        10.10.9.0/24 [110/2] via 10.10.0.2, 00:14:49, GigabitEthernet0/0/1
O        10.10.24.0/24 [110/2] via 10.10.0.3, 00:14:49, GigabitEthernet0/0/1
O        10.10.25.0/24 [110/2] via 10.10.0.3, 00:14:49, GigabitEthernet0/0/1


D1# show ip route ospf | begin Gateway
Gateway of last resort is not set


      10.0.0.0/8 is variably subnetted, 5 subnets, 2 masks
O        10.10.24.0/24 [110/2] via 10.10.0.3, 00:15:54, GigabitEthernet1/0/5
O        10.10.25.0/24 [110/2] via 10.10.0.3, 00:15:54, GigabitEthernet1/0/5
      192.168.1.0/26 is subnetted, 1 subnets
O        192.168.1.0 [110/2] via 10.10.0.1, 00:15:54, GigabitEthernet1/0/5


D2# show ip route ospf | begin Gateway
Gateway of last resort is not set


      10.0.0.0/8 is variably subnetted, 5 subnets, 2 masks
O        10.10.8.0/24 [110/2] via 10.10.0.2, 00:16:32, GigabitEthernet1/0/5
O        10.10.9.0/24 [110/2] via 10.10.0.2, 00:16:32, GigabitEthernet1/0/5
      192.168.1.0/26 is subnetted, 1 subnets
O        192.168.1.0 [110/2] via 10.10.0.1, 00:16:32, GigabitEthernet1/0/5
```

b. Use the show ip **ospf interface [brief]** command to verify which interfaces are enabled for OSPF, process ID, Area ID, and state. A missing interface could be the result of an incorrect **network** statement, IP addressing problem, or a disabled interface.

```
R1# show ip ospf interface brief
```

Interface	PID	Area	IP Address/Mask	Cost	State	Nbrs F/C
Lo1	123	0	192.168.1.1/26	1	P2P	0/0
Gi0/0/1	123	0	10.10.0.1/29	1	DROTH	2/2

```
D1# show ip ospf interface brief
```

Interface	PID	Area	IP Address/Mask	Cost	State	Nbrs F/C
Gi1/0/24	123	0	10.10.9.1/24	1	DR	0/0
Gi1/0/23	123	0	10.10.8.1/24	1	DR	0/0
Gi1/0/5	123	0	10.10.0.2/29	1	DR	2/2

```
D2# show ip ospf interface brief
```

Interface	PID	Area	IP Address/Mask	Cost	State	Nbrs F/C
Gi1/0/24	123	0	10.10.25.1/24	1	DR	0/0
Gi1/0/23	123	0	10.10.24.1/24	1	DR	0/0
Gi1/0/5	123	0	10.10.0.3/29	1	BDR	2/2

Note: Omitting the "**brief**" keyword displays detailed information about the OSPF enabled interfaces.

This State field defines the state of the link and can be:

- **DR** - This is the Designated Router on the multiaccess network (i.e., Ethernet) to which this interface is connected. The DR establishes OSPF adjacencies with all other routers on the network.

- **BDR** - This is the Backup Designated Router on the multiaccess network to which this interface is connected. Like the DR, the BDR establishes adjacencies with all other routers on the broadcast network.

- **DROTH** - This is a DROTHER. It is neither the DR nor the BDR on the multiaccess network. All non-DRs and BDRs on the broadcast network would be DROTHERs and establish adjacencies only with the DR and the BDR.

- **P2P** - This is an OSPF point-to-point interface and does not require a DR or BDR. In this state, the interface is fully functional and starts exchanging hello packets with all of its neighbors.

c. Use the **show ip ospf neighbor [detail]** command to verify which OSPF neighbor your device has established adjacencies with, the state, the next-hop IP address, and the exit interface to use. The reasons a neighbor may not be appearing include RIDs that are not unique, interconnecting interfaces that are not on a common subnet, MTU values that do not match, Area ID that is not correct, Hello and dead interval timers that do not match, or authentication type / credentials that do not match. The following output confirms that our devices have correctly established adjacencies. The output for R1 is shown below. Repeat the command for D1 and D2.

```
R1# show ip ospf neighbor

Neighbor ID     Pri   State       Dead Time   Address      Interface
2.2.2.2          1    FULL/DR     00:00:33    10.10.0.2    GigabitEthernet0/0/1
3.3.3.3          1    FULL/BDR    00:00:37    10.10.0.3    GigabitEthernet0/0/1
```

_____

_____

_____

_____

_____

d. Other OSPF validation commands include the **show ip ospf, show ip ospf topology-info, show ip ospf database** commands. Use these commands now and identify what types of information they generate which may be useful to know when troubleshooting a network.

Part 3: Configure and Verify the Advertising of a Default Route

In this part, you will configure a default static route to the internet on R1. R1 will then propagate the default route to other OSPF routers as an external Type 2 OSPF route (i.e., O*E2).

Propagating a default is the most efficient method to provide a consistent default gateway to all OSPF-enabled devices.

Step 1.   Configure default route advertisement on R1.

    a.   R1 will be the gateway of last resort for the OSPF internetwork. In our sample topology, the internet is simulated using the Lo0 interface. Configure a static default route out of the Lo0 interface on R1.

```
R1(config)# ip route 0.0.0.0 0.0.0.0 lo0
%Default route without gateway, if not a point-to-point interface, may impact
performance
```

Note: Disregard the informational message. In a production environment, a valid physical interface would be used to provide default gateway services.

    b.   Enter OSPF router configuration mode and use the **default-information originate** **[always]** [*metric metric-value*] [*metric-type type-value*] command to enable default route propagation. The **always** keyword advertises a default route even if a static default route does not exist while the route metric and metric type can be changed. R1 is configured to propagate the default route.

```
R1(config)# router ospf 123
R1(config-router)# default-information originate
R1(config-router)# end
```

Step 2.   Verify the default route advertisement.

    a.   Verify the routing table on R1.

```
R1# show ip route static | begin Gateway
Gateway of last resort is 0.0.0.0 to network 0.0.0.0

S*    0.0.0.0/0 is directly connected, Loopback0
```

    b.   Verify the routing table on D1 and D2.

```
D1# show ip route | include Gateway|0/0
Gateway of last resort is 10.10.0.1 to network 0.0.0.0
O*E2 0.0.0.0/0 [110/1] via 10.10.0.1, 00:06:55, GigabitEthernet1/0/5

D2# show ip route | include Gateway|0/0
Gateway of last resort is 10.10.0.1 to network 0.0.0.0
O*E2 0.0.0.0/0 [110/1] via 10.10.0.1, 00:09:36, GigabitEthernet1/0/5
```

Part 4: Implement OSPF Network Optimizing Features

In this part, you will configure OSPF optimizing features including:

- Passive interfaces
- Link costs
- Reference bandwidth
- Hello and Dead interval timers

Step 1.   Configure passive interfaces on R1, D1 and D2.

A passive interface does not send out OSPF messages or process any received OSPF packets. However, the passive interface network segment is still added to the link state database (LSDB) and advertised out of non-passive interfaces. For security reasons, LAN interfaces which are not connected to other OSPF routers should be passive.

There are two approaches to identify passive interfaces.

- Use the **passive-interface** *interface-id* router configuration command to make an interface passive. This is a good approach to use when there are only a few interfaces to make passive.

- Use the **passive-interface default** router config command to make all interfaces passive, and then make some interfaces not passive using the **no passive-interface** *interface-id* command. This is a good approach to use when there are many interfaces to make passive, but only a few interfaces that should not be passive.

a. R1 only needs the Lo1 interface to be passive. The first approach is used to make the Loopback 1 interface passive. Enter OSPF router configuration mode and make the Lo1 interface passive as shown.

```
R1(config-if)# router ospf 123
R1(config-router)# passive-interface lo1
R1(config-router)# end
```

b. Verify which interfaces are passive using the **show ip protocols** command.

```
R1# show ip protocols | section ospf
Routing Protocol is "ospf 123"
  Outgoing update filter list for all interfaces is not set
  Incoming update filter list for all interfaces is not set
  Router ID 1.1.1.1
  It is an autonomous system boundary router
 Redistributing External Routes from,
  Number of areas in this router is 1. 1 normal 0 stub 0 nssa
  Maximum path: 4
  Routing for Networks:
  Routing on Interfaces Configured Explicitly (Area 0):
    Loopback1
    GigabitEthernet0/0/1
  Passive Interface(s):
Loopback1
  Loopback1
  Routing Information Sources:
    Gateway         Distance      Last Update
    3.3.3.3              110      10:45:59
    2.2.2.2              110      10:45:59
    10.10.9.1            110      10:54:49
    10.10.25.1           110      10:49:26
  Distance: (default is 110)
```

c. A Layer 3 switch can potentially have many interfaces that should be passive. For example, assume that D1 and D2 only require their G1/0/5 interface to not be passive. However, all other interfaces should be passive. Using the first approach would be very time-consuming. For this reason, the second approach will be used. All active interfaces will be rendered passive and only interface G1/0/5 will be re-enabled.

```
D1(config)# router ospf 123
D1(config-router)# passive-interface default
D1(config-router)#
```

```
*Mar  1 12:30:42.637: %OSPF-5-ADJCHG: Process 123, Nbr 1.1.1.1 on
GigabitEthernet1/0/5 from FULL to DOWN, Neighbor Down: Interface down or
detached
*Mar  1 12:30:42.637: %OSPF-5-ADJCHG: Process 123, Nbr 3.3.3.3 on
GigabitEthernet1/0/5 from FULL to DOWN, Neighbor Down: Interface down or
detached

D1(config-router)# no passive-interface g1/0/5
D1(config-router)#
*Mar  1 12:31:35.880: %OSPF-5-ADJCHG: Process 123, Nbr 1.1.1.1 on
GigabitEthernet1/0/5 from LOADING to FULL, Loading Done
D1(config-router)#
*Mar  1 12:31:39.445: %OSPF-5-ADJCHG: Process 123, Nbr 3.3.3.3 on
GigabitEthernet1/0/5 from LOADING to FULL, Loading Done
D1(config-router)# end
```

Notice the information messages stating that the OSPF adjacency with R1 and D2 transitioned to the DOWN state. Disabling the passive feature on interface G1/0/5 re-enabled the OSPF adjacency.

    d. Repeat the process on D2.

_____

_____

_____

_____

_____

Step 2. Adjust OSPF link costs.

The OSPF path metric is based on the cumulative interface cost to the network. OSPF assigns the OSPF link cost using the formula **Cost = Reference Bandwidth / Interface Bandwidth**. The default reference bandwidth is 100 Mbps, therefore, the default formula is **Cost = 100,000,000/Interface Bandwidth**. For example, a FastEthernet interface would be assigned a cost of 1 (i.e., 100,000,000 / 100,000,000).

However, the default reference bandwidth does not differentiate interfaces faster than FastEthernet. Therefore, OSPF assigns the identical cost of "1" to FastEthernet, Gigabit Ethernet, and 10 GE interfaces. OSPF makes no distinction that the Gig and 10GE interfaces are faster.

Use the **auto-cost reference-bandwidth** *bandwidth-mbps* router configuration command to change the reference bandwidth as follows:

- **auto-cost reference-bandwidth 100** – Assigns the default reference bandwidth to 100 Mbps which is the default setting. With this setting, FastEthernet = 1, GigabitEthernet = 1, and 10GE = 1.

- **auto-cost reference-bandwidth 1000** – Assigns the default reference bandwidth to 1 Gbps. With this setting, FastEthernet = 10, GigabitEthernet = 1, and 10GE = 1.

- **auto-cost reference-bandwidth 10000** – Assigns the default reference bandwidth to 10 Gbps. With this setting, FastEthernet = 100, GigabitEthernet = 10, and 10GE = 1.

Note: The **auto-cost reference-bandwidth** must be the same on all routers in the area. Otherwise sub-optimal routing may occur.

a. On R1, change the reference bandwidth to account for the Gigabit interfaces as shown.

```
R1(config)# router ospf 123
R1(config-router)# auto-cost reference-bandwidth 1000
% OSPF: Reference bandwidth is changed.
      Please ensure reference bandwidth is consistent across all routers.
R1(config-router)# end
```

b. Verify that the reference bandwidth has changed to account for the Gigabit interfaces using the **show ip ospf** command.

```
R1# show ip ospf | include Ref
 Reference bandwidth unit is 1000 mbps
```

c. Repeat the steps on D1 and D2 to change the reference bandwidth to account for the Gigabit interfaces.

_____

_____

_____

_____

d. Verify the routing table on D2 to see if the route metrics have increased.

```
R1# show ip route ospf | begin Gateway
Gateway of last resort is 0.0.0.0 to network 0.0.0.0

      10.0.0.0/8 is variably subnetted, 6 subnets, 3 masks
O        10.10.8.0/24 [110/20] via 10.10.0.2, 00:05:57, GigabitEthernet0/0/1
O        10.10.9.0/24 [110/20] via 10.10.0.2, 00:05:57, GigabitEthernet0/0/1
O        10.10.24.0/24 [110/20] via 10.10.0.3, 00:03:31, GigabitEthcrnct0/0/1
O        10.10.25.0/24 [110/20] via 10.10.0.3, 00:03:31, GigabitEthernet0/0/1


D1# show ip route ospf | begin Gateway
Gateway of last resort is 10.10.0.1 to network 0.0.0.0

      10.0.0.0/8 is variably subnetted, 5 subnets, 2 masks
O        10.10.24.0/24 [110/20] via 10.10.0.3, 00:02:48, GigabitEthernet1/0/5
O        10.10.25.0/24 [110/20] via 10.10.0.3, 00:02:48, GigabitEthernet1/0/5
      192.168.1.0/26 is subnetted, 1 subnets
O        192.168.1.0 [110/11] via 10.10.0.1, 00:02:48, GigabitEthernet1/0/5
O*E2 0.0.0.0/0 [110/1] via 10.10.0.1, 00:02:48, GigabitEthernet1/0/5


D2# show ip route ospf | begin Gateway
Gateway of last resort is 10.10.0.1 to network 0.0.0.0

      10.0.0.0/8 is variably subnetted, 5 subnets, 2 masks
O        10.10.8.0/24 [110/20] via 10.10.0.2, 00:00:10, GigabitEthernet1/0/5
O        10.10.9.0/24 [110/20] via 10.10.0.2, 00:00:10, GigabitEthernet1/0/5
      192.168.1.0/26 is subnetted, 1 subnets
O        192.168.1.0 [110/11] via 10.10.0.1, 00:00:10, GigabitEthernet1/0/5
O*E2 0.0.0.0/0 [110/1] via 10.10.0.1, 00:00:10, GigabitEthernet1/0/5
```

Step 3.   Alter Hello and Dead interval timers.

OSPF Hello messages are exchanged to establish a neighbor relationship and to ensure that adjacent OSPF neighbors are still available. OSPF uses a hello timer and a dead interval timer which is four times the hello timer.

When a router receives a Hello packet, the dead interval resets and starts to decrement again. If subsequent hello packets are not received before the OSPF dead interval timer reaches 0, the neighbor state is changed to down. The router then sends out the appropriate topology change LSA to all other peers and the SPF algorithm must be recalculated on all routers in the area.

The default OSPF hello timer interval varies based on the OSPF network type. On broadcast and point-to-point links, the default hello timer interval is 10 seconds and dead timer interval is 40 seconds. On non-broadcast multiaccess (NBMA) and point-to-multipoint networks, the default hello interval is 30 seconds with a dead timer interval of 120 seconds.

You can alter the hello timer interval with values between 1 and 65,535 seconds using the **ip ospf hello-interval** *seconds* interface configuration command.

The dead interval can be modified using the **ip ospf dead-interval** *seconds* interface configuration command. However, the command is really not required because changing the hello timer interval automatically modifies the default dead interval.

a.   On R1, change the hello interval on the G0/0/1 interface to 5 seconds and a dead interval time to 20 seconds.

```
R1(config)# interface g0/0/1
R1(config-if)# ip ospf hello-interval 5
R1(config-if)# ip ospf dead-interval 20
R1(config-if)# end
*Dec 23 08:52:07.961: %OSPF-5-ADJCHG: Process 123, Nbr 2.2.2.2 on
GigabitEthernet0/0/1 from FULL to DOWN, Neighbor Down: Dead timer expired
*Dec 23 08:52:30.471: %OSPF-5-ADJCHG: Process 123, Nbr 3.3.3.3 on
GigabitEthernet0/0/1 from FULL to DOWN, Neighbor Down: Dead timer expired
```

Notice how R1 has received OSPF adjacency change messages for D1 and D2. The reason is because OSPF timers must match between interconnecting peers. Therefore, the D1 and D2 GigabitEthernet1/0/5 interface must also be configured with the identical timers.

b.   Verify that the timers have changed on G0/0/1 using the **show ip ospf interface** command.

```
R1# show ip ospf interface g0/0/1 | include Timer
  Timer intervals configured, Hello 5, Dead 20, Wait 20, Retransmit 5
```

c.   Configure D1 with the identical hello and dead interval timers on GigabitEthernet1/0/5 and verify.

```
*Mar  1 15:12:45.159: %OSPF-5-ADJCHG: Process 123, Nbr 1.1.1.1 on
GigabitEthernet0/5 from FULL to DOWN, Neighbor Down: Dead timer expired
D1(config)# interface g1/0/5
D1(config-if)# ip ospf hello-interval 5
D1(config-if)# ip ospf dead-interval 20
D1(config-if)#
*Mar  1 15:18:25.779: %OSPF-5-ADJCHG: Process 123, Nbr 1.1.1.1 on
GigabitEthernet1/0/5 from LOADING to FULL, Loading Done
D1(config-if)# end
```

```
D1#
*Mar  1 15:18:53.201: %OSPF-5-ADJCHG: Process 123, Nbr 3.3.3.3 on
GigabitEthernet1/0/5 from FULL to DOWN, Neighbor Down: Dead timer expired

D1# show ip ospf interface g1/0/5 | include Timer
   Timer intervals configured, Hello 5, Dead 20, Wait 20, Retransmit 5
```

Notice the first OSPF adjacency change message indicating that D1 had lost adjacency with R1. After the commands are entered, the next OSPF adjacency change message indicates that the adjacency with R1 has been re-established. However, the second adjacency change message indicates that the adjacency with D2 has been lost because its timers are not matching.

Also notice that the dead interval was automatically adjusted without having to configure the **ip ospf dead-interval 20** command on the interface.

d. Configure D2 with the identical hello and dead interval timers on GigabitEthernet1/0/5 and verify.

```
*Mar  1 15:12:34.045: %OSPF-5-ADJCHG: Process 123, Nbr 1.1.1.1 on
GigabitEthernet1/0/5 from FULL to DOWN, Neighbor Down: Dead timer expired
*Mar  1 15:19:24.717: %OSPF-5-ADJCHG: Process 123, Nbr 2.2.2.2 on
GigabitEthernet1/0/5 from FULL to DOWN, Neighbor Down: Dead timer expired
D2#
D2# conf t
D2(config)# interface g1/0/5
D2(config-if)# ip ospf hello-interval 5
D2(config-if)# ip ospf dead-interval 20
*Mar  1 15:38:48.158: %OSPF-5-ADJCHG: Process 123, Nbr 1.1.1.1 on
GigabitEthernet1/0/5 from LOADING to FULL, Loading Done
*Mar  1 15:38:52.965: %OSPF-5-ADJCHG: Process 123, Nbr 2.2.2.2 on
GigabitEthernet1/0/5 from LOADING to FULL, Loading Done
D2(config-if)# end

D2# show ip ospf interface g1/0/5 | include Timer
   Timer intervals configured, Hello 5, Dead 20, Wait 20, Retransmit 5
```

Again, notice the existing OSPF adjacency change messages indicating that D2 had lost adjacency with R1 and D1.

After the commands are entered, the next OSPF adjacency change messages indicate adjacencies with R1 and D1 have been re-established. And again, the dead interval was automatically adjusted without having to configure the **ip ospf dead-interval 20** command on the interface.

Part 5: DR and BDR Placement

In this part, you will configure OSPF DR and BDR placement on the multiaccess network.

By default, an OSPF router tries to establish neighbor adjacencies with all other OSPF routers. This is a concern with large multiaccess (i.e. Ethernet) networks. For instance, 10 routers interconnected to the same Layer 2 switch would require a total of 45 adjacencies to be established. This can cause excessive OSPF traffic and waste router resources.

For this reason, OSPF routers interconnected to the same multiaccess network elect a designated router (DR) and a backup designated router (BDR). All non-DR and BDR routers are referred to as DROTHERS and only form adjacencies with DR and BDR routers. This reduces the total number of adjacencies and improves network operations.

DR and BDR are automatically elected during the last phase of the 2-Way OSPF neighbor state, before the ExStart state.

DR and BDR elections are conducted as follows:

1. An OSPF router interface with a priority greater than 0 attempts to become BDR on the link.

2. If no BDR exists, then it elects itself the BDR. If there is a tie with another router, the highest router ID is used.

3. If there is no DR, the BDR promotes itself as DR.

4. The neighbor with the next highest priority is elected BDR.

The DR and BDR are the central focal points on a multiaccess network. In a large network, it is advantageous to choose which router should be DR and BDR.

When all OSPF routers have the same OSPF priority, the election is based on the higher router ID. Altering router ID to choose DR/BDR routers may not be convenient. A better alternative is to alter the interface priority.

By default, all OSPF routers on a multiaccess network have a priority of 1 assigned. An interface priority can be changed using the **ip ospf priority** *value* interface configuration command. The value can be between 0 and 255. Setting the value to 0 ensures the router will never become a DR or BDR. Setting the value greater than the default value of 1 makes the router a candidate to become the DR or BDR.

Note: It may be necessary to use the **clear ip ospf process** to ensure the proper devices are elected.

Step 1.   Verify current DR and BDR selection.

In the topology, R1, D1, and D2 are interconnected on the same Ethernet network. Therefore, a DR/BDR election has already transpired. The easiest way to determine the interface role is by viewing the OSPF interface with the **show ip ospf neighbor** command.

   a.   On R1, verify the current DR/BDR status using the **show ip ospf neighbor** command.

```
R1# show ip ospf neighbor


Neighbor ID     Pri   State           Dead Time   Address     Interface
2.2.2.2          1    FULL/DROTHER    00:00:19    10.10.0.2   GigabitEthernet0/0/1
3.3.3.3          1    FULL/DR         00:00:18    10.10.0.3   GigabitEthernet0/0/1
```

From the perspective of R1, D1 (i.e., 2.2.2.2) is a DROTHER and D2 (i.e., router ID 3.3.3.3) is the DR. We must then assume that R1 is the BDR.

   b.   Verify the current status of R1 using the **show ip ospf interface G0/0/1** command.

```
R1# show ip ospf interface g0/0/1
GigabitEthernet0/0/1 is up, line protocol is up
  Internet Address 10.10.0.1/29, Interface ID 7, Area 0
  Attached via Interface Enable
  Process ID 123, Router ID 1.1.1.1, Network Type BROADCAST, Cost: 10
  Topology-MTID   Cost    Disabled    Shutdown      Topology Name
       0           10       no          no             Base
  Enabled by interface config, including secondary ip addresses
  Transmit Delay is 1 sec, State BDR, Priority 1
```

```
Designated Router (ID) 3.3.3.3, Interface address 10.10.0.3
Backup Designated router (ID) 1.1.1.1, Interface address 10.10.0.1
Timer intervals configured, Hello 5, Dead 20, Wait 20, Retransmit 5
  oob-resync timeout 40
  Hello due in 00:00:02
Supports Link-local Signaling (LLS)
Cisco NSF helper support enabled
IETF NSF helper support enabled
Can be protected by per-prefix Loop-Free FastReroute
Can be used for per-prefix Loop-Free FastReroute repair paths
Not Protected by per-prefix TI-LFA
Index 1/2/2, flood queue length 0
Next 0x0(0)/0x0(0)/0x0(0)
Last flood scan length is 1, maximum is 2
Last flood scan time is 0 msec, maximum is 1 msec
Neighbor Count is 2, Adjacent neighbor count is 2
  Adjacent with neighbor 2.2.2.2
  Adjacent with neighbor 3.3.3.3   (Designated Router)
Suppress hello for 0 neighbor(s)
```

The output confirms that R1 is the BDR and that D2 (i.e., 3.3.3.3) is the DR.

    c. Verify the current DR/BDR status on D1 and D2 using the **show ip ospf neighbor** command.

Step 2. Change DR and BDR selection.

It is sometimes advantageous to choose which router is selected as DR and BDR. For example, we will change the DR and BDR assignment as follows:

- R1 is currently the BDR but should be the DR using a priority of 255.

- D1 is currently a DROTHER but should be the BDR using the default priority.

- D2 is currently DR but should never become DR or BDR using a priority of 0.

    a. Starting on D2, enter interface G1/0/5 and set the priority to 0 as shown.

```
D2(config)# interface g1/0/5
D2(config-if)# ip ospf priority 0
D2(config-if)#
*Mar  1 17:23:20.195: %OSPF-5-ADJCHG: Process 123, Nbr 2.2.2.2 on
GigabitEthernet1/0/5 from LOADING to FULL, Loading Done
D2(config-if)# end
```

Notice the OSPF message. The reason is because D1 (i.e., 2.2.2.2) just assumed either the DR or BDR role and has established an adjacency with D2.

    b. Verify the current DR / BDR placement.

```
D2# show ip ospf neighbor
Neighbor ID     Pri   State         Dead Time   Address     Interface
1.1.1.1          1    FULL/DR       00:00:18    10.10.0.1   GigabitEthernet1/0/5
2.2.2.2          1    FULL/BDR      00:00:18    10.10.0.2   GigabitEthernet1/0/5
```

The output confirms that R1 (i.e., 1.1.1.1) is now the DR and D1 (i.e., 2.2.2.2) is the BDR.

The reason R1 became the DR is because it had already been elected as BDR. When a DR fails, the elected BDR is automatically elected as DR to avoid network instability.

c. Although R1 is already the DR, change the interface priority to ensure it is always a candidate to be DR.

```
R1(config)# interface g0/0/1
R1(config-if)# ip ospf priority 255
R1(config-if)# end
```

d. Verify that R1 is now the DR.

```
R1# show ip ospf interface g0/0/1 | include State
  Transmit Delay is 1 sec, State DR, Priority 255
```

e. Verify the roles of D1 and D2.

```
R1# show ip ospf neighbor

Neighbor ID     Pri   State         Dead Time   Address     Interface
2.2.2.2           1   FULL/BDR      00:00:19    10.10.0.2   GigabitEthernet0/0/1
3.3.3.3           0   FULL/DROTHER  00:00:15    10.10.0.3   GigabitEthernet0/0/1
```

Router Interface Summary Table

Router Model	Ethernet Interface #1	Ethernet Interface #2	Serial Interface #1	Serial Interface #2
1800	Fast Ethernet 0/0 (F0/0)	Fast Ethernet 0/1 (F0/1)	Serial 0/0/0 (S0/0/0)	Serial 0/0/1 (S0/0/1)
1900	Gigabit Ethernet 0/0 (G0/0)	Gigabit Ethernet 0/1 (G0/1)	Serial 0/0/0 (S0/0/0)	Serial 0/0/1 (S0/0/1)
2801	Fast Ethernet 0/0 (F0/0)	Fast Ethernet 0/1 (F0/1)	Serial 0/1/0 (S0/1/0)	Serial 0/1/1 (S0/1/1)
2811	Fast Ethernet 0/0 (F0/0)	Fast Ethernet 0/1 (F0/1)	Serial 0/0/0 (S0/0/0)	Serial 0/0/1 (S0/0/1)
2900	Gigabit Ethernet 0/0 (G0/0)	Gigabit Ethernet 0/1 (G0/1)	Serial 0/0/0 (S0/0/0)	Serial 0/0/1 (S0/0/1)
4221	Gigabit Ethernet 0/0/0 (G0/0/0)	Gigabit Ethernet 0/0/1 (G0/0/1)	Serial 0/1/0 (S0/1/0)	Serial 0/1/1 (S0/1/1)
4300	Gigabit Ethernet 0/0/0 (G0/0/0)	Gigabit Ethernet 0/0/1 (G0/0/1)	Serial 0/1/0 (S0/1/0)	Serial 0/1/1 (S0/1/1)

Note: To find out how the router is configured, look at the interfaces to identify the type of router and how many interfaces the router has. There is no way to effectively list all the combinations of configurations for each router class. This table includes identifiers for the possible combinations of Ethernet and Serial interfaces in the device. The table does not include any other type of interface, even though a specific router may contain one. An example of this might be an ISDN BRI interface. The string in parenthesis is the legal abbreviation that can be used in Cisco IOS commands to represent the interface.

Device Config Final – Notes

CHAPTER 7

Advanced OSPF

7.1.2 Lab - Implement Multiarea OSPFv2

Topology

Addressing Table

Device	Interface	IPv4 Address
R1	G0/0/0	172.16.0.2/30
	G0/0/1	10.10.0.1/30
R2	Lo0	209.165.200.225/27
	G0/0/0	172.16.0.1/30
	G0/0/1	172.16.1.1/30
R3	G0/0/0	172.16.1.2/30
	G0/0/1	10.10.4.1/30

Device	Interface	IPv4 Address
D1	G1/0/11	10.10.0.2/30
	G1/0/23	10.10.1.1/24
D2	G1/0/11	10.10.4.2/30
	G1/0/23	10.10.5.1/24
PC1	NIC	10.10.1.10/24
PC2	NIC	10.10.5.10/24

Objectives

Part 1: Build the Network and Configure Basic Device Settings and Interface Addressing

Part 2: Configure and Verify Multiarea OSPF for IPv4 on R1, D1, and D2

Part 3: Exploring Link State Announcements

Background/Scenario

To make OSPF more efficient and scalable, OSPF supports hierarchical routing using the areas. An OSPF area is a group of routers that share the same link-state information in their link-state databases (LSDBs). When a large OSPF area is divided into smaller areas, it is called multiarea OSPF. Multiarea OSPF is useful in larger network deployments to reduce processing and memory overhead.

In this lab you will configure multiarea OSPF version 2 for IPv4. This lab was specifically designed to use three routers and two Layer 3 switches.

Note: This lab is an exercise in developing, deploying, and verifying how multiarea OSPF operates and does not reflect networking best practices.

Note: The router used with this CCNP hands-on lab is a Cisco 4221 and the two Layer 3 switches are Catalyst 3560 switches. Other routers and Layer 3 switches and Cisco IOS versions can be used. Depending on the model and Cisco IOS version, the commands available and the output produced might vary from what is shown in the labs.

Note: Ensure that the routers and switches have been erased and have no startup configurations. If you are unsure contact your instructor.

Required Resources

- 3 Routers (Cisco 4221 with Cisco IOS XE Release 16.9.4 universal image or comparable)
- 2 Switches (Cisco 3650 with Cisco IOS XE Release 16.9.4 universal image or comparable)
- 2 PCs (Windows with terminal emulation program, such as Tera Term)
- Console cables to configure the Cisco IOS devices via the console ports
- Ethernet cables as shown in the topology

Instructions

Part 1: Build the Network and Configure Basic Device Settings and Interface Addressing

In Part 1, you will set up the network topology and configure basic settings and interface addressing on the routers and Layer 3 switches.

Step 1.   Cable the network as shown in the topology.

Attach the devices as shown in the topology diagram, and cable as necessary.

Step 2.   Configure basic settings for the routers and switches.

    a.   Console into each router and Layer 3 switch, enter global configuration mode, and apply the basic settings and interface addressing using the following startup configurations for each device.

Router R1

```
hostname R1
no ip domain lookup
line con 0
 logging sync
 exec-time 0 0
 exit
banner motd # This is R1, Implement Multiarea OSPFv2 Lab #
interface g0/0/0
 ip add 172.16.0.2 255.255.255.252
 no shut
 exit
interface GigabitEthernet0/0/1
 ip address 10.10.0.1 255.255.255.252
 no shut
 exit
```

Router R2

```
hostname R2
no ip domain lookup
line con 0
logging sync
exec-time 0 0
exit
banner motd # This is R2, Implement Multiarea OSPFv2 Lab #
interface g0/0/0
 ip add 172.16.0.1 255.255.255.252
 no shut
 exit
interface GigabitEthernet0/0/1
 ip address 172.16.1.1 255.255.255.252
 no shut
 exit
interface lo0
 ip add 209.165.200.225 255.255.255.224
 exit
```

Router R3

```
hostname R3
no ip domain lookup
line con 0
 logging sync
 exec-time 0 0
 exit
banner motd # This is R3, Implement Multiarea OSPFv2 Lab #
interface g0/0/0
 ip add 172.16.1.2 255.255.255.252
 no shut
 exit
interface GigabitEthernet0/0/1
 ip address 10.10.4.1 255.255.255.252
 no shut
 exit
```

Switch D1

```
hostname D1
no ip domain lookup
line con 0
exec-timeout 0 0
logging synchronous
exit
banner motd # This is D1, Implement Multiarea OSPFv2 Lab #
interface g1/0/11
 no switchport
 ip address 10.10.0.2 255.255.255.252
 no shut
 exit
interface g1/0/23
 no switchport
 ip address 10.10.1.1 255.255.255.0
 no shut
 exit
```

Switch D2

```
hostname D2
no ip domain lookup
line con 0
 logging sync
 exec-time 0 0
 exit
banner motd # This is D2, Implement Multiarea OSPFv2 Lab #
interface g1/0/11
 no switchport
 ip address 10.10.4.2 255.255.255.252
 no shut
 exit
interface g1/0/23
 no switchport
```

```
ip address 10.10.5.1 255.255.255.0
no shut
exit
```

b. Save the running configuration to startup-config.

c. Verify the interface status using the **show ip interface brief** command.

```
R1# show ip interface brief | include manual
GigabitEthernet0/0/0   172.16.0.2      YES manual up            up
GigabitEthernet0/0/1   10.10.0.1       YES manual up            up


R2# show ip interface brief | include manual
GigabitEthernet0/0/0   172.16.0.1      YES manual up            up
GigabitEthernet0/0/1   172.16.1.1      YES manual up            up
Loopback0              209.165.200.225 YES manual up            up


R3# show ip interface brief | include manual
GigabitEthernet0/0/0   172.16.1.2      YES manual up            up
GigabitEthernet0/0/1   10.10.4.1       YES manual up            up


D1# show ip interface brief | include manual
GigabitEthernet1/0/11  10.10.0.2       YES manual up            up
GigabitEthernet1/0/23  10.10.1.1       YES manual up            up


D2# show ip interface brief | include manual
GigabitEthernet1/0/11  10.10.4.2       YES manual up            up
GigabitEthernet1/0/23  10.10.5.1       YES manual up            up
```

d. Verify direct connectivity between all five devices. R1 is shown as an example.

```
R1# ping 10.10.0.2
Type escape sequence to abort.
Sending 5, 100-byte ICMP Echos to 10.10.0.2, timeout is 2 seconds:
..!!!
Success rate is 60 percent (3/5), round-trip min/avg/max = 2/2/2 ms


R1# ping 172.16.0.1
Type escape sequence to abort.
Sending 5, 100-byte ICMP Echos to 172.16.0.1, timeout is 2 seconds:
.!!!!
Success rate is 80 percent (4/5), round-trip min/avg/max = 1/1/1 ms
```

All five devices should be able to reach the other directly connected networks. Troubleshoot if necessary.

Part 2: Configure Multiarea OSPFv2

In this part, you will implement multiarea OSPF. Multiarea OSPF defines a two-layer area hierarchy using a backbone area interconnecting regular areas. This is useful in larger network deployments to reduce processing and memory overhead.

In this topology, OSPF has the following three areas defined:

- **Area 0** – The backbone area. All regular areas should connect to the backbone area.

- **Area 1** and **Area 2** – Regular OSPF areas that connect to the backbone area.

The routers and switches in the topology are used in the following roles:

- **Internal routers** - R2 is an internal router in Area 0, D1 is internal in Area 1, and D2 is internal in Area 2.

- **Backbone routers** - R1, R2, and R3 are backbone routers as they all have interfaces in Area 0.

- **Area Border routers (ABRs)** – R1 and R3 are ABRs because they connect regular areas (i.e., Area 1 and Area 2) to the backbone Area 0.

- **Autonomous System Boundary router (ASBR)** – R2 is an ASBR because it connects to another non-OSPF network.

Recall that OSPF can be enabled using the traditional **network** router configuration command or by using the **ip ospf** *process-id* **area** *area-id* interface configuration command. Although the interface method is simpler, the OSPF routing configuration commands are applied to individual interfaces and not conveniently found in a central location. Therefore, in this lab, we will implement multiarea OSPF using wildcard masks.

You will now configure multiarea OSPF on all five devices starting with D1. You will also configure router IDs, reference bandwidths, and default route propagation.

Note: The verification output displayed in the following part assumes that the devices have been configured in the prescribed order. The output will vary if all devices are configured simultaneously.

Step 1. Implement OSPF on D1 using wildcard masks.

D1 will advertise its OSPF networks using the **network** router configuration command and wildcard masks. D1 is an internal router in Area 1. Therefore, both **network** commands will be configured for Area 1.

a. Layer 3 switches are not enabled to perform routing by default. Therefore, routing must be enabled using the **ip routing** global configuration command.

```
D1(config)# ip routing
```

b. Next, enter the OSPF router configuration mode using process ID 123, assign D1 the router ID 1.1.1.2. You will also set the reference bandwidth to distinguish between Gigabit Ethernet and FastEthernet interfaces. Changing the reference bandwidth to a higher value allows for a differentiation of cost between higher-speed interfaces.

```
D1(config)# router ospf 123
D1(config-router)# router-id 1.1.1.2
D1(config-router)# auto-cost reference-bandwidth 1000
% OSPF: Reference bandwidth is changed.
         Please ensure reference bandwidth is consistent across all routers.
```

Note: Setting the reference cost value too high may cause issues with low-bandwidth interfaces.

---

c. Configure D1 to advertise the FastEthernet 0/5 interface 10.10.0.0/30 network in OSPF Area 1. The wildcard mask can be calculated by deducting the subnet mask (i.e., /30 = 255.255.255.252) from 255.255.255.255, resulting in a wildcard mask of 0.0.0.3.

```
D1(config-router)# network 10.10.0.0 0.0.0.3 area 1
```

d. Next, configure D1 to advertise its Fa0/23 interface 10.10.1.0/24 network in OSPF Area 1 and return to privileged EXEC mode.

```
D1(config-router)# network 10.10.1.0 0.0.0.255 area 1
D1(config-router)# end
```

e. Verify the OSPF configuration on D1 using the **show ip protocols** command.

```
D1# show ip protocols
*** IP Routing is NSF aware ***

Routing Protocol is "ospf 123"
  Outgoing update filter list for all interfaces is not set
  Incoming update filter list for all interfaces is not set
  Router ID 1.1.1.2
  Number of areas in this router is 1. 1 normal 0 stub 0 nssa
  Maximum path: 4
  Routing for Networks:
    10.10.0.0 0.0.0.3 area 1
    10.10.1.0 0.0.0.255 area 1
  Routing Information Sources:
    Gateway         Distance      Last Update
  Distance: (default is 110)
```

The output confirms the router ID, and the number of areas, and the networks advertised. Notice there are no Routing Information Sources because there are no OSPF neighbors.

f. Verify the OSPF interfaces using the **show ip ospf interface brief** command.

```
D1# show ip ospf interface brief
Interface    PID   Area        IP Address/Mask     Cost  State Nbrs F/C
Gi1/0/23     123   1           10.10.1.1/24        10    DR    0/0
Gi1/0/11     123   1           10.10.0.2/30        10    DR    0/0
```

The output confirms that both G1/0/11 and G1/0/23 interfaces were correctly assigned to Area 1.

g. Finally, verify the OSPF routes in the routing table using the **show ip route ospf** command.

```
D1# show ip route ospf
```

Notice that no routes are displayed. This is because D1 does not yet have an OSPF neighbor.

Step 2. Implement OSPF on R1.

Next, configure R1. R1 is an ABR with an interface in Area 1 and the other interface in the backbone.

a. Enter the OSPF router configuration mode using process ID 123, assign R1 the router ID 1.1.1.1, and set the reference bandwidth to distinguish between Gigabit Ethernet and FastEthernet interfaces.

```
R1(config)# router ospf 123
R1(config-router)# router-id 1.1.1.1
R1(config-router)# auto-cost reference-bandwidth 1000
% OSPF: Reference bandwidth is changed.
        Please ensure reference bandwidth is consistent across all routers.
```

b. Configure R1 to advertise the G0/0/0 interface 172.16.0.0/30 network in OSPF Area 0.

```
R1(config-router)# network 172.16.0.0 0.0.0.3 area 0
```

c. Advertise the G0/0/1 interface 10.10.0.0/30 network in OSPF Area 1 and return to privileged EXEC mode.

```
R1(config-router)# network 10.10.0.0 0.0.0.3 area 1
R1(config-router)# end
*Jan  4 11:55:32.064: %OSPF-5-ADJCHG: Process 123, Nbr 1.1.1.2 on
GigabitEthernet0/0/1 from LOADING to FULL, Loading Done
```

Notice the informational message stating that a neighbor adjacency has been established with D1 (i.e., 1.1.1.2).

d. Verify the OSPF configuration on R1 using the **show ip protocols** command.

```
R1# show ip protocols | begin ospf
Routing Protocol is "ospf 123"
  Outgoing update filter list for all interfaces is not set
  Incoming update filter list for all interfaces is not set
  Router ID 1.1.1.1
  It is an area border router
  Number of areas in this router is 2. 2 normal 0 stub 0 nssa
  Maximum path: 4
  Routing for Networks:
    10.10.0.0 0.0.0.3 area 1
    172.16.0.0 0.0.0.3 area 0
  Routing Information Sources:
    Gateway         Distance      Last Update
    1.1.1.2              110       00:07:55
  Distance: (default is 110)
```

Like the previous output of D1, this output confirms the router ID, and the networks advertised. However, notice that it also explicitly states that R1 is an area border router (ABR), that it is in two areas, and that it has established an adjacency and exchanged routing information with D2 (i.e., 1.1.1.2).

e. Verify that the reference bandwidth has been changed using the **show ip ospf | begin Ref** command.

```
R1# show ip ospf | begin Ref
 Reference bandwidth unit is 1000 mbps
    Area BACKBONE(0) (Inactive)
        Number of interfaces in this area is 1
        Area has no authentication
        SPF algorithm last executed 00:14:41.606 ago
        SPF algorithm executed 2 times
        Area ranges are
        Number of LSA 3. Checksum Sum 0x01ABED
        Number of opaque link LSA 0. Checksum Sum 0x000000
        Number of DCbitless LSA 0
        Number of indication LSA 0
        Number of DoNotAge LSA 0
        Flood list length 0
    Area 1
        Number of interfaces in this area is 1
        Area has no authentication
        SPF algorithm last executed 00:15:07.141 ago
        SPF algorithm executed 5 times
```

```
                  Area ranges are
                  Number of LSA 4. Checksum Sum 0x0292B8
                  Number of opaque link LSA 0. Checksum Sum 0x000000
                  Number of DCbitless LSA 0
                  Number of indication LSA 0
                  Number of DoNotAge LSA 0
                  Flood list length 0
```

The output confirms that the reference bandwidth has been changed to distinguish GigabitEthernet interfaces. The output also confirms that R1 is in two areas and has two link-state databases (LSDBs).

Note: Area 0 is currently inactive because there are no other peers configured yet.

---

    f. Verify the active OSPF interfaces and assigned areas using the **show ip ospf interface brief** command.

```
R1# show ip ospf interface brief
Interface    PID    Area         IP Address/Mask    Cost    State Nbrs F/C
Gi0/0/0      123    0            172.16.0.2/30      1       DR    0/0
Gi0/0/1      123    1            10.10.0.1/30       1       BDR   1/1
```

The output confirms the interfaces, areas, and IP addresses.

    g. Verify which OSPF neighbors R1 has established an adjacency with using the **show ip ospf neighbor** command.

```
R1# show ip ospf neighbor

Neighbor ID    Pri    State        Dead Time    Address      Interface
1.1.1.2        1      FULL/DR      00:00:31     10.10.0.2    GigabitEthernet0/0/1
```

The output confirms that R1 has one neighbor (i.e., 1.1.1.2 = D1), they have a full adjacency established, the IP address of D1 is 10.10.0.2, and R1 can reach D1 using its G0/0/1 interface.

    h. Use the **show ip ospf neighbor detail** command to get additional information about neighbor adjacencies.

```
R1# show ip ospf neighbor detail
 Neighbor 1.1.1.2, interface address 10.10.0.2, interface-id 38
    In the area 1 via interface GigabitEthernet0/0/1
    Neighbor priority is 1, State is FULL, 6 state changes
    DR is 10.10.0.2 BDR is 10.10.0.1
    Options is 0x12 in Hello (E-bit, L-bit)
    Options is 0x52 in DBD (E-bit, L-bit, O-bit)
    LLS Options is 0x1 (LR)
    Dead timer due in 00:00:34
    Neighbor is up for 00:19:09
    Index 1/1/1, retransmission queue length 0, number of retransmission 0
    First 0x0(0)/0x0(0)/0x0(0) Next 0x0(0)/0x0(0)/0x0(0)
    Last retransmission scan length is 0, maximum is 0
    Last retransmission scan time is 0 msec, maximum is 0 msec
```

As shown, the output confirms various information about the OSPF neighbor including DR and BDR status.

i. Verify the OSPF routes in the routing table using the **show ip route ospf** command.

```
R1# show ip route ospf | begin Gateway
Gateway of last resort is not set


      10.0.0.0/8 is variably subnetted, 3 subnets, 3 masks
O        10.10.1.0/24 [110/11] via 10.10.0.2, 00:24:43, GigabitEthernet0/0/1
```

The output displays an entry for the D1 LAN. The O designation identifies this as an OSPF internal route. Network routes learned from other OSPF routers in the same area are known as *intra-area routes* and are identified in the IP routing table with an O.

j. Finally, get detailed information on how R1 learned about the OSPF entry using the **show ip route ospf 10.10.1.0** command.

```
R1# show ip route 10.10.1.0
Routing entry for 10.10.1.0/24
  Known via "ospf 123", distance 110, metric 11, type intra area
  Last update from 10.10.0.2 on GigabitEthernet0/0/1, 00:25:25 ago
  Routing Descriptor Blocks:
  * 10.10.0.2, from 1.1.1.2, 00:25:25 ago, via GigabitEthernet0/0/1
      Route metric is 11, traffic share count is 1
```

The output confirms that R1 learned about the intra-area route 10.10.1.0 from 10.10.0.2 with a router ID of 1.1.1.2 in OSPF 123.

Step 3.   Implement OSPF on R2.

Next, configure R2. R2 is an internal backbone router and will become an ASBR.

a. Enter the OSPF router configuration mode using process ID 123, assign R2 the router ID 2.2.2.1, and set the reference bandwidth to distinguish between Gigabit Ethernet and FastEthernet interfaces.

```
R2(config)# router ospf 123
R2(config-router)# router-id 2.2.2.1
R2(config-router)# auto-cost reference-bandwidth 1000
% OSPF: Reference bandwidth is changed.
       Please ensure reference bandwidth is consistent across all routers.
```

b. Configure R2 to advertise its two interfaces in OSPF Area 0.

```
R2(config-router)# network 172.16.0.0 0.0.0.3 area 0
R2(config-router)# network 172.16.1.0 0.0.0.3 area 0
*Dec 28 16:22:42.530: %OSPF-5-ADJCHG: Process 123, Nbr 1.1.1.1 on
GigabitEthernet0/0/0 from LOADING to FULL, Loading Done
```

The output confirms that an adjacency has been established with R1 (i.e., 1.1.1.1).

Note: Alternatively, the two **network** statements could be combined using **network 17.16.0.0 0.0.1.3 area 0.**

---

c. Configure R2 to propagate a default route to the internet. In our lab, the internet is represented as a loopback interface.

```
R2(config-router)# default-information originate
R2(config-router)# exit
R2(config)# ip route 0.0.0.0 0.0.0.0 lo0
%Default route without gateway, if not a point-to-point interface, may impact
performance
R2(config)# exit
```

d. Verify the OSPF configuration on R2 using the **show ip protocols** command.

```
R2# show ip protocols | begin ospf
Routing Protocol is "ospf 123"
  Outgoing update filter list for all interfaces is not set
  Incoming update filter list for all interfaces is not set
  Router ID 2.2.2.1
  It is an autonomous system boundary router
 Redistributing External Routes from,
  Number of areas in this router is 1. 1 normal 0 stub 0 nssa
  Maximum path: 4
  Routing for Networks:
    172.16.0.0 0.0.0.3 area 0
    172.16.1.0 0.0.0.3 area 0
  Routing Information Sources:
    Gateway         Distance      Last Update
    1.1.1.1               110      00:24:29
  Distance: (default is 110)
```

Again, this output confirms the router ID chosen, number of areas R2 is in (i.e., 1), networks advertised, and that it has established an adjacency and exchanged routing information with R1 (i.e., 1.1.1.1). R2 does not have an adjacency with D1 because it is in another area.

Notice as well, that it explicitly states that R2 is an autonomous system boundary router (ASBR). This is because it is now propagating a default route to all other routers in the OSPF domain.

e. Verify that the reference bandwidth has been changed using the **show ip ospf | begin Ref** command as shown.

```
R2# show ip ospf | begin Ref
 Reference bandwidth unit is 1000 mbps
    Area BACKBONE(0)
        Number of interfaces in this area is 2
        Area has no authentication
        SPF algorithm last executed 00:02:27.531 ago
        SPF algorithm executed 5 times
        Area ranges are
        Number of LSA 5. Checksum Sum 0x01C25B
        Number of opaque link LSA 0. Checksum Sum 0x000000
        Number of DCbitless LSA 0
        Number of indication LSA 0
        Number of DoNotAge LSA 0
        Flood list length 0
```

The output confirms that the reference bandwidth has been changed and also confirms that R2 has two interfaces in its link-state database (LSDB).

f. Verify the active OSPF interfaces and assigned areas using the **show ip ospf interface brief** command.

```
R2# show ip ospf interface brief
Interface    PID   Area         IP Address/Mask     Cost  State Nbrs F/C
Gi0/0/1      123   0            172.16.1.1/30       1     DR    0/0
Gi0/0/0      123   0            172.16.0.1/30       1     BDR   1/1
```

The output confirms that the two interfaces are in Area 0, their IP addresses, state, and neighbors.

g. Verify which OSPF neighbors R2 has established an adjacency with using the **show ip ospf neighbor** command.

```
R2# show ip ospf neighbor

Neighbor ID     Pri   State        Dead Time   Address       Interface
1.1.1.1           1   FULL/DR      00:00:36    172.16.0.2    GigabitEthernet0/0/0
```

The output confirms that R2 has one neighbor (i.e., 1.1.1.1 = R1) and they have a full adjacency established.

h. Use the **show ip ospf neighbor detail** command to get additional information about neighbor adjacencies.

```
R2# show ip ospf neigh detail
 Neighbor 1.1.1.1, interface address 172.16.0.2, interface-id 5
    In the area 0 via interface GigabitEthernet0/0/0
    Neighbor priority is 1, State is FULL, 6 state changes
    DR is 172.16.0.2 BDR is 172.16.0.1
    Options is 0x12 in Hello (E-bit, L-bit)
    Options is 0x52 in DBD (E-bit, L-bit, O-bit)
    LLS Options is 0x1 (LR)
    Dead timer due in 00:00:34
    Neighbor is up for 00:27:48
    Index 1/1/1, retransmission queue length 0, number of retransmission 0
    First 0x0(0)/0x0(0)/0x0(0) Next 0x0(0)/0x0(0)/0x0(0)
    Last retransmission scan length is 0, maximum is 0
    Last retransmission scan time is 0 msec, maximum is 0 msec
```

As shown, the output confirms various information about the OSPF neighbor including DR and BDR status.

i. Verify the OSPF routes in the routing table using the **show ip route ospf** command.

```
R2# show ip route ospf | begin Gateway

Gateway of last resort is 0.0.0.0 to network 0.0.0.0

      10.0.0.0/8 is variably subnetted, 2 subnets, 2 masks
O IA     10.10.0.0/30 [110/2] via 172.16.0.2, 00:28:19, GigabitEthernet0/0/0
O IA     10.10.1.0/24 [110/12] via 172.16.0.2, 00:28:19, GigabitEthernet0/0/0
```

The output displays that there is now a default gateway and two entries for the OSPF Area 1 networks. Notice how these routes are identified as **O IA** which means they are routes from another area. Network routes learned from OSPF routers in another area using an ABR are known as *interarea routes* as opposed to intra-area routes.

j. Verify the static route entry in the routing table.

```
R2# show ip route static | begin Gateway
Gateway of last resort is 0.0.0.0 to network 0.0.0.0

S*    0.0.0.0/0 is directly connected, Loopback0
```

k. Finally, get detailed information on how R2 learned about the OSPF entry using the **show ip route ospf 10.10.1.0** command.

```
R2# show ip route 10.10.1.0
Routing entry for 10.10.1.0/24
  Known via "ospf 123", distance 110, metric 12, type inter area
  Last update from 172.16.0.2 on GigabitEthernet0/0/0, 00:31:08 ago
  Routing Descriptor Blocks:
  * 172.16.0.2, from 1.1.1.1, 00:31:08 ago, via GigabitEthernet0/0/0
      Route metric is 12, traffic share count is 1
```

The output confirms that R2 learned about the interarea route 10.10.1.0 from OSPF 123 and specifically from R1, based on the router ID of 1.1.1.1.

Step 4. Implement OSPF on R3.

Next to configure is R3. Like R1, R3 is an ABR with an interface in Area 0 and one in Area 2.

a. Enter the OSPF router configuration mode using process ID 123, assign R3 the router ID 3.3.3.1, and set the reference bandwidth to distinguish between Gigabit Ethernet and FastEthernet interfaces.

```
R3(config)# router ospf 123
R3(config-router)# router-id 3.3.3.1
R3(config-router)# auto-cost reference-bandwidth 1000
% OSPF: Reference bandwidth is changed.
```

Note: Please ensure reference bandwidth is consistent across all routers.

b. Configure R3 to advertise its interfaces in OSPF Area 0 and Area 2 accordingly and then return to privileged EXEC mode.

```
R3(config-router)# network 172.16.1.0 0.0.0.3 area 0
R3(config-router)# network 10.10.4.0 0.0.0.3 area 2
R3(config-router)# end
R3#
*Jan  5 19:28:25.146: %OSPF-5-ADJCHG: Process 123, Nbr 2.2.2.1 on
GigabitEthernet0/0/0 from LOADING to FULL, Loading Done
```

c. Verify the OSPF configuration on R3 using the **show ip protocols** command.

```
R3# show ip protocols | begin ospf
Routing Protocol is "ospf 123"
  Outgoing update filter list for all interfaces is not set
  Incoming update filter list for all interfaces is not set
  Router ID 3.3.3.1
  It is an area border router
  Number of areas in this router is 2. 2 normal 0 stub 0 nssa
  Maximum path: 4
  Routing for Networks:
    10.10.4.0 0.0.0.3 area 2
    172.16.1.0 0.0.0.3 area 0
  Routing Information Sources:
    Gateway         Distance      Last Update
    1.1.1.1              110      00:01:43
    2.2.2.1              110      00:01:43
  Distance: (default is 110)
```

The output confirms the router ID, and that R3 is an ABR, it has interfaces in two areas, the networks it is advertising, and that R3 has R1 (i.e., 1.1.1.1) and R2 (i.e., 2.2.2.1) as sources of routing information.

d. Verify that the reference bandwidth has been changed using the **show ip ospf | begin Ref** command as shown.

```
R3# show ip ospf | begin Ref
 Reference bandwidth unit is 1000 mbps
    Area BACKBONE(0)
        Number of interfaces in this area is 1
        Area has no authentication
        SPF algorithm last executed 00:10:38.256 ago
        SPF algorithm executed 4 times
        Area ranges are
        Number of LSA 8. Checksum Sum 0x0396BA
        Number of opaque link LSA 0. Checksum Sum 0x000000
        Number of DCbitless LSA 0
        Number of indication LSA 0
        Number of DoNotAge LSA 0
        Flood list length 0
    Area 2
        Number of interfaces in this area is 1
        Area has no authentication
        SPF algorithm last executed 00:10:13.755 ago
        SPF algorithm executed 2 times
        Area ranges are
        Number of LSA 6. Checksum Sum 0x0362CF
        Number of opaque link LSA 0. Checksum Sum 0x000000
        Number of DCbitless LSA 0
        Number of indication LSA 0
        Number of DoNotAge LSA 0
        Flood list length 0
```

The output confirms that the reference bandwidth has been changed and also confirms that R2 has area information for Area 0 and Area 2.

e. Verify the active OSPF interfaces and assigned areas using the **show ip ospf interface brief** command.

```
R3# show ip ospf interface brief
Interface    PID    Area           IP Address/Mask     Cost   State  Nbrs F/C
Gi0/0/0      123    0              172.16.1.2/30       1      BDR    1/1
Gi0/0/1      123    2              10.10.4.1/30        1      DR     0/0
```

The output confirms the interfaces, process ID, areas, IP addresses, cost, state, and neighbors.

f. Verify which OSPF neighbors R2 has established an adjacency with using the **show ip ospf neighbor** command.

```
R3# show ip ospf neighbor
Neighbor ID    Pri    State        Dead Time    Address      Interface
2.2.2.1          1    FULL/DR      00:00:31     172.16.1.1   GigabitEthernet0/0/0
```

The output confirms that R3 has one neighbor (i.e., 2.2.2.1= R2) and they have a full adjacency established.

g. Use the **show ip ospf neighbor detail** command to get additional information about neighbor adjacencies.

```
R3# show ip ospf neighbor detail
 Neighbor 2.2.2.1, interface address 172.16.1.1, interface-id 6
    In the area 0 via interface GigabitEthernet0/0/0
    Neighbor priority is 1, State is FULL, 6 state changes
    DR is 172.16.1.1 BDR is 172.16.1.2
    Options is 0x12 in Hello (E-bit, L-bit)
    Options is 0x52 in DBD (E-bit, L-bit, O-bit)
    LLS Options is 0x1 (LR)
    Dead timer due in 00:00:37
    Neighbor is up for 00:21:50
    Index 1/1/1, retransmission queue length 0, number of retransmission 0
    First 0x0(0)/0x0(0)/0x0(0) Next 0x0(0)/0x0(0)/0x0(0)
    Last retransmission scan length is 0, maximum is 0
    Last retransmission scan time is 0 msec, maximum is 0 msec
```

As shown, the output confirms various information about the OSPF neighbor including DR and BDR status.

h. Verify the OSPF routes in the routing table using the **show ip route ospf** command.

```
R3# show ip route ospf | begin Gateway
Gateway of last resort is 172.16.1.1 to network 0.0.0.0

O*E2  0.0.0.0/0 [110/1] via 172.16.1.1, 00:26:08, GigabitEthernet0/0/0
        10.0.0.0/8 is variably subnetted, 4 subnets, 3 masks
O IA    10.10.0.0/30 [110/3] via 172.16.1.1, 00:26:08, GigabitEthernet0/0/0
O IA    10.10.1.0/24 [110/13] via 172.16.1.1, 00:26:08, GigabitEthernet0/0/0
        172.16.0.0/16 is variably subnetted, 3 subnets, 2 masks
O       172.16.0.0/30 [110/2] via 172.16.1.1, 00:26:08, GigabitEthernet0/0/0
```

The output verifies that R3 has received a default route from R2, two interarea routes (i.e., O IA routes) and one intra-area OSPF route (i.e., O routes). The O*E2 route indicates that this is an external route that did not originate in OSPF. The asterisk identifies this as a candidate default route.

i. Now get detailed information on how R3 learned about the O E2 and O IA routes.

```
R3# show ip route 0.0.0.0
Routing entry for 0.0.0.0/0, supernet
  Known via "ospf 123", distance 110, metric 1, candidate default path
  Tag 123, type extern 2, forward metric 1
  Last update from 172.16.1.1 on GigabitEthernet0/0/0, 00:28:41 ago
  Routing Descriptor Blocks:
  * 172.16.1.1, from 2.2.2.1, 00:28:41 ago, via GigabitEthernet0/0/0
      Route metric is 1, traffic share count is 1
      Route tag 123


R3# show ip route 10.10.1.0
Routing entry for 10.10.1.0/24
  Known via "ospf 123", distance 110, metric 13, type inter area
  Last update from 172.16.1.1 on GigabitEthernet0/0/0, 00:29:10 ago
```

```
Routing Descriptor Blocks:
* 172.16.1.1, from 1.1.1.1, 00:29:10 ago, via GigabitEthernet0/0/0
    Route metric is 13, traffic share count is 1
```

The output confirms that R3 learned about the default route from R2 (2.2.2.1) and inter-area routes from R1 (1.1.1.1) via OSPF.

Step 5.   Implement OSPF on D2.

Last to configure is D2. Like D1, D2 is an internal router in Area 2.

a.   Layer 3 switches are not enabled to perform routing by default. Therefore, routing must be enabled using the **ip routing** global configuration command.

```
D2(config)# ip routing
```

b.   Next, enter the OSPF router configuration mode using process ID 123, assign D2 the router ID 3.3.3.2, and set the reference bandwidth to distinguish between Gigabit Ethernet and FastEthernet interfaces.

```
D2(config)# router ospf 123
D2(config-router)# router-id 3.3.3.2
D2(config-router)# auto-cost reference-bandwidth 1000
% OSPF: Reference bandwidth is changed.
        Please ensure reference bandwidth is consistent across all routers.
```

c.   Configure D2 to advertise its interfaces in OSPF Area 2 and return to privileged EXEC mode.

```
D2(config-router)# network 10.10.4.0 0.0.0.3 area 2
D2(config-router)# network 10.10.5.0 0.0.0.255 area 2
D2(config-router)# end
D2#
*Mar  1 01:52:03.888: %OSPF-5-ADJCHG: Process 123, Nbr 3.3.3.1 on
GigabitEthernet1/0/11 from LOADING to FULL, Loading Done
```

d.   Verify the OSPF configuration on D2 using the **show ip protocols** command.

```
D2# show ip protocols
Routing Protocol is "ospf 123"
  Outgoing update filter list for all interfaces is not set
  Incoming update filter list for all interfaces is not set
  Router ID 3.3.3.2
  Number of areas in this router is 1. 1 normal 0 stub 0 nssa
  Maximum path: 4
  Routing for Networks:
    10.10.4.0 0.0.0.3 area 2
    10.10.5.0 0.0.0.255 area 2
  Routing Information Sources:
    Gateway         Distance      Last Update
    3.3.3.1              110      00:09:35
    2.2.2.1              110      00:04:38
  Distance: (default is 110)
```

As expected, we can verify the router ID, number of areas, networks being advertised, and routing sources. It may be surprising that R2 (i.e., 2.2.2.1) is displayed as a routing source. The reason is because it is the source of the default route.

e. Verify that the reference bandwidth has been changed using the **show ip ospf | begin Ref** command as shown.

```
D2# show ip ospf | begin Ref
 Reference bandwidth unit is 1000 mbps
    Area 2
        Number of interfaces in this area is 2
        Area has no authentication
        SPF algorithm last executed 00:13:01.877 ago
        SPF algorithm executed 3 times
<output omitted>
```

f. Verify the active OSPF interfaces and assigned areas using the **show ip ospf interface brief** command.

```
D2# show ip ospf interface brief
Interface   PID   Area              IP Address/Mask   Cost   State Nbrs F/C
Gi1/0/23    123   2                 10.10.5.1/24      10     DR    0/0
G11/0/11    123   2                 10.10.4.2/30      1      BDR   1/1
```

The output confirms that the two interfaces are in Area 2 and their IP addresses are correct.

g. Verify which OSPF neighbors D2 has established an adjacency with using the **show ip ospf neighbor** command.

```
D2# show ip ospf neighbor
Neighbor ID     Pri   State       Dead Time   Address     Interface
3.3.3.1          1    FULL/BDR    00:00:33    10.10.4.1   GigabitEthernet1/0/11
```

The output confirms that D2 has one neighbor (i.e., 3.3.3.1= R3) and they have a full adjacency established.

h. Verify the OSPF routes in the routing table using the **show ip route ospf | begin Gateway** command.

```
D2# show ip route ospf | begin Gateway
Gateway of last resort is 10.10.4.1 to network 0.0.0.0

O*E2  0.0.0.0/0 [110/1] via 10.10.4.1, 00:09:25, Gigabitthernet1/0/11
         10.0.0.0/8 is variably subnetted, 6 subnets, 3 masks
O IA     10.10.0.0/30 [110/4] via 10.10.4.1, 00:09:00, GigabitEthernet1/0/11
O IA     10.10.1.0/24 [110/14] via 10.10.4.1, 00:09:00, GigabitEthernet1/0/11
         172.16.0.0/30 is subnetted, 2 subnets
O IA     172.16.0.0 [110/3] via 10.10.4.1, 00:09:00, GigabitEthernet1/0/11
O IA     172.16.1.0 [110/2] via 10.10.4.1, 00:09:00, GigabitEthernet1/0/11
```

The output displays four interarea routes (i.e., O IA routes) and the OSPF external route from 2.2.2.1 (R2).

```
D2# show ip route 0.0.0.0
Routing entry for 0.0.0.0/0, supernet
  Known via "ospf 123", distance 110, metric 1, candidate default path
  Tag 123, type extern 2, forward metric 2
  Last update from 10.10.4.1 on GigabitEthernet1/0/11, 00:18:31 ago
  Routing Descriptor Blocks:
  * 10.10.4.1, from 2.2.2.1, 00:18:31 ago, via GigabitEthernet1/0/11
      Route metric is 1, traffic share count is 1
      Route tag 123
```

As we can see, the routing source for the default route is R2 (i.e., 2.2.2.1).

Step 6.   Verify end-to-end connectivity.

The multiarea OSPF network is now completely configured. We now need to verify the operation of OSPF.

a.   From PC1, verify that it has been assigned the correct IP address as listed in the Addressing Table using the **ipconfig** Windows command.

```
C:\Users\Student> ipconfig


Windows IP Configuration


Ethernet adapter Ethernet0:

   Connection-specific DNS Suffix  . :
   Link-local IPv6 Address . . . . . : fe80::7853:120b:ecdf:d718%6
   IPv4 Address. . . . . . . . . . . : 10.10.1.10
   Subnet Mask . . . . . . . . . . . : 255.255.255.0
   Default Gateway . . . . . . . . . : 10.10.1.1
```

b.   Verify end-to-end connectivity by pinging PC3.

```
C:\Users\Student> ping 10.10.5.10


Pinging 10.10.5.10 with 32 bytes of data:


Pinging 10.10.5.10 with 32 bytes of data:
Reply from 10.10.5.10: bytes=32 time=1ms TTL=123
Reply from 10.10.5.10: bytes=32 time=1ms TTL=123
Reply from 10.10.5.10: bytes=32 time=1ms TTL=123
Reply from 10.10.5.10: bytes=32 time=1ms TTL=123


Ping statistics for 10.10.5.10:
    Packets: Sent = 4, Received = 4, Lost = 0 (0% loss),
Approximate round trip times in milli-seconds:
    Minimum = 0ms, Maximum = 1ms, Average = 0ms
```

c.   Verify the route taken by doing a traceroute to PC3.

```
C:\Users\Student> tracert 10.10.5.10


Tracing route to DESKTOP-3FR7RKA [10.10.5.10]
over a maximum of 30 hops:

  1     1 ms     1 ms     1 ms   10.10.1.1
  2    <1 ms    <1 ms    <1 ms   10.10.0.1
  3     1 ms    <1 ms    <1 ms   172.16.0.1
  4     1 ms     1 ms    <1 ms   172.16.1.2
  5     1 ms     2 ms     2 ms   10.10.4.2
  6     1 ms    <1 ms    <1 ms   DESKTOP-3RF7RKA [10.10.5.10]


Trace complete.
```

This confirms end-to-end connectivity

Part 3: Exploring Link-State Announcements

In this part, you will verify that the network has converged and explore how link-state advertisements (LSAs) are used as the building blocks for the OSPF link-state database (LSDB).

OSPF routers create LSAs for every directly connected OSPF-enabled interface. It then sends those LSAs to OSPF peers to form adjacencies. Individually, LSAs are database records providing specific OSPF network details. Combined, they describe the entire topology of an OSPF area.

OSPF routers uses six LSA types for IPv4 routing:

- **Type 1, router LSA** – All OSPF-enabled routers create and send type 1 LSAs. The LSAs are immediately propagated within the area. An ABR does not forward the LSA outside the area.

- **Type 2, network LSA** – Only a DR generates and advertises a type 2 LSA. The type 2 network LSA lists each of the attached routers that make up the transit network, including the DR itself, and the subnet mask that is used on the link. The DR floods the LSA to all OSPF routers (i.e., 224.0.0.5) on the multiaccess network. The content of the displayed type 2 LSA describes the network segment listing the DR address, the attached routers, and the used subnet mask. This information is used by each router participating in OSPF to build the exact picture of the described multiaccess segment, which cannot be fully described with just type 1 LSAs.

- **Type 3, summary LSA** – ABRs do not forward type 1 or type 2 LSAs into other areas. ABRs flood type 3 LSAs to propagate network information to other areas. Type 3 summary LSAs describe networks that are in an area to the rest of the areas in the OSPF autonomous system.

- **Type 4, ASBR summary LSA** – When there is an ASBR in the OSPF domain, it advertises itself using a special type 1 LSA. When an ABR receives this type 1 LSA, it builds a type 4 LSA to advertise the existence of the ASBR and floods it to other areas. Subsequent ABRs regenerate a type 4 LSA and flood it into their areas.

- **Type 5, AS external LSA** – ASBRs generate a type 5 external LSAs to advertise external OSPF routes to the OSPF domain. Type 5 LSAs are originated by the ASBR and are flooded to the entire autonomous system.

- **Type 7, NSSA external LSA** – This is a special LSA generated by a not-so-stubby (NSSA) ASBR to advertise external OSPF networks to an OSPF domain. The ABR converts the type 7 LSA to a type 5 LSA and propagates it to other areas. An NSSA network is a special-case area type used to reduce the amount of flooding, the LSDB size, and the routing table size in routers within the area.

Note: Other LSAs also exist but are out of scope of this lab.

The focus of this section will be on LSA types 1, 2, and 3 which are used to identify intra-area and interarea routes.

Step 1. Verifying OSPF and Exploring LSAs on D1.

D1 is an internal router and generates type 1 LSAs. It is also the DR on the link connecting to R1 and therefore generates type 2 LSAs.

a. On D1, display the list of neighbors using the **show ip ospf neighbors** command.

```
D1# show ip ospf neighbor

Neighbor ID     Pri   State       Dead Time   Address      Interface
1.1.1.1           1   FULL/BDR    00:00:32    10.10.0.1    GigabitEthernet1/0/11
```

The output confirms that R1 (i.e., 1.1.1.1) is a neighbor and is the BDR on the link. Therefore, D1 must be the DR.

b. Verify the OSPF routing table using the **show ip router ospf | begin Gateway** command.

```
D1# show ip route ospf | begin Gateway

Gateway of last resort is 10.10.0.1 to network 0.0.0.0

O*E2  0.0.0.0/0 [110/1] via 10.10.0.1, 01:15:48, GigabitEthernet1/0/11
         10.0.0.0/8 is variably subnetted, 6 subnets, 3 masks
O IA     10.10.4.0/30 [110/4] via 10.10.0.1, 00:52:50, GigabitEthernet1/0/11
O IA     10.10.5.0/24 [110/14] via 10.10.0.1, 00:24:49, GigabitEthernet1/0/11
         172.16.0.0/30 is subnetted, 2 subnets
O IA     172.16.0.0 [110/2] via 10.10.0.1, 02:05:06, GigabitEthernet1/0/11
O IA     172.16.1.0 [110/3] via 10.10.0.1, 01:18:11, GigabitEthernet1/0/11
```

The routing table lists the four interarea networks and one external OSPF network.

c. D1 learned about these networks from LSAs. A router maintains an LSDB for each area it has interfaces in. Because D1 is an internal OSPF router, it will only have entries for Area 1. To display the contents of the LSDB of D1, use the **show ip ospf database** command.

```
D1# show ip ospf database

            OSPF Router with ID (1.1.1.2) (Process ID 123)

                Router Link States (Area 1)

Link ID         ADV Router      Age         Seq#        Checksum Link count
1.1.1.1         1.1.1.1         1806        0x80000005 0x00DC15 1
1.1.1.2         1.1.1.2         167         0x80000005 0x001AA6 2

                Net Link States (Area 1)

Link ID         ADV Router      Age         Seq#        Checksum
10.10.0.2       1.1.1.2         167         0x80000003 0x00B462

                Summary Net Link States (Area 1)

Link ID         ADV Router      Age         Seq#        Checksum
10.10.4.0       1.1.1.1         1806        0x80000002 0x00A86E
10.10.5.0       1.1.1.1         1807        0x80000002 0x0014F4
172.16.0.0      1.1.1.1         1807        0x80000002 0x00DBA1
172.16.1.0      1.1.1.1         1807        0x80000002 0x00DAA0

                Summary ASB Link States (Area 1)

Link ID         ADV Router      Age         Seq#        Checksum
2.2.2.1         1.1.1.1         1807        0x80000002 0x00131C
```

```
                        Type-5 AS External Link States


Link ID          ADV Router       Age       Seq#       Checksum Tag
0.0.0.0          2.2.2.1          1939      0x80000002 0x009F90 123
```

Notice how the command output is divided into the following five sections:

- **Router Link States** – These are the type 1 LSAs received by D1 and they identify the routers (i.e., 1.1.1.1 = R1, 1.1.1.2 = D1) in Area 1 that sent them, and the number of links that the routers have in the area. Therefore, R1 only has one interface in Area 1 and D1 has 2 interfaces in Area 1.

- **Net Link States** – These are the type 2 LSAs generated by the DR. In our example, the DR is 1.1.1.2 (i.e., D1) on the link 10.10.0.2.

- **Summary Net Link States** – These are the type 3 LSAs describing remote networks (i.e., our O IA networks or interarea routes) and the router that advertised them to D1.

- **Summary ASB Link States** – This is a type 4 LSA sent by the ABR (i.e., 1.1.1.1 = R1) advertising that there is an ASBR in the network (i.e., 2.2.2.1).

- **Type-5 AS External Link States** – This is a type 5 LSA advertising a default route (i.e., 0.0.0.0) and the router that is advertising it (i.e., 2.2.2.1).

d. Additional information about the Router Link States type 1 LSA can be gathered using the **show ip ospf database router** command.

```
D1# show ip ospf database router


                OSPF Router with ID (1.1.1.2) (Process ID 123)


                        Router Link States (Area 1)


Routing Bit Set on this LSA in topology Base with MTID 0
LS age: 843
Options: (No TOS-capability, DC)
LS Type: Router Links
Link State ID: 1.1.1.1
Advertising Router: 1.1.1.1
LS Seq Number: 80000007
Checksum: 0xD817
Length: 36
Area Border Router
Number of Links: 1

  Link connected to: a Transit Network
    (Link ID) Designated Router address: 10.10.0.2
    (Link Data) Router Interface address: 10.10.0.1
     Number of TOS metrics: 0
      TOS 0 Metrics: 1



LS age: 1196
Options: (No TOS-capability, DC)
```

```
        LS Type: Router Links
        Link State ID: 1.1.1.2
        Advertising Router: 1.1.1.2
        LS Seq Number: 80000006
        Checksum: 0x18A7
        Length: 48
        Number of Links: 2

    Link connected to: a Transit Network
         (Link ID) Designated Router address: 10.10.0.2
         (Link Data) Router Interface address: 10.10.0.2
          Number of MTID metrics: 0
           TOS 0 Metrics: 1

      Link connected to: a Stub Network
        (Link ID) Network/subnet number: 10.10.1.0
        (Link Data) Network Mask: 255.255.255.0
         Number of MTID metrics: 0
          TOS 0 Metrics: 10
```

The output provides more information about the type 1 LSAs. The first router link (i.e., type 1) LSA is from R1 (i.e., 1.1.1.1). It is an ABR with only 1 link in Area 1 which is the transit network connecting to D1. The second router link portion identifies the transit network connecting to R1 and the stub network of D1 (i.e., 10.10.1.0/24).

An OSPF link can be connected to a stub, to another router (point-to-point), or to a transit network. The transit network usually describes an Ethernet segment which can include two or more routers. If the link is connected to a transit network, the LSA also includes the IP address of the DR.

e. To learn more about type 2 network LSAs, use the **show ip ospf database network** command.

D1# **show ip ospf database network**

```
                OSPF Router with ID (1.1.1.2) (Process ID 123)


                    Net Link States (Area 1)

    LS age: 845
    Options: (No TOS-capability, DC)
    LS Type: Network Links
    Link State ID: 10.10.0.2 (address of Designated Router)
    Advertising Router: 1.1.1.2
    LS Seq Number: 80000005
    Checksum: 0xB064
    Length: 32
    Network Mask: /30
            Attached Router: 1.1.1.2
            Attached Router: 1.1.1.1
```

The content of the type 2 LSA describes the network segment listing the DR address, the attached routers, and subnet mask using CIDR notation. This information is used by each router in the area to build the exact picture of the described multiaccess segment, which cannot be fully described with just type 1 LSAs.

f. To learn more about type 3 summary LSAs, use the **show ip ospf database summary** command.

```
D1# show ip ospf database summary

             OSPF Router with ID (1.1.1.2) (Process ID 123)

                 Summary Net Link States (Area 1)

  LS age: 987
  Options: (No TOS-capability, DC, Upward)
  LS Type: Summary Links(Network)
  Link State ID: 10.10.4.0 (summary Network Number)
  Advertising Router: 1.1.1.1
  LS Seq Number: 80000005
  Checksum: 0xA271
  Length: 28
  Network Mask: /30
        MTID: 0        Metric: 3

  LS age: 987
  Options: (No TOS-capability, DC, Upward)
  LS Type: Summary Links(Network)
  Link State ID: 10.10.5.0 (summary Network Number)
  Advertising Router: 1.1.1.1
  LS Seq Number: 80000005
  Checksum: 0xEF7
  Length: 28
  Network Mask: /24
        MTID: 0        Metric: 13

  LS age: 988
  Options: (No TOS-capability, DC, Upward)
  LS Type: Summary Links(Network)
  Link State ID: 172.16.0.0 (summary Network Number)
  Advertising Router: 1.1.1.1
  LS Seq Number: 80000005
  Checksum: 0xD5A4
  Length: 28
  Network Mask: /30
        MTID: 0        Metric: 1

  LS age: 989
  Options: (No TOS-capability, DC, Upward)
  LS Type: Summary Links(Network)
  Link State ID: 172.16.1.0 (summary Network Number)
  Advertising Router: 1.1.1.1
```

```
LS Seq Number: 80000005
Checksum: 0xD4A3
Length: 28
Network Mask: /30
      MTID: 0        Metric: 2
```

The output lists four type 3 LSAs. The LSAs identify the interarea networks, which ABR advertised, and the network mask using CIDR notation.

g. To learn more about type 4 summary LSAs, use the **show ip ospf database asbr-summary** command.

```
D1# show ip ospf database asbr-summary

              OSPF Router with ID (1.1.1.2) (Process ID 123)


              Summary ASB Link States (Area 1)


    LS age: 591
    Options: (No TOS-capability, DC, Upward)
    LS Type: Summary Links(AS Boundary Router)
    Link State ID: 2.2.2.1 (AS Boundary Router address)
    Advertising Router: 1.1.1.1
    LS Seq Number: 80000006
    Checksum: 0xB20
    Length: 28
    Network Mask: /0
          MTID: 0        Metric: 1
```

The output lists one type 4 LSA advertised by R1 identifying 2.2.2.1 as an ASBR.

h. Finally, to learn more about type 5 AS external link LSAs, use the **show ip ospf database external** command.

```
D1# show ip ospf database external

              OSPF Router with ID (1.1.1.2) (Process ID 123)


              Type-5 AS External Link States


    LS age: 1024
    Options: (No TOS-capability, DC, Upward)
    LS Type: AS External Link
    Link State ID: 0.0.0.0 (External Network Number )
    Advertising Router: 2.2.2.1
    LS Seq Number: 80000006
    Checksum: 0x9794
    Length: 36
    Network Mask: /0
          Metric Type: 2 (Larger than any link state path)
          TOS: 0
          Metric: 1
          Forward Address: 0.0.0.0
          External Route Tag: 123
```

The output lists one type 5 LSA identifying that 0.0.0.0/0 is available from 2.2.2.1 (i.e., R2).

Step 2. Verifying OSPF and exploring LSAs on an ABR R1.

R1 is an ABR with interfaces in Area 1 and Area 0. Therefore, R1 will have two LSDBs. Display the LSDB on R1.

```
R1# show ip ospf database

                OSPF Router with ID (1.1.1.1) (Process ID 123)


                Router Link States (Area 0)

Link ID          ADV Router       Age         Seq#        Checksum Link count
1.1.1.1          1.1.1.1          1250        0x80000009 0x001E87 1
2.2.2.1          2.2.2.1          1284        0x8000000C 0x00A06E 2
3.3.3.1          3.3.3.1          1220        0x80000008 0x00BDDA 1


                Net Link States (Area 0)

Link ID          ADV Router       Age         Seq#        Checksum
172.16.0.2       1.1.1.1          1284        0x80000006 0x002A3E
172.16.1.1       2.2.2.1          1284        0x80000006 0x0067F9


                Summary Net Link States (Area 0)

Link ID          ADV Router       Age         Seq#        Checksum
10.10.0.0        1.1.1.1          1250        0x80000008 0x00B462
10.10.1.0        1.1.1.1          1250        0x80000006 0x0024E6
10.10.4.0        3.3.3.1          1220        0x80000008 0x0058B4
10.10.5.0        3.3.3.1          1220        0x80000006 0x00C739


                Router Link States (Area 1)

Link ID          ADV Router       Age         Seq#        Checksum Link count
1.1.1.1          1.1.1.1          1250        0x80000009 0x00D419 1
1.1.1.2          1.1.1.2          1632        0x80000008 0x0014A9 2


                Net Link States (Area 1)

Link ID          ADV Router       Age         Seq#        Checksum
10.10.0.2        1.1.1.2          1632        0x80000006 0x00AE65


                Summary Net Link States (Area 1)

Link ID          ADV Router       Age         Seq#        Checksum
10.10.4.0        1.1.1.1          1250        0x80000006 0x00A072
10.10.5.0        1.1.1.1          1250        0x80000006 0x000CF8
172.16.0.0       1.1.1.1          1250        0x80000006 0x00D3A5
172.16.1.0       1.1.1.1          1250        0x80000006 0x00D2A4
```

```
                        Summary ASB Link States (Area 1)

       Link ID          ADV Router      Age        Seq#        Checksum
       2.2.2.1          1.1.1.1         1250       0x80000006 0x000B20


                        Type-5 AS External Link States

       Link ID          ADV Router      Age        Seq#        Checksum Tag
       0.0.0.0          2.2.2.1         1284       0x80000006 0x009794 123
```

The output displays the type 1, 2, and 3 LSAs in Area 0, and then lists the type 1, 2, 3, and 4 LSAs in Area 1. The last section displays the type 5 LSAs.

To learn more about each LSA type, use the following commands:

- **show ip ospf database router**

- **show ip ospf database network**

- **show ip ospf database summary**

- **show ip ospf database asbr-summary**

- **show ip ospf database external**

Step 3.   Verifying OSPF and exploring LSAs on the ASBR R2.

R2 is an ASBR with interfaces in Area 0 and an external non-OSPF network.

Display the LSDB on R2.

R2# **show ip ospf database**

```
                   OSPF Router with ID (2.2.2.1) (Process ID 123)

                        Router Link States (Area 0)

       Link ID          ADV Router      Age        Seq#        Checksum Link count
       1.1.1.1          1.1.1.1         1790       0x80000009 0x001E87 1
       2.2.2.1          2.2.2.1         1822       0x8000000C 0x00A06E 2
       3.3.3.1          3.3.3.1         1759       0x80000008 0x00BDDA 1


                        Net Link States (Area 0)

       Link ID          ADV Router      Age        Seq#        Checksum
       172.16.0.2       1.1.1.1         1822       0x80000006 0x002A3E
       172.16.1.1       2.2.2.1         1822       0x80000006 0x0067F9


                        Summary Net Link States (Area 0)

       Link ID          ADV Router      Age        Seq#        Checksum
       10.10.0.0        1.1.1.1         1790       0x80000008 0x00B462
       10.10.1.0        1.1.1.1         1790       0x80000006 0x0024E6
       10.10.4.0        3.3.3.1         1759       0x80000008 0x0058B4
       10.10.5.0        3.3.3.1         1759       0x80000006 0x00C739
```

```
               Type-5 AS External Link States

Link ID        ADV Router      Age        Seq#       Checksum Tag
0.0.0.0        2.2.2.1         1822       0x80000006 0x009794 123
```

The output displays the type 1, 2, 3 and 5 LSAs in Area 0. Notice that there is no type 4 LSA because R2 is the ASBR and only an ABR can generate an LSA4.

Router Interface Summary Table

Router Model	Ethernet Interface #1	Ethernet Interface #2	Serial Interface #1	Serial Interface #2
1800	Fast Ethernet 0/0 (F0/0)	Fast Ethernet 0/1 (F0/1)	Serial 0/0/0 (S0/0/0)	Serial 0/0/1 (S0/0/1)
1900	Gigabit Ethernet 0/0 (G0/0)	Gigabit Ethernet 0/1 (G0/1)	Serial 0/0/0 (S0/0/0)	Serial 0/0/1 (S0/0/1)
2801	Fast Ethernet 0/0 (F0/0)	Fast Ethernet 0/1 (F0/1)	Serial 0/1/0 (S0/1/0)	Serial 0/1/1 (S0/1/1)
2811	Fast Ethernet 0/0 (F0/0)	Fast Ethernet 0/1 (F0/1)	Serial 0/0/0 (S0/0/0)	Serial 0/0/1 (S0/0/1)
2900	Gigabit Ethernet 0/0 (G0/0)	Gigabit Ethernet 0/1 (G0/1)	Serial 0/0/0 (S0/0/0)	Serial 0/0/1 (S0/0/1)
4221	Gigabit Ethernet 0/0/0 (G0/0/0)	Gigabit Ethernet 0/0/1 (G0/0/1)	Serial 0/1/0 (S0/1/0)	Serial 0/1/1 (S0/1/1)
4300	Gigabit Ethernet 0/0/0 (G0/0/0)	Gigabit Ethernet 0/0/1 (G0/0/1)	Serial 0/1/0 (S0/1/0)	Serial 0/1/1 (S0/1/1)

Note: To find out how the router is configured, look at the interfaces to identify the type of router and how many interfaces the router has. There is no way to effectively list all the combinations of configurations for each router class. This table includes identifiers for the possible combinations of Ethernet and Serial interfaces in the device. The table does not include any other type of interface, even though a specific router may contain one. An example of this might be an ISDN BRI interface. The string in parenthesis is the legal abbreviation that can be used in Cisco IOS commands to represent the interface.

Device Config Final – Notes

7.1.3 Lab - OSPFv2 Route Summarization and Filtering

Topology

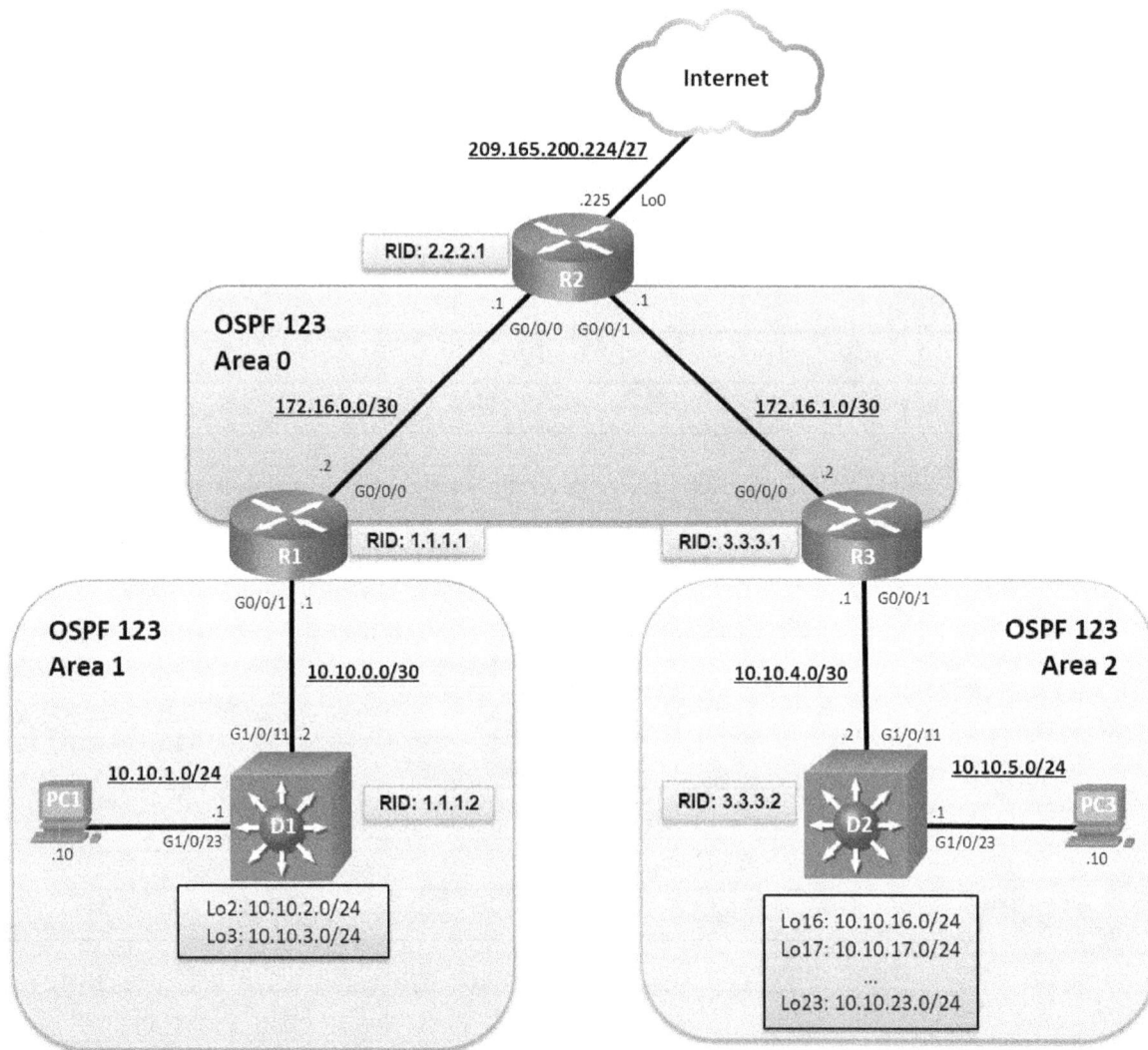

Addressing Table

Device	Interface	IPv4 Address
R1	G0/0/0	172.16.0.2/30
	G0/0/1	10.10.0.1/30
	Lo0	209.165.200.225/27
R2	G0/0/0	172.16.0.1/30
	G0/0/1	172.16.1.1/30
R3	G0/0/0	172.16.1.2/30
	G0/0/1	10.10.4.1/30

Device	Interface	IPv4 Address
D1	G1/0/11	10.10.0.2/30
	G1/0/23	10.10.1.1/24
	Lo2	10.10.2.1/24
	Lo3	10.10.3.1/24
D2	G1/0/11	10.10.4.2/30
	G1/0/23	10.10.5.1/24
	Lo16	10.10.16.1/24
	Lo17	10.10.17.1/24
	Lo18	10.10.18.1/24
	Lo19	10.10.19.1/24
	Lo20	10.10.20.1/24
	Lo21	10.10.21.1/24
	Lo22	10.10.22.1/24
	Lo23	10.10.23.1/24
PC1	NIC	10.10.1.10/24
PC2	NIC	10.10.5.10/24

Objectives

Part 1: Build the Network, Configure Basic Device Settings and Routing

Part 2: OSPFv2 Route Summarization

Part 3: OSPFv2 Route Filtering

Background/Scenario

Areas make OSPF more scalable and increase efficiency. Consider a large multinational organization with a thousand OSPF routers. If all routers were in a single area, the information contained in their LSDB would be overwhelming. Segmenting the OSPF domain into multiple areas reduces the size of the LSDB for each area, making SPF tree calculations faster, and decreasing LSDB flooding between routers when a link flaps.

To make OSPF even more scalable and efficient, network routes can be summarized and advertised in other areas. As well, specific route filtering can be used to provide more precise control on route propagation.

In this lab, you will configure route summarization and route filtering in a multiarea OSPF version 2 network. This lab was specifically designed to use three routers and two Layer 3 switches. To help visualize the potential of summarization and route filtering, additional loopback interfaces will be configured to simulate LANs and create larger routing tables.

Note: This lab is an exercise in developing, deploying, and verifying how OSPF route summarization and filtering operates and does not reflect networking best practices.

Note: The router used with this CCNP hands-on lab is a Cisco 4221 and the two Layer 3 switches are Catalyst 3560 switches. Other routers and Layer 3 switches and Cisco IOS versions can be used. Depending on the model and Cisco IOS version, the commands available and the output produced might vary from what is shown in the labs.

Note: Ensure that the routers and switches have been erased and have no startup configurations. If you are unsure contact your instructor.

Required Resources

- 3 Routers (Cisco 4221 with Cisco IOS XE Release 16.9.4 universal image or comparable)

- 2 Switches (Cisco 3650 with Cisco IOS XE Release 16.9.4 universal image or comparable)

- 2 PCs (Windows with terminal emulation program, such as Tera Term)

- Console cables to configure the Cisco IOS devices via the console ports

- Ethernet cables as shown in the topology

Instructions

Part 1: Build the Network, Configure Basic Device Settings and Routing

In Part 1, you will set up the network topology and configure basic settings and interface addressing on the router and Layer 3 switches. You will also configure multiarea OSPFv2 on the OSPF backbone routers R1, R2, and R3. You will manually configure OSPFv2 on D1 and D2.

Step 1.   Cable the network as shown in the topology.

Attach the devices as shown in the topology diagram, and cable as necessary.

Step 2.   Configure basic settings for the routers.

    a.   Console into each router, enter global configuration mode, and apply the basic settings, interface addressing, and OSPFv2 configuration. The configuration for each device is provided for you below.

Note: Routers were configured with OSPFv2.

Router R1

```
hostname R1
no ip domain lookup
line con 0
 logging sync
 exec-time 0 0
 exit
banner motd # This is R1, OSPFv2 Route Summarization and Filtering Lab #
interface g0/0/0
 ip add 172.16.0.2 255.255.255.252
 no shut
```

```
 exit
interface GigabitEthernet0/0/1
 ip address 10.10.0.1 255.255.255.252
 no shut
 exit
router ospf 123
 router-id 1.1.1.1
 auto-cost reference-bandwidth 1000
 network 10.10.0.0 0.0.0.3 area 1
 network 172.16.0.0 0.0.0.3 area 0
 exit
```

Router R2

```
hostname R2
no ip domain lookup
line con 0
 logging sync
 exec-time 0 0
 exit
banner motd # This is R2, OSPFv2 Route Summarization and Filtering Lab #
interface g0/0/0
 ip add 172.16.0.1 255.255.255.252
 no shut
 exit
interface GigabitEthernet0/0/1
 ip address 172.16.1.1 255.255.255.252
 no shut
 exit
int lo0
 ip add 209.165.200.225 255.255.255.224
 exit
ip route 0.0.0.0 0.0.0.0 Loopback0
router ospf 123
 router-id 2.2.2.1
 auto-cost reference-bandwidth 1000
 network 172.16.0.0 0.0.0.3 area 0
 network 172.16.1.0 0.0.0.3 area 0
 default-information originate
 exit
```

Router R3

```
hostname R3
no ip domain lookup
line con 0
 logging sync
 exec-time 0 0
 exit
banner motd # This is R3, OSPFv2 Route Summarization and Filtering Lab #
interface g0/0/0
 ip add 172.16.1.2 255.255.255.252
 no shut
```

```
    exit
  interface GigabitEthernet0/0/1
    ip address 10.10.4.1 255.255.255.252
    no shut
    exit
  router ospf 123
    router-id 3.3.3.1
    auto-cost reference-bandwidth 1000
    network 10.10.4.0 0.0.0.3 area 2
    network 172.16.1.0 0.0.0.3 area 0
    exit
```

b. Save the running configuration to startup-config.

Step 3. Configure basic settings for the switches.

a. Console into the switch, enter global configuration mode, and apply the basic settings and interface addressing. A command list for each switch is provided below.

Note: OSPF routing will be manually configured.

Switch D1

```
hostname D1
no ip domain lookup
line con 0
 exec-timeout 0 0
 logging synchronous
 exit
banner motd # This is D1, OSPFv2 Route Summarization and Filtering Lab #
interface g1/0/11
 no switchport
 ip address 10.10.0.2 255.255.255.252
 no shut
 exit
interface g1/0/23
 no switchport
 ip address 10.10.1.1 255.255.255.0
 no shut
 exit
int Lo2
 ip add 10.10.2.1 255.255.255.0
 ip ospf network point-to-point
 exit
int Lo3
 ip add 10.10.3.1 255.255.255.0
 ip ospf network point-to-point
exit
```

Switch D2

```
hostname D2
no ip domain lookup
line con 0
 logging sync
```

```
 exec-time 0 0
 exit
banner motd # This is D2, OSPFv2 Route Summarization and Filtering Lab #
interface g1/0/11
 no switchport
 ip address 10.10.4.2 255.255.255.252
 no shut
 exit
interface g1/0/23
 no switchport
 ip address 10.10.5.1 255.255.255.0
 no shut
 exit
int Lo16
 ip add 10.10.16.1 255.255.255.0
 ip ospf network point-to-point
 exit
int Lo17
 ip add 10.10.17.1 255.255.255.0
 ip ospf network point-to-point
 exit
int Lo18
 ip add 10.10.18.1 255.255.255.0
 ip ospf network point-to-point
 exit
int Lo19
 ip add 10.10.19.1 255.255.255.0
 ip ospf network point-to-point
 exit
int Lo20
 ip add 10.10.20.1 255.255.255.0
 ip ospf network point-to-point
 exit
int Lo21
 ip add 10.10.21.1 255.255.255.0
 ip ospf network point-to-point
 exit
int Lo22
 ip add 10.10.22.1 255.255.255.0
 ip ospf network point-to-point
 exit
int Lo23
 ip add 10.10.23.1 255.255.255.0
 ip ospf network point-to-point
 exit
```

b. Save the running configuration to startup-config.

c. Verify the interfaces configured on D1.

```
D1# show ip interface brief | include manual
GigabitEthernet1/0/11  10.10.0.2      YES manual up              up
GigabitEthernet1/0/23  10.10.1.1      YES manual up              up
Loopback2              10.10.2.1      YES manual up              up
Loopback3              10.10.3.1      YES manual up              up
```

Notice the loopback interfaces configured on D1. Theses interfaces were configured for lab purposes to simulate other LANs.

Note: Loopback interfaces were numbered based on the network address (e.g., Lo2 = 10.10.2.0/24) for convenience only.

---

d. Verify the interfaces configured on D2.

```
D2# show ip interface brief | include manual
GigabitEthernet1/0/11  10.10.4.2      YES manual up              up
GigabitEthernet1/0/23  10.10.5.1      YES manual up              up
Loopback16             10.10.16.1     YES manual up              up
Loopback17             10.10.17.1     YES manual up              up
Loopback18             10.10.18.1     YES manual up              up
Loopback19             10.10.19.1     YES manual up              up
Loopback20             10.10.20.1     YES manual up              up
Loopback21             10.10.21.1     YES manual up              up
Loopback22             10.10.22.1     YES manual up              up
Loopback23             10.10.23.1     YES manual up              up
```

Again, notice the loopback interfaces configured on D1. Theses interfaces were configured for lab purposes to simulate other LANs.

Step 4.   Verify routing on R1, R2, and R3.

a. Verify the routing table of R1 using the **show ip route ospf** command.

```
R1# show ip route ospf | begin Gateway
Gateway of last resort is 172.16.0.1 to network 0.0.0.0

O*E2  0.0.0.0/0 [110/1] via 172.16.0.1, 00:22:17, GigabitEthernet0/0/0
        10.0.0.0/8 is variably subnetted, 3 subnets, 2 masks
O IA    10.10.4.0/30 [110/3] via 172.16.0.1, 00:21:43, GigabitEthernet0/0/0
        172.16.0.0/16 is variably subnetted, 3 subnets, 2 masks
O       172.16.1.0/30 [110/2] via 172.16.0.1, 00:22:11, GigabitEthernet0/0/0
```

The R1 routing table contains an OSPF internal or intra-area route, an interarea route, and an external route to the default gateway.

b. Verify the routing table of R2 using the **show ip route ospf** command.

```
R2# show ip route ospf | begin Gateway
Gateway of last resort is 0.0.0.0 to network 0.0.0.0

        10.0.0.0/30 is subnetted, 2 subnets
O IA    10.10.0.0 [110/2] via 172.16.0.2, 00:19:40, GigabitEthernet0/0/0
O IA    10.10.4.0 [110/2] via 172.16.1.2, 00:19:07, GigabitEthernet0/0/1
```

R2 is propagating the static default route and therefore does not have an external type 2 OSPF route (i.e., O* E2) in the routing table like R1 and R3.

c. Verify the routing table of R3 using the **show ip route ospf** command.

```
R3# show ip route ospf | begin Gateway
Gateway of last resort is 172.16.1.1 to network 0.0.0.0

O*E2  0.0.0.0/0 [110/1] via 172.16.1.1, 00:20:00, GigabitEthernet0/0/0
         10.0.0.0/8 is variably subnetted, 3 subnets, 2 masks
O IA     10.10.0.0/30 [110/3] via 172.16.1.1, 00:20:00, GigabitEthernet0/0/0
         172.16.0.0/16 is variably subnetted, 3 subnets, 2 masks
O        172.16.0.0/30 [110/2] via 172.16.1.1, 00:20:00, GigabitEthernet0/0/0
```

Like R1, R3 has an internal route (LSA 2), an interarea route (LSA 3), and an external route (LSA 5).

The LANs connected to D1 and D2 are not yet advertised.

Step 5. Enable OSPFv2 on D1.

a. On D1, enable IP routing using the **ip routing global configuration** command.

```
D1(config)# ip routing
```

b. Next, enter the OSPF router configuration mode using process ID 123, assign D1 the router ID 1.1.1.2 and set the reference bandwidth to distinguish between Gigabit Ethernet and FastEthernet interfaces.

```
D1(config)# router ospf 123
D1(config-router)# router-id 1.1.1.2
D1(config-router)# auto-cost reference-bandwidth 1000
% OSPF: Reference bandwidth is changed.
        Please ensure reference bandwidth is consistent across all routers.
```

Note: Setting the reference cost value too high may cause issues with low-bandwidth interfaces.

c. Next, we need to have D1 advertise all four of its directly connected interfaces. Although this could be accomplished using four separate **network** statements, we will use the wildcard mask to advertise all four interfaces using one **network** statement.

```
D1(config-router)# network 10.10.0.0 0.0.3.255 area 1
D1(config-router)# end
*Mar  1 01:01:22.540: %OSPF-5-ADJCHG: Process 123, Nbr 1.1.1.1 on
GigabitEthernet1/0/11 from LOADING to FULL, Loading Done
```

d. Verify the OSPF routing table on D1.

```
D1# show ip route ospf | begin Gateway
Gateway of last resort is 10.10.0.1 to network 0.0.0.0

O*E2  0.0.0.0/0 [110/1] via 10.10.0.1, 00:05:20, GigabitEthernet1/0/11
         10.0.0.0/8 is variably subnetted, 9 subnets, 3 masks
O IA     10.10.4.0/30 [110/4] via 10.10.0.1, 00:05:20, GigabitEthernet1/0/11
         172.16.0.0/30 is subnetted, 2 subnets
O IA     172.16.0.0 [110/2] via 10.10.0.1, 00:05:20, GigabitEthernet1/0/11
O IA     172.16.1.0 [110/3] via 10.10.0.1, 00:05:20, GigabitEthernet1/0/11
```

e. Verify the routing table of R2 using the **show ip route ospf** command.

```
R2# show ip route ospf | begin Gateway
Gateway of last resort is 0.0.0.0 to network 0.0.0.0


      10.0.0.0/8 is variably subnetted, 5 subnets, 2 masks
O IA     10.10.0.0/30 [110/2] via 172.16.0.2, 00:40:29, GigabitEthernet0/0/0
O IA     10.10.1.0/24 [110/12] via 172.16.0.2, 00:06:56, GigabitEthernet0/0/0
O IA     10.10.2.0/24 [110/3] via 172.16.0.2, 00:06:56, GigabitEthernet0/0/0
O IA     10.10.3.0/24 [110/3] via 172.16.0.2, 00:06:56, GigabitEthernet0/0/0
O IA     10.10.4.0/30 [110/2] via 172.16.1.2, 00:39:56, GigabitEthernet0/0/1
```

Notice how its routing table now includes routes to the D1 LANs. Notice also how this has increased the number of routing entries.

Step 6.   Enable OSPFv2 on D2.

a. On D2, enable IP routing using the **ip routing** global configuration command.

```
D2(config)# ip routing
```

b. Next, enter the OSPF router configuration mode using process ID 123, assign D2 the router ID 3.3.3.2 and set the reference bandwidth to distinguish between Gigabit Ethernet and FastEthernet interfaces.

```
D2(config)# router ospf 123
D2(config-router)# router-id 3.3.3.2
D2(config-router)# auto-cost reference-bandwidth 1000
% OSPF: Reference bandwidth is changed.
        Please ensure reference bandwidth is consistent across all routers.
```

Note: Setting the reference cost value too high may cause issues with low-bandwidth interfaces.

c. Advertise the 10.10.4.0/30 and 10.10.5.0 /255 networks. Again, this could be accomplished using separate **network** statements. However, the wildcard mask can be used to advertise both interfaces using one **network** statement as shown.

```
D2(config-router)# network 10.10.4.0 0.0.1.255 area 2
D2(config-router)#
*Mar  1 01:15:02.643: %OSPF-5-ADJCHG: Process 123, Nbr 3.3.3.1 on
GigabitEthernet1/0/11 from LOADING to FULL, Loading Done
```

Note: The wildcard mask 0.0.1.255 matches both networks 10.10.4.0/30 and 10.10.5.0/24

d. Next, advertise the 10.10.16.0/24 through to 10.10.23.0/24 loopback interface networks. Traditionally, this would require 8 **network** statements. But again, the wildcard mask can be used to advertise all 8 interfaces using one **network** statement as shown.

```
D2(config-router)# network 10.10.16.0 0.0.7.255 area 2
D2(config-router)# end
```

Note: The wildcard mask 0.0.7.255 matches networks 10.10.16.0/24 through to 10.10.23.0/24.

Step 7.   Verify Routing.

a. Verify the routing table of D2 using the **show ip route ospf** command.

```
D2# show ip route ospf | begin Gateway
Gateway of last resort is 10.10.4.1 to network 0.0.0.0
```

```
O*E2  0.0.0.0/0 [110/1] via 10.10.4.1, 00:02:19, GigabitEthernet1/0/11
        10.0.0.0/8 is variably subnetted, 24 subnets, 3 masks
O IA      10.10.0.0/30 [110/4] via 10.10.4.1, 00:02:19, GigabitEthernet1/0/11
O IA      10.10.1.0/24 [110/14] via 10.10.4.1, 00:02:19, GigabitEthernet1/0/11
O IA      10.10.2.0/24 [110/5] via 10.10.4.1, 00:02:19, GigabitEthernet1/0/11
O IA      10.10.3.0/24 [110/5] via 10.10.4.1, 00:02:19, GigabitEthernet1/0/11
        172.16.0.0/30 is subnetted, 2 subnets
O IA      172.16.0.0 [110/3] via 10.10.4.1, 00:02:19, GigabitEthernet1/0/11
O IA      172.16.1.0 [110/2] via 10.10.4.1, 00:02:19, GigabitEthernet1/0/11
```

D2 has OSPF route entries for:

- One external OSPF route to the gateway of last resort.

- The four D1 LANs (i.e., 10.10.0.0/30 through 10.10.3.0/24)

- The two Area 0 networks (i.e., 172.16.0.0/30 and 172.16.1.0/30)

b. From R2, verify the routing table using the **show ip route ospf** command.

```
R2# show ip route ospf | begin Gateway
Gateway of last resort is 0.0.0.0 to network 0.0.0.0


        10.0.0.0/8 is variably subnetted, 14 subnets, 2 masks
O IA      10.10.0.0/30 [110/2] via 172.16.0.2, 01:00:10, GigabitEthernet0/0/0
O IA      10.10.1.0/24 [110/12] via 172.16.0.2, 00:26:37, GigabitEthernet0/0/0
O IA      10.10.2.0/24 [110/3] via 172.16.0.2, 00:26:37, GigabitEthernet0/0/0
O IA      10.10.3.0/24 [110/3] via 172.16.0.2, 00:26:37, GigabitEthernet0/0/0
O IA      10.10.4.0/30 [110/2] via 172.16.1.2, 00:59:37, GigabitEthernet0/0/1
O IA      10.10.5.0/24 [110/12] via 172.16.1.2, 00:09:55, GigabitEthernet0/0/1
O IA      10.10.16.0/24 [110/3] via 172.16.1.2, 00:00:13, GigabitEthernet0/0/1
O IA      10.10.17.0/24 [110/3] via 172.16.1.2, 00:00:13, GigabitEthernet0/0/1
O IA      10.10.18.0/24 [110/3] via 172.16.1.2, 00:00:13, GigabitEthernet0/0/1
O IA      10.10.19.0/24 [110/3] via 172.16.1.2, 00:00:13, GigabitEthernet0/0/1
O IA      10.10.20.0/24 [110/3] via 172.16.1.2, 00:00:13, GigabitEthernet0/0/1
O IA      10.10.21.0/24 [110/3] via 172.16.1.2, 00:00:13, GigabitEthernet0/0/1
O IA      10.10.22.0/24 [110/3] via 172.16.1.2, 00:00:13, GigabitEthernet0/0/1
O IA      10.10.23.0/24 [110/3] via 172.16.1.2, 00:00:13, GigabitEthernet0/0/1
```

Notice how the routing table of R2 now includes routes to the D1 and D2 LANs. And again, notice how this has increased the number of routing entries.

c. From D1, verify the routing table using the **show ip route ospf** command.

```
D1# show ip route ospf | begin Gateway
Gateway of last resort is 10.10.0.1 to network 0.0.0.0


O*E2  0.0.0.0/0 [110/1] via 10.10.0.1, 00:18:43, GigabitEthernet1/0/11
        10.0.0.0/8 is variably subnetted, 18 subnets, 3 masks
O IA      10.10.4.0/30 [110/4] via 10.10.0.1, 00:18:43, GigabitEthernet1/0/11
O IA      10.10.5.0/24 [110/5] via 10.10.0.1, 00:09:56, GigabitEthernet1/0/11
O IA      10.10.16.0/24 [110/5] via 10.10.0.1, 00:08:27, GigabitEthernet1/0/11
O IA      10.10.17.0/24 [110/5] via 10.10.0.1, 00:08:27, GigabitEthernet1/0/11
O IA      10.10.18.0/24 [110/5] via 10.10.0.1, 00:08:27, GigabitEthernet1/0/11
O IA      10.10.19.0/24 [110/5] via 10.10.0.1, 00:08:27, GigabitEthernet1/0/11
O IA      10.10.20.0/24 [110/5] via 10.10.0.1, 00:08:27, GigabitEthernet1/0/11
```

```
O IA     10.10.21.0/24 [110/5] via 10.10.0.1, 00:08:27, GigabitEthernet1/0/11
O IA     10.10.22.0/24 [110/5] via 10.10.0.1, 00:08:27, GigabitEthernet1/0/11
O IA     10.10.23.0/24 [110/5] via 10.10.0.1, 00:08:27, GigabitEthernet1/0/11
         172.16.0.0/30 is subnetted, 2 subnets
O IA     172.16.0.0 [110/2] via 10.10.0.1, 00:18:43, GigabitEthernet1/0/11
O IA     172.16.1.0 [110/3] via 10.10.0.1, 00:18:43, GigabitEthernet1/0/11
```

Notice the OSPF routing table now includes the additional interarea routes from D2.

Part 2: OSPFv2 Route Summarization

As shown in Part 1, routing tables increase in the number of entries as more and more networks are connected to the OSPF domain.

To reduce the size of the routing table and LSDB, network prefixes must be summarized. Route summarization improves OSPF performance as fewer network entries are required.

Route summarization involves consolidating multiple routes into a single advertisement. Proper route summarization reduces the bandwidth, memory, and CPU resources consumed by the OSPF process.

OSPF routes can only be summarized between areas. Interarea route summarization is configured on ABRs using the **area** *area-id* **range** *network subnet-mask* [**advertise** | **not-advertise**] [**cost** *metric*] router configuration command.

Parameter	Description
area *area-id*	■ Identifies the area subject to route summarization.
address	■ The summary address designated for a range of addresses.
mask	■ The IP subnet mask used for the summary route.
advertise	■ Enabled by default, it sets the address range status to advertise and generate a type 3 summary LSA.
not-advertise	■ (Optional) Sets the address range status to DoNotAdvertise. ■ Can be used for route filtering as the type 3 summary LSA is suppressed, and the component networks remain hidden from other networks.
cost *cost*	■ (Optional) Metric or cost for this summary route, which is used during the OSPF SPF calculation to determine the shortest paths to the destination. ■ The value can be 0 to 16777215.

In this part, you will learn how to reduce the number of routing entries without compromising access to any networks.

Step 1.   Configure interarea route summarization on R1.

Area 1 consists of networks 10.10.0.0/30, 10.10.1.0/24, 10.10.2.0/24, and 10.10.3.0/24. To calculate the summary address of these networks:

1) List the networks in binary format.

2) Count the number of left-most matching bits to determine the mask.

3) Copy the matching bits and add zero bits to determine the network address.

The four networks are listed in binary format.

Network	1st Octet	2nd Octet	3rd Octet	4th Octet
10.10.0.0	0000 1010	0000 1010	0000 0000	0000 0000
10.10.1.0	0000 1010	0000 1010	0000 0001	0000 0000
10.10.2.0	0000 1010	0000 1010	0000 0010	0000 0000
10.10.3.0	0000 1010	0000 1010	0000 0011	0000 0000

There are 22 left-most bits that match. Octet 1 and 2 match for a sum of 16 bits. There are 6 left-most bits that match in the 3rd octet which results in a total of 22 bits that match.

A /22 subnet converts to 255.255.252.0.

Therefore, the summary network address of networks 10.10.0.0/30, 10.10.1.0/24, 10.10.2.0/24, and 10.10.3.0/24 is 10.10.0.0 255.255.252.0.

a. On the Area 1 ABR router R1, enter OSFP router config mode.

```
R1(config)# router ospf 123
```

b. Summarize the D1 LANs using the **area 1 range 10.10.0.0 255.255.252.0** router configuration command.

```
R1(config-router)# area 1 range 10.10.0.0 255.255.252.0
R1(config-router)# end
```

Step 2. Verify the interarea route summarization.

a. Verify the routing table of R1 using the **show ip route ospf** command.

```
R1# show ip route ospf | begin Gateway
Gateway of last resort is 172.16.0.1 to network 0.0.0.0

O*E2  0.0.0.0/0 [110/1] via 172.16.0.1, 00:40:51, GigabitEthernet0/0/0
         10.0.0.0/8 is variably subnetted, 16 subnets, 4 masks
O        10.10.0.0/22 is a summary, 00:40:51, Null0
O        10.10.1.0/24 [110/11] via 10.10.0.2, 00:40:51, GigabitEthernet0/0/1
O        10.10.2.0/24 [110/2] via 10.10.0.2, 00:40:51, GigabitEthernet0/0/1
O        10.10.3.0/24 [110/2] via 10.10.0.2, 00:40:51, GigabitEthernet0/0/1
O IA     10.10.4.0/30 [110/4] via 172.16.0.1, 00:40:51, GigabitEthernet0/0/0
O IA     10.10.5.0/24 [110/4] via 172.16.0.1, 00:40:51, GigabitEthernet0/0/0
O IA     10.10.16.0/24 [110/4] via 172.16.0.1, 00:40:51, GigabitEthernet0/0/0
O IA     10.10.17.0/24 [110/4] via 172.16.0.1, 00:40:51, GigabitEthernet0/0/0
O IA     10.10.18.0/24 [110/4] via 172.16.0.1, 00:40:51, GigabitEthernet0/0/0
O IA     10.10.19.0/24 [110/4] via 172.16.0.1, 00:40:51, GigabitEthernet0/0/0
O IA     10.10.20.0/24 [110/4] via 172.16.0.1, 00:40:51, GigabitEthernet0/0/0
O IA     10.10.21.0/24 [110/4] via 172.16.0.1, 00:40:51, GigabitEthernet0/0/0
O IA     10.10.22.0/24 [110/4] via 172.16.0.1, 00:40:51, GigabitEthernet0/0/0
O IA     10.10.23.0/24 [110/4] via 172.16.0.1, 00:40:51, GigabitEthernet0/0/0
         172.16.0.0/16 is variably subnetted, 3 subnets, 2 masks
O        172.16.1.0/30 [110/2] via 172.16.0.1, 00:40:51, GigabitEthernet0/0/0
```

The routing table of the ABR using the summary address adds a discard route entry to the Null 0 interface to prevent routing loops. The portions of the summarized network range will not have a more specific route in the routing table.

b. Verify the routing table of R2 using the **show ip route ospf** command.

```
R2# show ip route ospf | begin Gateway
Gateway of last resort is 0.0.0.0 to network 0.0.0.0

      10.0.0.0/8 is variably subnetted, 11 subnets, 3 masks
O IA     10.10.0.0/22 [110/2] via 172.16.0.2, 00:36:38, GigabitEthernet0/0/0
O IA     10.10.4.0/30 [110/2] via 172.16.1.2, 02:20:03, GigabitEthernet0/0/1
O IA     10.10.5.0/24 [110/12] via 172.16.1.2, 01:30:21, GigabitEthernet0/0/1
O IA     10.10.16.0/24 [110/3] via 172.16.1.2, 01:20:39, GigabitEthernet0/0/1
O IA     10.10.17.0/24 [110/3] via 172.16.1.2, 01:20:39, GigabitEthernet0/0/1
O IA     10.10.18.0/24 [110/3] via 172.16.1.2, 01:20:39, GigabitEthernet0/0/1
O IA     10.10.19.0/24 [110/3] via 172.16.1.2, 01:20:39, GigabitEthernet0/0/1
O IA     10.10.20.0/24 [110/3] via 172.16.1.2, 01:20:39, GigabitEthernet0/0/1
O IA     10.10.21.0/24 [110/3] via 172.16.1.2, 01:20:39, GigabitEthernet0/0/1
O IA     10.10.22.0/24 [110/3] via 172.16.1.2, 01:20:39, GigabitEthernet0/0/1
O IA     10.10.23.0/24 [110/3] via 172.16.1.2, 01:20:39, GigabitEthernet0/0/1
```

Notice how the previous four route entries for network 10.10.0.0 to 10.0.3.0 have now been summarized onto one route.

Step 3. Configure interarea route summarization on R3.

Area 2 consists of networks 10.10.4.0/30 and 10.10.5.0/24. It also consists of LANs 10.10.16.0/24 through to 10.10.23.0/24. These networks are not contiguous and cannot easily be summarized. For this reason, two summary commands will be configured on R3.

a. On R3 enter OSFP router config mode.

```
R3(config)# router ospf 123
```

b. The first summary advertisement will be for the 10.10.4.0/30 and 10.10.5.0/24 networks. To summarize they are listed in binary format.

Network	1st Octet	2nd Octet	3rd Octet	4th Octet
10.10.4.0	0000 1010	0000 1010	0000 0100	0000 0000
10.10.5.0	0000 1010	0000 1010	0000 0101	0000 0000

There are 23 left-most bits that match. Octet 1 and 2 match for a sum of 16 bits. There are 7 left-most bits that match in the 3rd octet which results in a total of 23 bits that match.

A /23 subnet converts to 255.255.254.0. Therefore, the summary network address of networks 10.10.4.0/30 and 10.10.5.0/24 is 10.10.4.0 255.255.254.0.

Summarize the D2 LANs using the **area 2 range 10.10.4.0 255.255.254.0** router configuration command.

```
R3(config-router)# area 2 range 10.10.4.0 255.255.254.0
```

c. The second summary advertisement will be for the 10.10.16.0/24 through to 10.10.23.0/24 networks. Although all eight networks could be listed in binary format, it is possible to discover the summary addresses by only listing the first network and last network in binary format.

Network	1st Octet	2nd Octet	3rd Octet	4th Octet
10.10.16.0	0000 1010	0000 1010	0001 0000	0000 0000
10.10.23.0	0000 1010	0000 1010	0001 0111	0000 0000

There are 21 left-most bits that match. Octet 1 and 2 match for a sum of 16 bits. There are 5 left-most bits that match in the 3rd octet which results in a total of 21 bits that match.

A /21 subnet converts to 255.255.248.0. Therefore, the summary network address of networks 10.10.16.0/24 through to 10.10.23.0/24 is 10.10.16.0 255.255.248.0.

A cost can also be assigned to a summary route by using the **cost** keyword.

Summarize the D2 LANs and assign them a cost of 65 using the **area 2 range 10.10.16.0 255.255.248.0 cost 65** router configuration command.

```
R3(config-router)# area 2 range 10.10.16.0 255.255.248.0 cost 65
```

Step 4.   Verify the interarea route summarization

a.   Verify the routing table of R1 using the **show ip route ospf** command.

```
R3# show ip route ospf | begin Gateway
Gateway of last resort is 172.16.1.1 to network 0.0.0.0

O*E2  0.0.0.0/0 [110/1] via 172.16.1.1, 00:01:14, GigabitEthernet0/0/0
      10.0.0.0/8 is variably subnetted, 14 subnets, 6 masks
O IA    10.10.0.0/22 [110/3] via 172.16.1.1, 00:01:14, GigabitEthernet0/0/0
O       10.10.4.0/23 is a summary, 00:01:14, Null0
O       10.10.5.0/24 [110/2] via 10.10.4.2, 00:01:14, GigabitEthernet0/0/1
O       10.10.16.0/21 is a summary, 00:01:14, Null0
O       10.10.16.0/24 [110/2] via 10.10.4.2, 00:01:14, GigabitEthernet0/0/1
O       10.10.17.0/24 [110/2] via 10.10.4.2, 00:01:14, GigabitEthernet0/0/1
O       10.10.18.0/24 [110/2] via 10.10.4.2, 00:01:14, GigabitEthernet0/0/1
O       10.10.19.0/24 [110/2] via 10.10.4.2, 00:01:14, GigabitEthernet0/0/1
O       10.10.20.0/24 [110/2] via 10.10.4.2, 00:01:14, GigabitEthernet0/0/1
O       10.10.21.0/24 [110/2] via 10.10.4.2, 00:01:14, GigabitEthernet0/0/1
O       10.10.22.0/24 [110/2] via 10.10.4.2, 00:01:14, GigabitEthernet0/0/1
O       10.10.23.0/24 [110/2] via 10.10.4.2, 00:01:14, GigabitEthernet0/0/1
      172.16.0.0/16 is variably subnetted, 3 subnets, 2 masks
O       172.16.0.0/30 [110/2] via 172.16.1.1, 00:01:14, GigabitEthernet0/0/0
```

R3 added two discard route entries to the Null 0 interface to prevent routing loops.

b.   Verify the routing table of R2 using the **show ip route ospf** command.

```
R2# show ip route ospf | begin Gateway
Gateway of last resort is 0.0.0.0 to network 0.0.0.0

      10.0.0.0/8 is variably subnetted, 3 subnets, 3 masks
O IA    10.10.0.0/22 [110/2] via 172.16.0.2, 00:01:25, GigabitEthernet0/0/0
O IA    10.10.4.0/23 [110/2] via 172.16.1.2, 00:01:25, GigabitEthernet0/0/1
O IA    10.10.16.0/21 [110/66] via 172.16.1.2, 00:00:04, GigabitEthernet0/0/1
```

Notice how R2 now has only two route entries for the D2 LANs. Previous to route summarization on R3, R2 had 10 route entries for the D2 LANs. Also notice the cost of the 10.10.16.0/21 route has been influenced.

c.   Verify the routing table of D1 using the **show ip route ospf** command.

```
D1# show ip route ospf | begin Gateway
Gateway of last resort is 10.10.0.1 to network 0.0.0.0

O*E2  0.0.0.0/0 [110/1] via 10.10.0.1, 01:19:46, GigabitEthernet1/0/11
          10.0.0.0/8 is variably subnetted, 10 subnets, 5 masks
O IA     10.10.4.0/23 [110/4] via 10.10.0.1, 00:10:04, GigabitEthernet1/0/11
O IA     10.10.16.0/21 [110/68] via 10.10.0.1, 00:09:23, GigabitEthernet1/0/11
          172.16.0.0/30 is subnetted, 2 subnets
O IA     172.16.0.0 [110/2] via 10.10.0.1, 01:19:46, GigabitEthernet1/0/11
O IA     172.16.1.0 [110/3] via 10.10.0.1, 01:19:46, GigabitEthernet1/0/11
```

Notice how its routing table is smaller.

d.   Verify connectivity to D2.

```
D1# ping 10.10.23.1 source 10.10.1.1

Type escape sequence to abort.
Sending 5, 100-byte ICMP Echos to 10.10.23.1, timeout is 2 seconds:
Packet sent with a source address of 10.10.1.1
!!!!!
Success rate is 100 percent (5/5), round-trip min/avg/max = 1/2/8 ms
```

Part 3: OSPFv2 Route Filtering

In this part, you will learn about OSPF route filtering. Route filtering is a method for selectively identifying routes that are advertised or received from neighbor routers. Route filtering may be used to manipulate traffic flows, reduce memory utilization, or improve security.

Filtering of routes with vector-based routing protocols is straightforward. This is because the routes are filtered as routing updates and are advertised to downstream neighbors. However, with link-state routing protocols such as OSPF, every router in an area shares a complete copy of the link-state database. Therefore, filtering of routes generally occurs as routes enter the area on the ABR.

The following sections describe three techniques for filtering routes with OSPF.

- **Filtering with Summarization** – An easy router filtering method is to use the **area** *area-id* **range** *network subnet-mask* **not-advertise** router config command. However, it is limited in its ability to filter.

- **Area Filtering** – OSPF area filtering is accomplished by using the **area** *area-id* **filter-list prefix** *prefix-list-name* {**in** | **out**} router config command on the ABR.

- **Local OSPF Filtering** – To enable a route to exist in the OSPF LSDB and prevent it from being installed in the local routing table, use the distribute list feature.

Step 1.   Filter with summarization.

a.   As an example of filtering with summarization, we will remove the last route summarization command configured on R3.

```
R3(config-router)# no area 2 range 10.10.16.0 255.255.248.0
```

b. On D1, verify that all of the 1010.16.0/24 through 10.10.23.0/24 networks are in the routing table.

```
D1# show ip route ospf | begin Gateway
Gateway of last resort is 10.10.0.1 to network 0.0.0.0

O*E2  0.0.0.0/0 [110/1] via 10.10.0.1, 01:25:49, GigabitEthernet1/0/11
        10.0.0.0/8 is variably subnetted, 17 subnets, 4 masks
O IA    10.10.4.0/23 [110/4] via 10.10.0.1, 00:16:07, GigabitEthernet1/0/11
O IA    10.10.16.0/24 [110/5] via 10.10.0.1, 00:00:07, GigabitEthernet1/0/11
O IA    10.10.17.0/24 [110/5] via 10.10.0.1, 00:00:07, GigabitEthernet1/0/11
O IA    10.10.18.0/24 [110/5] via 10.10.0.1, 00:00:07, GigabitEthernet1/0/11
O IA    10.10.19.0/24 [110/5] via 10.10.0.1, 00:00:07, GigabitEthernet1/0/11
O IA    10.10.20.0/24 [110/5] via 10.10.0.1, 00:00:07, GigabitEthernet1/0/11
O IA    10.10.21.0/24 [110/5] via 10.10.0.1, 00:00:07, GigabitEthernet1/0/11
O IA    10.10.22.0/24 [110/5] via 10.10.0.1, 00:00:07, GigabitEthernet1/0/11
O IA    10.10.23.0/24 [110/5] via 10.10.0.1, 00:00:07, GigabitEthernet1/0/11
        172.16.0.0/30 is subnetted, 2 subnets
O IA    172.16.0.0 [110/2] via 10.10.0.1, 01:31:12, GigabitEthernet1/0/11
O IA    172.16.1.0 [110/3] via 10.10.0.1, 01:31:12, GigabitEthernet1/0/11
```

The D2 LANs are in the routing table of D1.

c. Now, on R3, filter the 10.10.18.0/24 network from being advertised to another area using the **not-advertise** keyword.

```
R3(config-router)# area 2 range 10.10.18.0 255.255.255.0 not-advertise
```

d. On D1, verify the routing table.

```
D1# show ip route ospf | begin Gateway
Gateway of last resort is 10.10.0.1 to network 0.0.0.0

O*E2  0.0.0.0/0 [110/1] via 10.10.0.1, 01:31:12, GigabitEthernet1/0/11
        10.0.0.0/8 is variably subnetted, 16 subnets, 4 masks
O IA    10.10.4.0/23 [110/4] via 10.10.0.1, 00:21:30, GigabitEthernet1/0/11
O IA    10.10.16.0/24 [110/5] via 10.10.0.1, 00:05:30, GigabitEthernet1/0/11
O IA    10.10.17.0/24 [110/5] via 10.10.0.1, 00:05:30, GigabitEthernet1/0/11
O IA    10.10.19.0/24 [110/5] via 10.10.0.1, 00:05:30, GigabitEthernet1/0/11
O IA    10.10.20.0/24 [110/5] via 10.10.0.1, 00:05:30, GigabitEthernet1/0/11
O IA    10.10.21.0/24 [110/5] via 10.10.0.1, 00:05:30, GigabitEthernet1/0/11
O IA    10.10.22.0/24 [110/5] via 10.10.0.1, 00:05:30, GigabitEthernet1/0/11
O IA    10.10.23.0/24 [110/5] via 10.10.0.1, 00:05:30, GigabitEthernet1/0/11
        172.16.0.0/30 is subnetted, 2 subnets
O IA    172.16.0.0 [110/2] via 10.10.0.1, 01:31:12, GigabitEthernet1/0/11
O IA    172.16.1.0 [110/3] via 10.10.0.1, 01:31:12, GigabitEthernet1/0/11
```

Notice that the 10.10.18.0/24 prefix is no longer in the routing table of D1.

Step 2.   Use area filtering.

On R1, filter the 10.10.2.0/24 network from being advertised into OSPF Area 0 by creating a prefix list and then referencing the list in the **area** *area-id* **filter-list prefix** *prefix-list-name* {**in** | **out**} command on R1. You will then filter the 10.10.3.0 network from being propagated into Area 2.

a. O R1, remove the route summarization command that was configured in Part 2.

```
R1(config-router)# no area 1 range 10.10.0.0 255.255.252.0
R1(config-router)# exit
```

b. Verify that the routing table of R2 has the 4 entries from Area 1.

```
R2# show ip route ospf | include 0/0/0
O IA     10.10.0.0/30 [110/2] via 172.16.0.2, 00:03:46, GigabitEthernet0/0/0
O IA     10.10.1.0/24 [110/12] via 172.16.0.2, 00:03:46, GigabitEthernet0/0/0
O IA     10.10.2.0/24 [110/3] via 172.16.0.2, 00:03:46, GigabitEthernet0/0/0
O IA     10.10.3.0/24 [110/3] via 172.16.0.2, 00:03:46, GigabitEthernet0/0/0
```

c. Create the following prefix list on R1 to deny 10.10.2.0/24 but permit everything else.

```
R1(config)# ip prefix-list FILTER-1 deny 10.10.2.0/24
R1(config)# ip prefix-list FILTER-1 permit 0.0.0.0/0 le 32
```

d. Enter OSPF router configuration mode and assign the prefix filter incoming in Area 0.

```
R1(config)# router ospf 123
R1(config-router)# area 0 filter-list prefix FILTER-1 in
```

e. Verify that 10.10.2.0 is not in the routing table of R2.

```
R2# show ip route ospf | include 0/0/0
O IA     10.10.0.0/30 [110/2] via 172.16.0.2, 00:08:59, GigabitEthernet0/0/0
O IA     10.10.1.0/24 [110/12] via 172.16.0.2, 00:08:59, GigabitEthernet0/0/0
O IA     10.10.3.0/24 [110/3] via 172.16.0.2, 00:08:59, GigabitEthernet0/0/0
```

Notice that the 10.10.2.0/24 prefix has been filtered from Area 0 and is no longer in the R2 routing table.

f. Verify that D2 has a route entry for 10.10.3.0/24.

```
D2# show ip route | inc 10.10.3.0
O IA     10.10.3.0/24 [110/5] via 10.10.4.1, 00:13:22, GigabitEthernet1/0/11
```

g. On R3, create the following prefix list to deny 10.10.3.0/24 but permit everything else.

```
R3(config)# ip prefix-list FILTER-1 deny 10.10.3.0/24
R3(config)# ip prefix-list FILTER-1 permit 0.0.0.0/0 le 32
```

h. On R3, enter OSPF router configuration mode and assign the prefix filter outgoing from Area 0.

```
R3(config)# router ospf 123
R3(config-router)# area 0 filter-list prefix FILTER-1 out
```

i. Verify that 10.10.3.0 is not in the routing table of D2.

```
D2# show ip route | inc 10.10.3.0
```

Step 3. Use local OSPF filtering.

A distribute list should not be used for filtering prefixes between areas. A distribute list is configured using the **distribute-list** {*acl-number* | *acl*-name | **prefix** *prefix-list-name* | **route-map** *route-map-name*} **in** router configuration command.

In this step, we will filter the 10.10.20.0/24 network from entering the R2 routing table.

a. On R2, verify that 10.10.20.0 is in the routing table.

```
R2# show ip route | include 10.10.20.0
O IA     10.10.20.0/24 [110/3] via 172.16.1.2, 01:32:39, GigabitEthernet0/0/1
```

b. Next enter an ACL called OSPF-FILTER that denies 10.10.20.0/24 from entering the R2 routing table.

```
R2(config)# ip access-list standard OSPF-FILTER
R2(config-std-nacl)# deny 10.10.20.0 0.0.0.255
R2(config-std-nacl)# permit any
R2(config-std-nacl)# exit
```

c. On R2, enter OSPF router configuration mode and assign the distribute list filter.

```
R2(config)# router ospf 123
R2(config-router)# distribute-list OSPF-FILTER in
R2(config-router)# end
```

d. Verify that 10.10.20.0 prefix is not in the routing table of R2.

```
R2# show ip route | include 10.10.20.0
```

e. Verify that the 10.10.20.0 prefix is still being propagated in the area. Verify the routing table of R1.

```
R1# show ip route | include 10.10.20.0
O IA     10.10.20.0/24 [110/4] via 172.16.0.1, 00:45:23, GigabitEthernet0/0/0
```

The 10.10.20.0/24 prefix still appears in the routing table of R1. The distribute list only filtered the route from entering the routing table on R2 but is still in the LSDB for Area 0.

Router Interface Summary Table

Router Model	Ethernet Interface #1	Ethernet Interface #2	Serial Interface #1	Serial Interface #2
1800	Fast Ethernet 0/0 (F0/0)	Fast Ethernet 0/1 (F0/1)	Serial 0/0/0 (S0/0/0)	Serial 0/0/1 (S0/0/1)
1900	Gigabit Ethernet 0/0 (G0/0)	Gigabit Ethernet 0/1 (G0/1)	Serial 0/0/0 (S0/0/0)	Serial 0/0/1 (S0/0/1)
2801	Fast Ethernet 0/0 (F0/0)	Fast Ethernet 0/1 (F0/1)	Serial 0/1/0 (S0/1/0)	Serial 0/1/1 (S0/1/1)
2811	Fast Ethernet 0/0 (F0/0)	Fast Ethernet 0/1 (F0/1)	Serial 0/0/0 (S0/0/0)	Serial 0/0/1 (S0/0/1)
2900	Gigabit Ethernet 0/0 (G0/0)	Gigabit Ethernet 0/1 (G0/1)	Serial 0/0/0 (S0/0/0)	Serial 0/0/1 (S0/0/1)
4221	Gigabit Ethernet 0/0/0 (G0/0/0)	Gigabit Ethernet 0/0/1 (G0/0/1)	Serial 0/1/0 (S0/1/0)	Serial 0/1/1 (S0/1/1)
4300	Gigabit Ethernet 0/0/0 (G0/0/0)	Gigabit Ethernet 0/0/1 (G0/0/1)	Serial 0/1/0 (S0/1/0)	Serial 0/1/1 (S0/1/1)

Note: To find out how the router is configured, look at the interfaces to identify the type of router and how many interfaces the router has. There is no way to effectively list all the combinations of configurations for each router class. This table includes identifiers for the possible combinations of Ethernet and Serial interfaces in the device. The table does not include any other type of interface, even though a specific router may contain one. An example of this might be an ISDN BRI interface. The string in parenthesis is the legal abbreviation that can be used in Cisco IOS commands to represent the interface.

Device Config Final – Notes

8.1.2 Lab - Troubleshoot OSPFv2

Topology

Addressing Table

Device	Interface	IPv4 Address	Subnet Mask
R1	G0/0/0	10.2.0.1	255.255.255.0
	G0/0/1	10.0.1.1	255.255.255.0
R2	G0/0/0	10.2.0.2	255.255.255.0
	Lo1	10.2.1.1	255.255.255.0
R3	G0/0/1	10.0.3.1	255.255.255.0
	Lo1	10.3.1.1	255.255.255.0
D1	G1/0/1 (Po1)	10.0.2.1	255.255.255.0
	G1/0/2 (Po1)	10.0.2.1	255.255.255.0
	G1/0/3 (Po2)	10.1.1.1	255.255.255.0

Device	Interface	IPv4 Address	Subnet Mask
	G1/0/4 (Po2)	10.1.1.1	255.255.255.0
	G1/0/11	10.0.1.2	255.255.255.0
	G1/0/23	10.1.2.1	255.255.255.0
	G1/0/24	10.1.3.1	255.255.255.0
D2	G1/0/1 (Po1)	10.0.2.2	255.255.255.0
	G1/0/2 (Po1)	10.0.2.2	255.255.255.0
	G1/0/3 (Po2)	10.1.1.2	255.255.255.0
	G1/0/4 (Po2)	10.1.1.2	255.255.255.0
	G1/0/11	10.0.3.2	255.255.255.0
PC1	NIC	10.1.3.100/24	Default Gateway 10.1.3.1

Objectives

Troubleshoot network issues related to the configuration and operation of OSPFv2.

Background/Scenario

Your University network is using multiarea OSPFv2. The 10.3.1.0/24 network has a single end device which connects to an observatory. Reading the network design documentation, you notice two key items regarding this connection:

- The connection from R3 to the observatory is a microwave link. Because this link can be unstable, it is required to be in its own OSPFv2 area.

- The data from the observatory is sent to an astronomy research team on the 10.1.3.0/24 network (PC1).

    Although the topology has a limited number of routers, you should use the appropriate troubleshooting commands to help find and solve the problems in the three trouble tickets as if this were a much more complex topology with many more routers and networks.

    You will be loading configurations with intentional errors onto the network. Your tasks are to FIND the error(s), document your findings and the command(s) or method(s) used to fix them, FIX the issue(s) presented here and then test the network to ensure both of the following conditions are met:

    1) the complaint received in the ticket is resolved

    2) full reachability is restored

Note: The routers used with CCNP hands-on labs are Cisco 4221 with Cisco IOS XE Release 16.9.4 (universalk9 image). The switches used in the labs are Cisco Catalyst 3650 with Cisco IOS XE Release 16.9.4 (universalk9 image) and Cisco Catalyst 2960s with Cisco IOS Release 15.2(2) (lanbasek9 image). Other routers, switches, and Cisco IOS versions can be used. Depending on the model and Cisco IOS version, the commands available and the output produced might vary from what is shown in the labs. Refer to the Router Interface Summary Table at the end of the lab for the correct interface identifiers.

Note: Make sure that the switches have been erased and have no startup configurations. If you are unsure, contact your instructor.

Required Resources

- 3 Routers (Cisco 4221 with Cisco IOS XE Release 16.9.4 universal image or comparable)

- 2 Switches (Cisco 3560 with Cisco IOS XE Release 16.9.4 universal image or comparable)

- 1 PC (choice of operating system with terminal emulation program installed)

- Console cables to configure the Cisco IOS devices via the console ports

- Ethernet cables as shown in the topology

Instructions

Part 1: Trouble Ticket 8.1.2.1

Scenario:

Changes were made to minimize the number of routes in area 2. The intention was that all internal OSPF routers in area 2 would only receive a default route via OSPF, and they would not receive any interarea OSPF routes. However, after the change was made users began to indicate that they no longer have connectivity to devices on the 10.2.1.0/24 network.

Use the commands listed below to load the configuration files for this trouble ticket:

Device	Command
R1	copy flash:/enarsi/8.1.2.1-r1-config.txt run
R2	copy flash:/enarsi/8.1.2.1-r2-config.txt run
R3	copy flash:/enarsi/8.1.2.1-r3-config.txt run
D1	copy flash:/enarsi/8.1.2.1-d1-config.txt run
D2	copy flash:/enarsi/8.1.2.1-d2-config.txt run

- All routers should have the 10.2.1.0/24 network in their routing tables.

- All area 2 internal OSPF routers should only receive a default route via OSPF, they should not have any interarea routes.

- PC1 should be able to ping 10.2.1.1/24.

- R3, D1, and D2 should see the 10.2.1.0/24 network in their routing table.

- R2 should only have a single OSPF route in its routing table, and that should be a default route learned via OSPF.

- When you have fixed the ticket, change the MOTD on EACH DEVICE using the following command:

  banner motd # This is $(hostname) FIXED from ticket <ticket number> #

- Then save the configuration by issuing the **wri** command (on each device).

- Inform your instructor that you are ready for the next ticket.

- After the instructor approves your solution for this ticket, issue the **reset.now** privileged EXEC command This script will clear your configurations and reload the devices.

Part 2: Trouble Ticket 8.1.2.2

Scenario:

Recently Layer 3 switch D2 was replaced. However, the backup config file was corrupt and unable to be used. One of the network technicians configured D2 manually using the network documentation including the topology. After D2 was brought online, D2's routing table only sees the 10.3.1.0/24 network via OSPF.

Use the commands listed below to load the configuration files for this trouble ticket:

Device	Command
R1	`copy flash:/enarsi/8.1.2.2-r1-config.txt run`
R2	`copy flash:/enarsi/8.1.2.2-r2-config.txt run`
R3	`copy flash:/enarsi/8.1.2.2-r3-config.txt run`
D1	`copy flash:/enarsi/8.1.2.2-d1-config.txt run`
D2	`copy flash:/enarsi/8.1.2.2-d2-config.txt run`

- D2 should see all OSPF networks.

- When you have fixed the ticket, change the MOTD on EACH DEVICE using the following command:

  banner motd # This is $(hostname) FIXED from ticket <ticket number> #

- Then save the configuration by issuing the **wri** command (on each device).

- Inform your instructor that you are ready for the next ticket.

- After the instructor approves your solution for this ticket, issue the **reset.now** privileged EXEC command. This script will clear your configurations and reload the devices.

Part 3: Trouble Ticket 8.1.2.3

Topology Update:

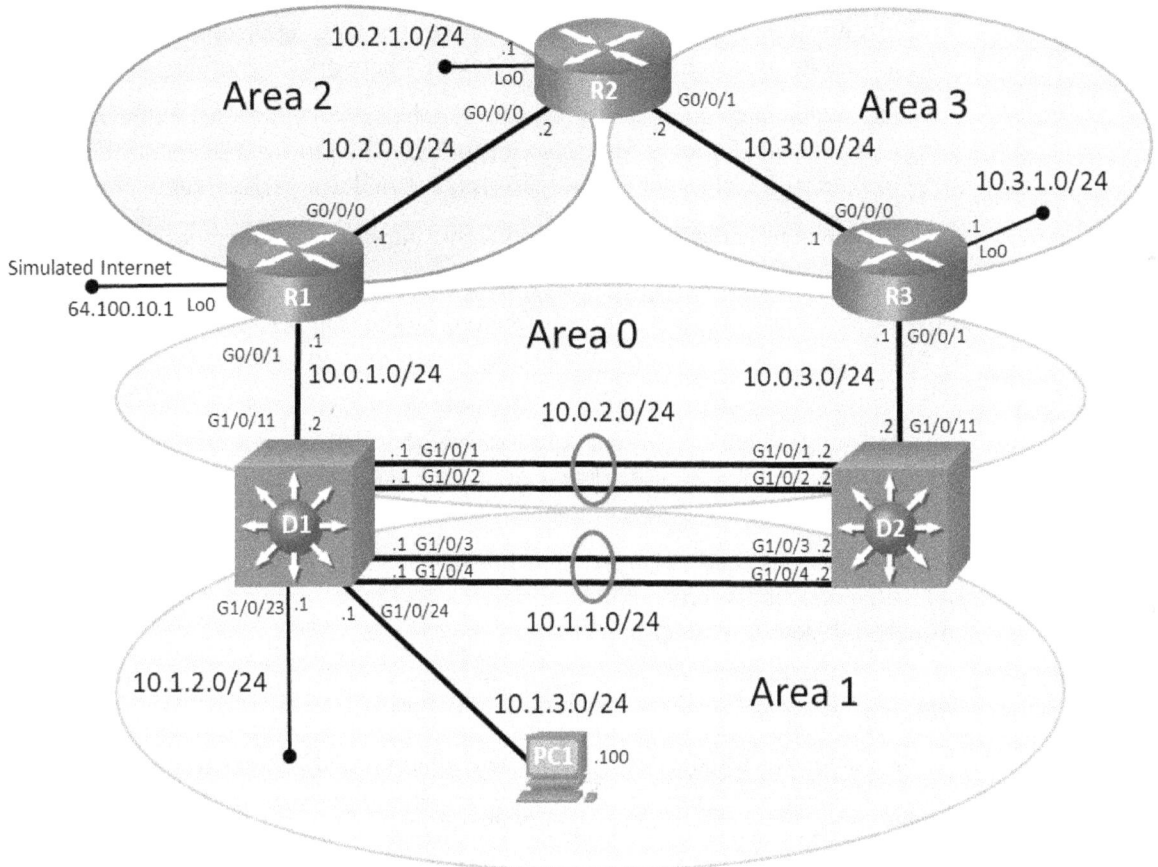

Addressing Table Update:

Device	Interface	IPv4 Address	Subnet Mask
R2	G0/0/0	10.2.0.2	255.255.255.0
	G0/0/1	10.3.0.2	255.255.255.0
	Lo1	10.2.1.1	255.255.255.0
R3	G0/0/0	10.3.0.1	255.255.255.0
	G0/0/1	10.0.3.1	255.255.255.0
	Lo1	10.3.1.1	255.255.255.0

Scenario:

The 10.3.1.0/24 network has a single end device which connects to an observatory. Reading the network design documentation, you notice three key items:

- The connection from R3 to the observatory is a microwave link. Because this link can be unstable, it is required to be in its own OSPFv2 area. The data from the observatory is sent to an astronomy research team on the 10.1.3.0/24 network (PC5).

- For redundancy purposes the design team had installed a new link between R2 and R3. This link results in an additional path from R1 to PC5. The primary path is through D2 to R3. The new backup path is through R2 to R3. This is verified using traceroute.

- Although redundancy is helpful, the most important requirement is that 101.3.0/24 is in its own OSPF area.

A troubleshooting ticket has just been received by the help desk. During construction of a new building on campus, the link between R3 and D2 was cut. It may take four or five days before this link can be restored.

The astronomy team has informed your manager that they are not receiving the data from the observatory and they cannot wait more than a day for the data from the observatory. Your manager is confused about why the data is not being forwarded using the redundant connection via R2.

Your task is to diagnose this problem and resolve it. Connectivity between the 10.3.1.0/24 and 10.1.3.0/24 networks is critical. In addition, the 10.3.1.0/24 network must be in its own area so the unstable link does not affect other areas.

Note: OSPF area 2 and area 3 are both normal OSPF areas.

Use the commands listed below to load the configuration files for this trouble ticket:

Device	Command
R1	`copy flash:/enarsi/8.1.2.3-r1-config.txt run`
R2	`copy flash:/enarsi/8.1.2.3-r2-config.txt run`
R3	`copy flash:/enarsi/8.1.2.3-r3-config.txt run`
D1	`copy flash:/enarsi/8.1.2.3-d1-config.txt run`
D2	`copy flash:/enarsi/8.1.2.3-d2-config.txt run`

Note: To simulate the link being cut, shut down the G0/0/1 interface on R3:

```
R3(config)# inter g 0/0/1
R3(config-if)# shutdown
```

- PC1 should be able to ping the 10.3.1.1.

- Network 10.3.1.0/24 must be in its own area.

- When you have fixed the ticket, change the MOTD on EACH DEVICE using the following command:

  banner motd # This is $(hostname) FIXED from ticket <ticket number> #

- Then save the configuration by issuing the **wri** command (on each device).

- Inform your instructor that you are ready for the next ticket.

- After the instructor approves your solution for this ticket, issue the **reset.now** privileged EXEC command. This script will clear your configurations and reload the devices.

Router Interface Summary Table

Router Model	Ethernet Interface #1	Ethernet Interface #2	Serial Interface #1	Serial Interface #2
1800	Fast Ethernet 0/0 (F0/0)	Fast Ethernet 0/1 (F0/1)	Serial 0/0/0 (S0/0/0)	Serial 0/0/1 (S0/0/1)
1900	Gigabit Ethernet 0/0 (G0/0)	Gigabit Ethernet 0/1 (G0/1)	Serial 0/0/0 (S0/0/0)	Serial 0/0/1 (S0/0/1)
2801	Fast Ethernet 0/0 (F0/0)	Fast Ethernet 0/1 (F0/1)	Serial 0/1/0 (S0/1/0)	Serial 0/1/1 (S0/1/1)
2811	Fast Ethernet 0/0 (F0/0)	Fast Ethernet 0/1 (F0/1)	Serial 0/0/0 (S0/0/0)	Serial 0/0/1 (S0/0/1)
2900	Gigabit Ethernet 0/0 (G0/0)	Gigabit Ethernet 0/1 (G0/1)	Serial 0/0/0 (S0/0/0)	Serial 0/0/1 (S0/0/1)
4221	Gigabit Ethernet 0/0/0 (G0/0/0)	Gigabit Ethernet 0/0/1 (G0/0/1)	Serial 0/1/0 (S0/1/0)	Serial 0/1/1 (S0/1/1)
4300	Gigabit Ethernet 0/0/0 (G0/0/0)	Gigabit Ethernet 0/0/1 (G0/0/1)	Serial 0/1/0 (S0/1/0)	Serial 0/1/1 (S0/1/1)

Note: To find out how the router is configured, look at the interfaces to identify the type of router and how many interfaces the router has. There is no way to effectively list all the combinations of configurations for each router class. This table includes identifiers for the possible combinations of Ethernet and Serial interfaces in the device. The table does not include any other type of interface, even though a specific router may contain one. An example of this might be an ISDN BRI interface. The string in parenthesis is the legal abbreviation that can be used in Cisco IOS commands to represent the interface.

Device Config Final – Notes

9.1.2 Lab - Implement Multiarea OSPFv3

Topology

Addressing Table

Device	Interface	IPv4 Address	IPv6 Address	IPv6 Link-Local
R1	G0/0/0	172.16.0.2/30	2001:db8:acad:a001::2/64	fe80::1:2
	G0/0/1	10.10.0.1/30	2001:db8:acad:1001::1/64	fe80::1:1
R2	Lo0	209.165.200.225/27	2001:db8:feed:209::1/64	fe80::2:3
	G0/0/0	172.16.0.1/30	2001:db8:acad:a001::1/64	fe80::2:1
	G0/0/1	172.16.1.1/30	2001:db8:acad:a002::1/64	fe80::2:2
R3	G0/0/0	172.16.1.2/30	2001:db8:acad:a002::2/64	fe80::3:2
	G0/0/1	10.10.4.1/30	2001:db8:acad:2001::1/64	fe80::3:1
D1	G1/0/11	10.10.0.2/30	2001:db8:acad:1001::2/64	fe80::d1:2
	G1/0/23	10.10.1.0/24	2001:db8:acad:1002::1/64	fe80::d1:1
D2	G1/0/11	10.10.4.2/30	2001:db8:acad:2001::2/64	fe80::d2:2
	G1/0/23	10.10.5.1/24	2001:db8:acad:2002::1/64	fe80::d2:1

Objectives

Part 1: Build the Topology and Configure Basic Device Settings and IP Addressing

Part 2: Configure Traditional OSPFv3 for IPv6 on D1

Part 3: Configure OSPFv3 for Address Families (AF) IPv4 and AF IPv6

Part 4: Verify OSPFv3 AF

Part 5: Tune OSPFv3 AF

Background/Scenario

In this lab, you will configure the network with multiarea OSPFv3 routing using the AF feature for both IPv4 and IPv6 in OSPF areas 0, 1 and 2. This lab was specifically designed to use three routers and two Layer 3 switches that support OSPFv3 using AF.

It should be noted that OSPFv3 runs on top of IPv6 and uses IPv6 link local addresses for OSPFv3 control packets. Therefore, it is required that IPv6 be enabled on an OSPFv3 link, although the link may not be participating in any IPv6 AFs. Additionally, OSPFv3 AF for IPv4 unicast is not backwards compatible with OSPFv2.

Note: The routers used with CCNP hands-on labs are Cisco 4221 with Cisco IOS XE Release 16.9.4 (universalk9 image). Other routers and Cisco IOS versions can be used. Depending on the model and Cisco IOS version, the commands available and the output produced might vary from what is shown in the labs.

Note: The switches used with CCNP hands-on labs are Cisco Catalyst 3650s with Cisco IOS XE Release 16.9.4 (universalk9 image). Other switches and Cisco IOS versions can be used. Depending on the model and Cisco IOS version, the commands available and output produced might vary from what is shown in the labs.

Note: Ensure that the routers and switches have been erased and have no startup configurations. If you are unsure contact your instructor.

Required Resources

- 3 Routers (Cisco 4221 with Cisco IOS XE Release 16.9.4 universal image or comparable)
- 2 Switches (Cisco 3650 with Cisco IOS XE Release 16.9.4 universal image or comparable)
- Console cables to configure the Cisco IOS devices via the console ports
- Ethernet cables as shown in the topology

Instructions

Part 1: Build the Network and Configure Basic Device Settings and Interface Addressing

In Part 1, you will set up the network topology and configure basic settings and interface addressing on routers and switches.

Step 1.   Cable the network as shown in the topology.

Attach the devices as shown in the topology diagram, and cable as necessary.

Step 2.   Configure basic settings for each router.

a.   Console into each device, enter global configuration mode, and apply the basic settings and interface addressing using the following startup configurations for each device.

Router R1

```
hostname R1
no ip domain lookup
line con 0
 logging sync
 exec-time 0 0
 exit
interface g0/0/0
 ip add 172.16.0.2 255.255.255.252
 ipv6 add 2001:db8:acad:a001::2/64
 ipv6 add fe80::1:2 link-local
 no shut
 exit
interface GigabitEthernet0/0/1
 ipv6 add 2001:db8:acad:1001::1/64
 ipv6 add fe80::1:1 link-local
 no shut
 exit
```

Router R2

```
hostname R2
no ip domain lookup
line con 0
 logging sync
 exec-time 0 0
 exit
interface g0/0/0
 ip add 172.16.0.1 255.255.255.252
 ipv6 add 2001:db8:acad:a001::1/64
 ipv6 add fe80::2:1 link-local
 no shut
 exit
interface GigabitEthernet0/0/1
 ip address 172.16.1.1 255.255.255.252
 ipv6 add 2001:db8:acad:a002::1/64
 ipv6 add fe80::2:2 link-local
 no shut
 exit
int lo0
 ip add 209.165.200.225 255.255.255.224
 ipv6 add 2001:db8:feed:209::1/64
 ipv6 add fe80::2:3 link-local
 exit
```

Router R3

```
hostname R3
no ip domain lookup
line con 0
 logging sync
 exec-time 0 0
 exit
interface g0/0/0
 ip add 172.16.1.2 255.255.255.252
 ipv6 add 2001:db8:acad:a002::2/64
 ipv6 add fe80::3:2 link-local
 no shut
 exit
interface GigabitEthernet0/0/1
 ip address 10.10.4.1 255.255.255.252
 ipv6 add 2001:db8:acad:2001::1/64
 ipv6 add fe80::3:1 link-local
 no shut
 exit
```

Switch D1

```
hostname D1
no ip domain lookup
line con 0
 exec-timeout 0 0
 logging synchronous
 exit
interface g1/0/11
 no switchport
 ipv6 add 2001:db8:acad:1001::2/64
 ipv6 add fe80::d1:2 link-local
 no shutdown
 exit
interface g1/0/23
 no switchport
 ipv6 add 2001:db8:acad:1002::1/64
 ipv6 add fe80::d1:1 link-local
 no shutdown
 exit
```

Switch D2

```
host D2
no ip domain lookup
line con 0
 logging sync
 exec-time 0 0
 exit
interface gi1/0/11
 no switchport
 ip address 10.10.4.2 255.255.255.252
```

```
ipv6 add 2001:db8:acad:2001::2/64
ipv6 add fe80::d2:2 link-local
no shut
exit
interface gi1/0/23
no switchport
ip address 10.10.5.1 255.255.255.0
ipv6 add 2001:db8:acad:2002::1/64
ipv6 add fe80::d2:1 link-local
no shut
exit
```

b. Save the running configuration to startup-config.

Part 2: Configure Traditional OSPFv3 for IPv6 on D1

Step 1. Configure traditional OSPFv3 on D1.

Traditional OSPFv3 implements OSPF routing for IPv6. In this part of the lab, you will configure traditional OSPFv3 for routing IPv6 on D1, which is in the IPv6-only area.

a. OSPFv3 messages are sourced from the router's IPv6 link-local address. Earlier in this lab, IPv6 GUA and link-local addresses were statically configured on each router's interface. The link-local addresses were statically configured to make these addresses more recognizable than being automatically created using EUI-64. Issue the **show ipv6 interface brief** command to verify the GUA and link-local addresses on the router's interfaces.

```
D1# show ipv6 interface brief
<output omitted>
GigabitEthernet1/0/11  [up/up]
    FE80::D1:2
    2001:DB8:ACAD:1001::2
<output omitted>
GigabitEthernet1/0/23  [up/up]
    FE80::D1:1
    2001:DB8:ACAD:1002::1
<output omitted>
```

b. IPv6 routing is disabled by default. Enable IPv6 routing using the **ipv6 unicast-routing** command in global configuration mode.

```
D1(config)# ipv6 unicast-routing
```

c. Most Cisco IOS versions have IPv6 CEF enabled by default when IPv6 routing is enabled. Use the **show ipv6 cef** command to verify whether IPv6 CEF is enabled. If you need to enable IPv6 CEF, use the **ipv6 cef** command. If IPv6 CEF is disabled you will see an IOS message similar to "%IPv6 CEF not running".

```
D1# show ipv6 cef
::/0
  no route
::/127
  discard
```

```
2001:DB8:ACAD:1001::/64
  attached to GigabitEthernet1/0/11
2001:DB8:ACAD:1001::2/128
  receive for GigabitEthernet1/0/11
2001:DB8:ACAD:1002::/64
  attached to GigabitEthernet1/0/23
2001:DB8:ACAD:1002::1/128
  receive for GigabitEthernet1/0/23
FE80::/10
  receive for Null0
FF00::/8
  multicast
FF02::/16
  receive
```

d. Configure the OSPFv3 process on D1. Similar to OSPFv2, the process ID does not have to match other routers to form neighbor adjacencies, although that is considered best practice. Configure the 32-bit OSPFv3 router ID on each router. Enable OSPFv3 directly on the interfaces using the interface **ipv6 ospf** *pid* **area** *area* command.

```
D1(config)# ipv6 unicast-routing
D1(config)# ipv6 router ospf 123
D1(config-rtr)# router-id 1.1.1.2
D1(config-rtr)# exit
D1(config)# interface g1/0/11
D1(config-if)# ipv6 ospf 123 area 1
D1(config-if)# exit
D1(config)# interface g1/0/23
D1(config-if)# ipv6 ospf 123 area 1
D1(config-if)# exit
```

e. The **show ipv6 ospf** command can be used to verify the OSPF router ID. If the OSPFv3 router ID is uses a 32-bit value other than the one specified by the **router-id** command, you can reset the router ID by using the **clear ipv6 ospf** *pid* **process** command and re-verify using the command **show ipv6 ospf.**

```
D1# show ipv6 ospf
 Routing Process "ospfv3 123" with ID 1.1.1.2
 Supports NSSA (compatible with RFC 3101)
 Supports Database Exchange Summary List Optimization (RFC 5243)
 Event-log enabled, Maximum number of events: 1000, Mode: cyclic
 Router is not originating router-LSAs with maximum metric
 Initial SPF schedule delay 50 msecs
 Minimum hold time between two consecutive SPFs 200 msecs
 Maximum wait time between two consecutive SPFs 5000 msecs
 Initial LSA throttle delay 50 msecs
 Minimum hold time for LSA throttle 200 msecs
 Maximum wait time for LSA throttle 5000 msecs
 Minimum LSA arrival 100 msecs
 LSA group pacing timer 240 secs
 Interface flood pacing timer 33 msecs
 Retransmission pacing timer 66 msecs
```

```
Retransmission limit dc 24 non-dc 24
EXCHANGE/LOADING adjacency limit: initial 300, process maximum 300
Number of external LSA 0. Checksum Sum 0x000000
Number of areas in this router is 1. 1 normal 0 stub 0 nssa
Graceful restart helper support enabled
Reference bandwidth unit is 100 mbps
RFC1583 compatibility enabled
    Area 1
        Number of interfaces in this area is 2
        SPF algorithm executed 5 times
        Number of LSA 12. Checksum Sum 0x0486C1
        Number of DCbitless LSA 0
        Number of indication LSA 0
        Number of DoNotAge LSA 0
        Flood list length 0
```

f. The **show ipv6 protocols** command can be used to verify general OSPFv3 information such as areas and enabled interfaces.

```
D1# show ipv6 protocols
IPv6 Routing Protocol is "connected"
IPv6 Routing Protocol is "ND"
IPv6 Routing Protocol is "ospf 123"
  Router ID 1.1.1.2
  Number of areas: 1 normal, 0 stub, 0 nssa
  Interfaces (Area 1):
    GigabitEthernet1/0/23
    GigabitEthernet1/0/11
  Redistribution:
None
```

Part 3: Configure OSPFv3 for AF IPv4 and AF IPv6

OSPFv3 with the address family (AF) unifies OSPF configuration for both IPv4 and IPv6. Each OSPFv3 AF is a single process, so you may have two processes per interface, but only one process per AF. OSPFv3 messages are sent over IPv6 which requires that IPv6 routing is enabled and that the interface has a link-local IPv6 address. This is the requirement even if only the IPv4 AF is configured.

In this section you will configure OSPFv3 with AF for the IPv4 and IPv6 address families on R1, R2, R3, D1 and D2.

Step 1. Configure OSPFv3 with AF on R1.

a. After enabling IPv6 unicast routing, configure OSPFv3 with AF on R1 using the **router ospfv3** *pid* command. Use the ? to see the address families available.

```
R1(config)# ipv6 unicast-routing
R1(config)# router ospfv3 123
R1(config-router)# address-family ?
  ipv4  Address family
  ipv6  Address family
```

b. Next, specify the AF for IPv4 and use the ? to see the available options.

```
R1(config-router)# address-family ipv4 ?
  unicast  Address Family modifier
  vrf      Specify parameters for a VPN Routing/Forwarding instance
  <cr>
```

c. Enter the AF for IPv4 unicast using the command **address-family ipv4 unicast**. Use the ? to examine the options in AF configuration mode. Some of the more common configuration commands are highlighted. Use the **router-id** command to configure the router ID for the IPv4 AF.

```
R1(config-router)# address-family ipv4 unicast
R1(config-router-af)# ?
Router Address Family configuration commands:
    adjacency                 Control adjacency formation
    area                      OSPF area parameters
    authentication            Authentication parameters
    auto-cost                 Calculate OSPF interface cost according to bandwidth
    auto-cost-determination   Calculate OSPF interface cost according to bandwidth
    bfd                       BFD configuration commands
    compatible                Compatibility list
    default                   Set a command to its defaults
    default-information       Control distribution of default information
    default-metric            Set metric of redistributed routes
    discard-route             Enable or disable discard-route installation
    distance                  Define an administrative distance
    distribute-list           Filter networks in routing updates
    event-log                 Event Logging
    exit-address-family       Exit from Address Family configuration mode
    graceful-restart          Graceful-restart options
    help                      Description of the interactive help system
    ignore                    Do not complain about specific event
    interface-id              Source of the interface ID
    limit                     Limit a specific OSPF feature
    local-rib-criteria        Enable or disable usage of local RIB as route
                              criteria
    log-adjacency-changes     Log changes in adjacency state
    manet                     Specify MANET OSPF parameters
    max-lsa                   Maximum number of non self-generated LSAs to accept
    max-metric                Set maximum metric
    maximum-paths             Forward packets over multiple paths
    mpls                      MPLS Traffic Engineering configs
    no                        Negate a command or set its defaults
    passive-interface         Suppress routing updates on an interface
    prefix-suppression        Enable prefix suppression
    process-min-time          Percentage of quantum to be used before releasing
                              CPU
    queue-depth               Hello/Router process queue depth
    redistribute              Redistribute information from another routing
                              protocol
    router-id                 router-id for this OSPF process
    shutdown                  Shutdown the router process
```

```
snmp                    Modify snmp parameters
statistics              Enable or disable OSPF statistics options
summary-address         Configure IP address summaries
summary-prefix          Configure IP address summaries
timers                  Adjust routing timers
R1(config-router-af)#
R1(config-router-af)# router-id 1.1.1.1
```

d. Exit the IPv4 AF configuration mode and enter the AF IPv6 configuration mode. The **exit-address-family** (or a shorter version of **exit**) command is used to exit address family configuration mode. Issue the **address-family ipv6 unicast** command to enter the IPv6 AF. For the IPv6 AF, use the **router-id** command to configure the router ID. It isn't necessary to configure a different router ID for IPv6 AF but it is a valid option. The **exit** command is used to return to global configuration mode.

```
R1(config-router-af)# exit-address-family
R1(config-router)# address-family ipv6 unicast
R1(config-router-af)# router-id 1.1.1.1
R1(config-router-af)# exit-address-family
R1(config-router)# exit
```

e. OSPFv3 is enabled directly on the interfaces for both IPv4 and IPv6 AFs using the **ospfv3** *pid* [**ipv4** | **ipv6**] **area** *area-id* interface command. Use this command to enable OSPFv3 on both of R1's interfaces.

```
R1(config)# interface g0/0/0
R1(config-if)# ospfv3 123 ipv4 area 0
R1(config-if)# ospfv3 123 ipv6 area 0
R1(config-if)# exit
R1(config)# interface g0/0/1
R1(config-if)# ospfv3 123 ipv4 area 1
R1(config-if)# ospfv3 123 ipv6 area 1
```

Step 2. Configure OSPFv3 with AF IPv4 and AF IPv6 on R2.

Enable IPv6 unicast routing and configure the OSPFv3 with AF for both IPv4 and IPv6 on R2, similar to the configuration for R1.

```
R2(config)# ipv6 unicast-routing

R2(config)# router ospfv3 123
R2(config-router)# address-family ipv4 unicast
R2(config-router-af)# router-id 2.2.2.1
R2(config-router-af)# exit-address-family
R2(config-router)# address-family ipv6 unicast
R2(config-router-af)# router-id 2.2.2.1
R2(config-router-af)# exit-address-family
R2(config-router)# exit

R2(config)# interface g0/0/0
R2(config-if)# ospfv3 123 ipv4 area 0
R2(config-if)# ospfv3 123 ipv6 area 0
R2(config-if)# exit
R2(config)# interface g0/0/1
R2(config-if)# ospfv3 123 ipv4 area 0
R2(config-if)# ospfv3 123 ipv6 area 0
```

Step 3.   Configure OSPFv3 with IPv4 AF and IPv6 AF on R3.

Enable IPv6 unicast routing and configure the OSPFv3 with AF for both IPv4 and IPv6 on R3, similar to the configurations for R1 and R2. On R3, set the router ID for both IPv4 AF and IPv6 AF with a single command as shown.

```
R3(config)# ipv6 unicast-routing

R3(config)# router ospfv3 123
R3(config-router)# router-id 3.3.3.1
R3(config-router)# address-family ipv4 unicast
R3(config-router-af)# exit-address-family
R3(config-router)# address-family ipv6 unicast
R3(config-router-af)# exit-address-family
R3(config-router)# exit

R3(config)# interface g0/0/0
R3(config-if)# ospfv3 123 ipv4 area 0
R3(config-if)# ospfv3 123 ipv6 area 0
R3(config-if)# exit
R3(config)# interface g0/0/1
R3(config-if)# ospfv3 123 ipv4 area 2
R3(config-if)# ospfv3 123 ipv6 area 2
```

Step 4.   Configure OSPFv3 with AF on D2.

a. Enter the following command to enable routing for IPv4. (This may not be required depending on model and IOS.)

```
D2(config)# ip routing
```

b. Enter the following command to enable routing for IPv6. (This may not be required depending on model and IOS.)

```
D2(config)# ipv6 unicast-routing
```

Note: By default, the 3650 supports IPv6 interface configuration.

---

c. Configure the OSPFv3 with AF for both IPv4 and IPv6 on D2, similar to the configurations for R1, R2 and R3.

```
D2(config)# router ospfv3 123
D2(config-router)# address-family ipv4 unicast
D2(config-router-af)# router-id 3.3.3.2
D2(config-router-af)# exit-address-family
D2(config-router)# address-family ipv6 unicast
D2(config-router-af)# router-id 3.3.3.2
D2(config-router-af)# exit-address-family
D2(config-router)# exit

D2(config)# interface g1/0/11
D2(config-if)# ospfv3 123 ipv4 area 2
D2(config-if)# ospfv3 123 ipv6 area 2
D2(config-if)# exit
D2(config)# interface g 1/0/23
D2(config-if)# ospfv3 123 ipv4 area 2
D2(config-if)# ospfv3 123 ipv6 area 2
```

Part 4: Verify OSPFv3

The commands to verify traditional OSPFv3 and OSPFv3 with AF may differ. This is because OSPFv3 with AF commands include information for both IPv4 and IPv6 address families, whereas traditional OSPFv3 is for IPv6 only.

Step 1.   Verifying neighbor adjacencies.

    a.   Use the **show ipv6 ospf neighbor** command on D1 to display OSPFv3 neighbors. This is a command used for routers configured with traditional OSPFv3. The equivalent command for OSPFv2 would be **show ip ospf neighbor**.

```
D1# show ipv6 ospf neighbor


          OSPFv3 Router with ID (1.1.1.2) (Process ID 123)


Neighbor ID     Pri   State       Dead Time   Interface ID   Interface
1.1.1.1           1   FULL/DR     00:00:39    6              GigabitEthernet1/0/11
```

    b.   This same command on a router running OSPFv3 with AF would generate similar output. For example, on R1 issue the same **show ipv6 ospf neighbor** command. Notice the output is only OSPFv3 for the IPv6 AF.

```
R1# show ipv6 ospf neighbor


          OSPFv3 Router with ID (1.1.1.1) (Process ID 123)


Neighbor ID     Pri   State       Dead Time   Interface ID   Interface
2.2.2.1           1   FULL/BDR    00:00:31    5              GigabitEthernet0/0/0
1.1.1.2           1   FULL/BDR    00:00:38    471            GigabitEthernet0/0/1
```

    c.   Now, issue the **show ospfv3 neighbor** command on R1. This is a command used for routers configured for OSPFv3 with AF. Notice the output includes neighbors for both IPv4 and IPv6 address families.

```
R1# show ospfv3 neighbor


          OSPFv3 123 address-family ipv4 (router-id 1.1.1.1)


Neighbor ID     Pri   State       Dead Time   Interface ID   Interface
2.2.2.1           1   FULL/BDR    00:00:38    5              GigabitEthernet0/0/0


          OSPFv3 123 address-family ipv6 (router-id 1.1.1.1)


Neighbor ID     Pri   State       Dead Time   Interface ID   Interface
2.2.2.1           1   FULL/BDR    00:00:32    5              GigabitEthernet0/0/0
1.1.1.2           1   FULL/BDR    00:00:30    471            GigabitEthernet0/0/1
```

    Traditional OSPFv3 commands are similar to those for OSPFv2, except **ipv6** is used as an argument instead of **ip**, for example **show ip ospf neighbor** and **show ipv6 ospf neighbor**. OSPFv3 with AF uses the argument ospfv3 which includes both OSPF for IPv4 and IPv6 AFs. For example, **show ospfv3 neighbor**.

    Traditional OSPFv3 commands can be used when a router is configured for OSPFv3 with AF, but the OSPFv3 AF router will only show OSPF for IPv6 AF information. OSPFv3 with AF commands cannot be used on routers configured with traditional OSPFv3.

To summarize the **show** command arguments:

- OSPFv2: Use **show ip ospf** (IPv4 only)

- Traditional OSPFv3: Use **show ipv6 ospf** (IPv6 only)

- OSPFv3 with AF: Use **show ospfv3** (IPv4 and IPv6 AF) or **show ipv6 ospf** (IPv6 only)

Question:

Why does the **show ipv6 ospf neighbor** command only display OSPFv3 neighbors in the IPv6 AF?

_____

_____

_____

Step 2. Examining the IP routing tables.

a. Use the **show ipv6 route ospf** command on D1 to display OSPFv3 routing entries in the IPv6 routing table.

```
D1# show ipv6 route ospf
IPv6 Routing Table - default - 9 entries
Codes: C - Connected, L - Local, S - Static, U - Per-user Static route
       B - BGP, R - RIP, H - NHRP, I1 - ISIS L1
       I2 - ISIS L2, IA - ISIS interarea, IS - ISIS summary, D - EIGRP
       EX - EIGRP external, ND - ND Default, NDp - ND Prefix, DCE - Destination
       NDr - Redirect, RL - RPL, O - OSPF Intra, OI - OSPF Inter
       OE1 - OSPF ext 1, OE2 - OSPF ext 2, ON1 - OSPF NSSA ext 1
       ON2 - OSPF NSSA ext 2, la - LISP alt, lr - LISP site-registrations
       ld - LISP dyn-eid, lA - LISP away, le - LISP extranet-policy
OI  2001:DB8:ACAD:2001::/64 [110/4]
     via FE80::1:1, GigabitEthernet1/0/11
OI  2001:DB8:ACAD:2002::/64 [110/5]
     via FE80::1:1, GigabitEthernet1/0/11
OI  2001:DB8:ACAD:A001::/64 [110/2]
     via FE80::1:1, GigabitEthernet1/0/11
OI  2001:DB8:ACAD:A002::/64 [110/3]
     via FE80::1:1, GigabitEthernet1/0/11
```

Question:

Display the routes using the **show ip route ospf**. Why are there no routes displayed using this command?

_____

_____

b. Understanding the difference between commands associated with OSPFv2 and OSPFv3 can seem challenging at times. The **show ip route ospfv3** command is used to view OSPFv3 routes in the IPv4 routing table. The **show ipv6 route ospf** command is used to view OSPFv3 routes in the IPv6 routing table. The **show ipv6 route ospf** command is the same command used with traditional OSPFv3 for IPv6.

```
R1# show ip route ospf

R1# show ip route ospfv3
Codes: L - local, C - connected, S - static, R - RIP, M - mobile, B - BGP
       D - EIGRP, EX - EIGRP external, O - OSPF, IA - OSPF inter area
       N1 - OSPF NSSA external type 1, N2 - OSPF NSSA external type 2
       E1 - OSPF external type 1, E2 - OSPF external type 2
       i - IS-IS, su - IS-IS summary, L1 - IS-IS level-1, L2 - IS-IS level-2
       ia - IS-IS inter area, * - candidate default, U - per-user static route
       o - ODR, P - periodic downloaded static route, H - NHRP, l - LISP
       a - application route
       + - replicated route, % - next hop override, p - overrides from PfR


Gateway of last resort is not set

      10.0.0.0/8 is variably subnetted, 4 subnets, 3 masks
O IA     10.10.4.0/30 [110/3] via 172.16.0.1, 00:17:34, GigabitEthernet0/0/0
O IA     10.10.5.0/24 [110/4] via 172.16.0.1, 00:17:34, GigabitEthernet0/0/0
      172.16.0.0/16 is variably subnetted, 3 subnets, 2 masks
O        172.16.1.0/30 [110/2] via 172.16.0.1, 00:17:34, GigabitEthernet0/0/0

R1# show ipv6 route ospfv3
                     ^
% Invalid input detected at '^' marker.

R1# show ipv6 route ospf
IPv6 Routing Table - default - 9 entries
Codes: C - Connected, L - Local, S - Static, U - Per-user Static route
       B - BGP, R - RIP, H - NHRP, I1    ISIS L1
       I2 - ISIS L2, IA - ISIS interarea, IS - ISIS summary, D - EIGRP
       EX - EIGRP external, ND - ND Default, NDp - ND Prefix, DCE - Destination
       NDr - Redirect, RL - RPL, O - OSPF Intra, OI - OSPF Inter
       OE1 - OSPF ext 1, OE2 - OSPF ext 2, ON1 - OSPF NSSA ext 1
       ON2 - OSPF NSSA ext 2, a - Application
O   2001:DB8:ACAD:1002::/64 [110/2]
     via FE80::D1:2, GigabitEthernet0/0/1
OI  2001:DB8:ACAD:2001::/64 [110/3]
     via FE80::2:1, GigabitEthernet0/0/0
OI  2001:DB8:ACAD:2002::/64 [110/4]
     via FE80::2:1, GigabitEthernet0/0/0
O   2001:DB8:ACAD:A002::/64 [110/2]
     via FE80::2:1, GigabitEthernet0/0/0
```

Question:

Why doesn't the **show ip route ospf** command display any routes on R1?

_____

_____

Step 3. Examining the OSPF LSDB.

a. D1 is running traditional OSPFv3. The **show ipv6 ospf database** command is used to display a summary of the OSPFv3 LSDB.

```
D1# show ipv6 ospf database

            OSPFv3 Router with ID (1.1.1.2) (Process ID 123)

               Router Link States (Area 1)

ADV Router      Age        Seq#        Fragment ID  Link count  Bits
1.1.1.1         1096       0x80000009  0            1           B
1.1.1.2         1110       0x80000005  0            1           None


               Net Link States (Area 1)

ADV Router      Age        Seq#        Link ID    Rtr count
1.1.1.1         1152       0x80000001  6          2


            Inter Area Prefix Link States (Area 1)

ADV Router      Age        Seq#        Prefix
1.1.1.1         1096       0x80000003  2001:DB8:ACAD:A001::/64
1.1.1.1         1096       0x80000003  2001:DB8:ACAD:A002::/64
1.1.1.1         833        0x80000005  2001:DB8:ACAD:2001::/64
1.1.1.1         1497       0x80000002  2001:DB8:ACAD:2002::/64


               Link (Type-8) Link States (Area 1)

ADV Router      Age        Seq#        Link ID    Interface
1.1.1.2         1150       0x80000001  39         Gi1/0/23
1.1.1.1         1096       0x80000006  6          Gi1/0/11
1.1.1.2         1151       0x80000001  38         Gi1/0/11


            Intra Area Prefix Link States (Area 1)

ADV Router      Age        Seq#        Link ID    Ref-lstype  Ref-LSID
1.1.1.1         1152       0x80000001  6144       0x2002      6
1.1.1.2         1150       0x80000003  0          0x2001      0
```

b. R1 is running OSPFv3 with AF. The **show ospfv3 database** command is used to display a summary of the OSPFv3 LSDB for both the IPv4 and IPv6 AFs.

```
R1# show ospfv3 database

            OSPFv3 123 address-family ipv4 (router-id 1.1.1.1)


               Router Link States (Area 0)
```

```
ADV Router        Age       Seq#          Fragment ID  Link count  Bits
 1.1.1.1          532       0x80000005    0            1           None
 2.2.2.1          508       0x80000008    0            2           None
 3.3.3.1          507       0x80000006    0            1           B


                  Net Link States (Area 0)


ADV Router        Age       Seq#          Link ID    Rtr count
 2.2.2.1          539       0x80000001    5          2
 3.3.3.1          512       0x80000001    5          2


                  Inter Area Prefix Link States (Area 0)


ADV Router        Age       Seq#          Prefix
 3.3.3.1          553       0x80000001    10.10.4.0/30
 3.3.3.1          513       0x80000001    10.10.5.0/24


                  Link (Type-8) Link States (Area 0)


ADV Router        Age       Seq#          Link ID    Interface
 1.1.1.1          579       0x80000001    5          Gi0/0/0
 2.2.2.1          579       0x80000001    5          Gi0/0/0


                  Intra Area Prefix Link States (Area 0)


ADV Router        Age       Seq#          Link ID    Ref-lstype  Ref-LSID
 2.2.2.1          539       0x80000001    5120       0x2002      5
 3.3.3.1          512       0x80000001    5120       0x2002      5


                  Router Link States (Area 1)


ADV Router        Age       Seq#          Fragment ID  Link count  Bits
 1.1.1.1          602       0x80000001    0            0           None


          OSPFv3 123 address-family ipv6 (router-id 1.1.1.1)


                  Router Link States (Area 0)


ADV Router        Age       Seq#          Fragment ID  Link count  Bits
 1.1.1.1          530       0x80000005    0            1           B
 2.2.2.1          508       0x80000009    0            2           None
 3.3.3.1          508       0x80000006    0            1           B


                  Net Link States (Area 0)


ADV Router        Age       Seq#          Link ID    Rtr count
 2.2.2.1          539       0x80000001    5          2
 3.3.3.1          511       0x80000001    5          2
```

```
                      Inter Area Prefix Link States (Area 0)

ADV Router        Age           Seq#          Prefix
 1.1.1.1          579           0x80000001    2001:DB8:ACAD:1001::/64
 1.1.1.1          559           0x80000001    2001:DB8:ACAD:1002::/64
 3.3.3.1          551           0x80000001    2001:DB8:ACAD:2001::/64
 3.3.3.1          512           0x80000001    2001:DB8:ACAD:2002::/64

                      Link (Type-8) Link States (Area 0)

ADV Router        Age           Seq#          Link ID    Interface
 1.1.1.1          578           0x80000002    5          Gi0/0/0
 2.2.2.1          578           0x80000002    5          Gi0/0/0

                      Intra Area Prefix Link States (Area 0)

ADV Router        Age           Seq#          Link ID    Ref-lstype   Ref-LSID
 2.2.2.1          539           0x80000001    5120       0x2002       5
 3.3.3.1          511           0x80000001    5120       0x2002       5

                      Router Link States (Area 1)

ADV Router        Age           Seq#          Fragment ID  Link count  Bits
 1.1.1.1          553           0x80000006    0            1           B
 1.1.1.2          552           0x80000025    0            1           None

                      Net Link States (Area 1)

ADV Router        Age           Seq#          Link ID    Rtr count
 1.1.1.2          560           0x80000001    38         2

                      Inter Area Prefix Link States (Area 1)

ADV Router        Age           Seq#          Prefix
 1.1.1.1          578           0x80000001    2001:DB8:ACAD:A001::/64
 1.1.1.1          538           0x80000001    2001:DB8:ACAD:A002::/64
 1.1.1.1          506           0x80000001    2001:DB8:ACAD:2002::/64
 1.1.1.1          506           0x80000001    2001:DB8:ACAD:2001::/64

                      Link (Type-8) Link States (Area 1)

ADV Router        Age           Seq#          Link ID    Interface
 1.1.1.1          559           0x8000000C    6          Gi0/0/1
 1.1.1.2          598           0x80000002    38         Gi0/0/1

                      Intra Area Prefix Link States (Area 1)

ADV Router        Age           Seq#          Link ID    Ref-lstype   Ref-LSID
 1.1.1.2          481           0x80000016    0          0x2001       0
 1.1.1.2          560           0x80000001    38912      0x2002       38
```

Question:

What would the **show ipv6 route database** command display on R1, if anything?

_____

_____

Part 5: Tune OSPFv3

Step 1.   Configuring a passive interface.

a.   To configure a passive interface in traditional OSPFv3, use the **passive-interface** command in OSPFv3 router mode.

```
D1(config)# ipv6 router ospf 123
D1(config-rtr)# passive-interface g1/0/23
```

b.   To configure a passive interface in OSPFv3 with AF, you can use the **passive-interface** command in OSPFv3 router mode to configure the passive interface for both IPv4 and IPv6 AFs.

```
D2(config)# router ospfv3 123
D2(config-router)# passive-interface g1/0/23
```

c.   As an alternative, you can use the **passive-interface** command within AF configuration mode to configure the passive interface for a specific AFs.

```
D2(config-router)# no passive-interface g1/0/23
D2(config-router)# address-family ipv4 unicast
D2(config-router-af)# passive-interface g1/0/23
D2(config-router-af)# exit-address-family
D2(config-router)# address-family ipv6 unicast
D2(config-router-af)# passive-interface g1/0/23
D2(config-router-af)# exit-address-family
```

Step 2.   Configuring summarization.

a.   The **area** *area* **range** *ipv6-summary-address* command is used to summarize prefixes from one area into another. The area is the area from which the prefixes are summarized.

```
R1(config)# router ospfv3 123
R1(config-router)# address-family ipv6 unicast
R1(config-router-af)# area 1 range 2001:db8:acad:1000::/52

R3(config)# router ospfv3 123
R3(config-router)# address-family ipv6 unicast
R3(config-router-af)# area 2 range 2001:db8:acad:2000::/52
```

b.   Notice that R2 is now receiving the summarized prefixes.

```
R2# show ipv6 route ospf
<output omitted>
OI  2001:DB8:ACAD:1000::/52 [110/3]
     via FE80::1:2, GigabitEthernet0/0/0
OI  2001:DB8:ACAD:2000::/52 [110/3]
     via FE80::3:2, GigabitEthernet0/0/1
```

Question:

Why is prefix summarization considered desirable? How does it stabilize routing?

_____

_____

_____

Step 3.   Modifying the network type.

    a.   OSPFv3 supports the same network types as OSPFv2. Notice that the Ethernet interfaces between R2 and R1, and R2 and R3, elect a DR and a BDR. This is because Ethernet is a multiaccess network. However, these are point-to-point links and there is no need for a DR or BDR.

```
R2# show ospfv3 interface brief
```

Interface	PID	Area	AF	Cost	State	Nbrs F/C
Gi0/0/1	123	0	ipv4	1	BDR	1/1
Gi0/0/0	123	0	ipv4	1	DR	1/1
Gi0/0/1	123	0	ipv6	1	BDR	1/1
Gi0/0/0	123	0	ipv6	1	DR	1/1

    b.   These connections can be changed to point-to-point using the **ospfv3 network point-to-point** interface command. This command needs to be configured on both sides of the point to point interface.

```
R2(config)# interface g0/0/1
R2(config-if)# ospfv3 network point-to-point
R2(config-if)# exit
R2(config)# interface g0/0/0
R2(config-if)# ospfv3 network point-to-point


R1(config)# interface g0/0/0
R1(config-if)# ospfv3 network point-to-point


R3(config)# interface g0/0/0
R3(config-if)# ospfv3 network point-to-point
```

    c.   Notice that the links have now change to P2P.

```
R2# show ospfv3 interface brief
```

Interface	PID	Area	AF	Cost	State	Nbrs F/C
Gi0/0/1	123	0	ipv4	1	P2P	1/1
Gi0/0/0	123	0	ipv4	1	P2P	1/1
Gi0/0/1	123	0	ipv6	1	P2P	1/1
Gi0/0/0	123	0	ipv6	1	P2P	1/1

Question:

What is the effect on the state of the interface when changing a broadcast network to point-to-point?

_____

_____

Step 4. Advertising a default route.

    a. Similar to OSPFv2, an ASBR in OSPFv3 advertises using the **default-information** command. Configure a static default route for IPv4 and IPv6 on R2.

Note: Without a default route in the routing table, OSPF would require the **default-information originate always** command to advertise a default route.

```
R2(config)# ipv6 route ::/0 lo0
R2(config)# ip route 0.0.0.0 0.0.0.0 lo0

R2(config)# router ospfv3 123
R2(config-router)# address-family ipv6 unicast
R2(config-router-af)# default-information originate
R2(config-router-af)# exit
R2(config-router)# address-family ipv4 unicast
R2(config-router-af)# default-information originate
R2(config-router-af)# exit
```

    b. Verify D1 is receiving an IPv6 default route via OSPFv3.

```
D1# show ipv6 route ospf
<output omitted>
OE2 ::/0 [110/1], tag 123
    via FE80::1:1, GigabitEthernet1/0/11
OI  2001:DB8:ACAD:2000::/52 [110/5]
    via FE80::1:1, GigabitEthernet1/0/11
OI  2001:DB8:ACAD:A001::/64 [110/2]
    via FE80::1:1, GigabitEthernet1/0/11
OI  2001:DB8:ACAD:A002::/64 [110/3]
    via FE80::1:1, GigabitEthernet1/0/11
```

    c. Verify D2 is receiving an IPv4 default route via OSPFv3.

```
D2# show ip route ospfv3
<output omitted>
Gateway of last resort is 10.10.4.1 to network 0.0.0.0

O*E2  0.0.0.0/0 [110/1] via 10.10.4.1, 00:01:13, GigabitEthernet1/0/11
      172.16.0.0/30 is subnetted, 2 subnets
O IA     172.16.0.0 [110/3] via 10.10.4.1, 00:02:55, GigabitEthernet1/0/11
O IA     172.16.1.0 [110/2] via 10.10.4.1, 00:20:22, GigabitEthernet1/0/11
```

Router Interface Summary Table

Router Model	Ethernet Interface #1	Ethernet Interface #2	Serial Interface #1	Serial Interface #2
1800	Fast Ethernet 0/0 (F0/0)	Fast Ethernet 0/1 (F0/1)	Serial 0/0/0 (S0/0/0)	Serial 0/0/1 (S0/0/1)
1900	Gigabit Ethernet 0/0 (G0/0)	Gigabit Ethernet 0/1 (G0/1)	Serial 0/0/0 (S0/0/0)	Serial 0/0/1 (S0/0/1)
2801	Fast Ethernet 0/0 (F0/0)	Fast Ethernet 0/1 (F0/1)	Serial 0/1/0 (S0/1/0)	Serial 0/1/1 (S0/1/1)

Router Model	Ethernet Interface #1	Ethernet Interface #2	Serial Interface #1	Serial Interface #2
2811	Fast Ethernet 0/0 (F0/0)	Fast Ethernet 0/1 (F0/1)	Serial 0/0/0 (S0/0/0)	Serial 0/0/1 (S0/0/1)
2900	Gigabit Ethernet 0/0 (G0/0)	Gigabit Ethernet 0/1 (G0/1)	Serial 0/0/0 (S0/0/0)	Serial 0/0/1 (S0/0/1)
4221	Gigabit Ethernet 0/0/0 (G0/0/0)	Gigabit Ethernet 0/0/1 (G0/0/1)	Serial 0/1/0 (S0/1/0)	Serial 0/1/1 (S0/1/1)
4300	Gigabit Ethernet 0/0/0 (G0/0/0)	Gigabit Ethernet 0/0/1 (G0/0/1)	Serial 0/1/0 (S0/1/0)	Serial 0/1/1 (S0/1/1)

Note: To find out how the router is configured, look at the interfaces to identify the type of router and how many interfaces the router has. There is no way to effectively list all the combinations of configurations for each router class. This table includes identifiers for the possible combinations of Ethernet and Serial interfaces in the device. The table does not include any other type of interface, even though a specific router may contain one. An example of this might be an ISDN BRI interface. The string in parenthesis is the legal abbreviation that can be used in Cisco IOS commands to represent the interface.

Device Config Final – Notes

_____

_____

_____

_____

_____

_____

_____

_____

_____

_____

_____

_____

_____

_____

_____

_____

_____

_____

_____

_____

CHAPTER 10

Troubleshooting OSPFv3

10.1.2 Lab - Troubleshoot OSPFv3

Topology

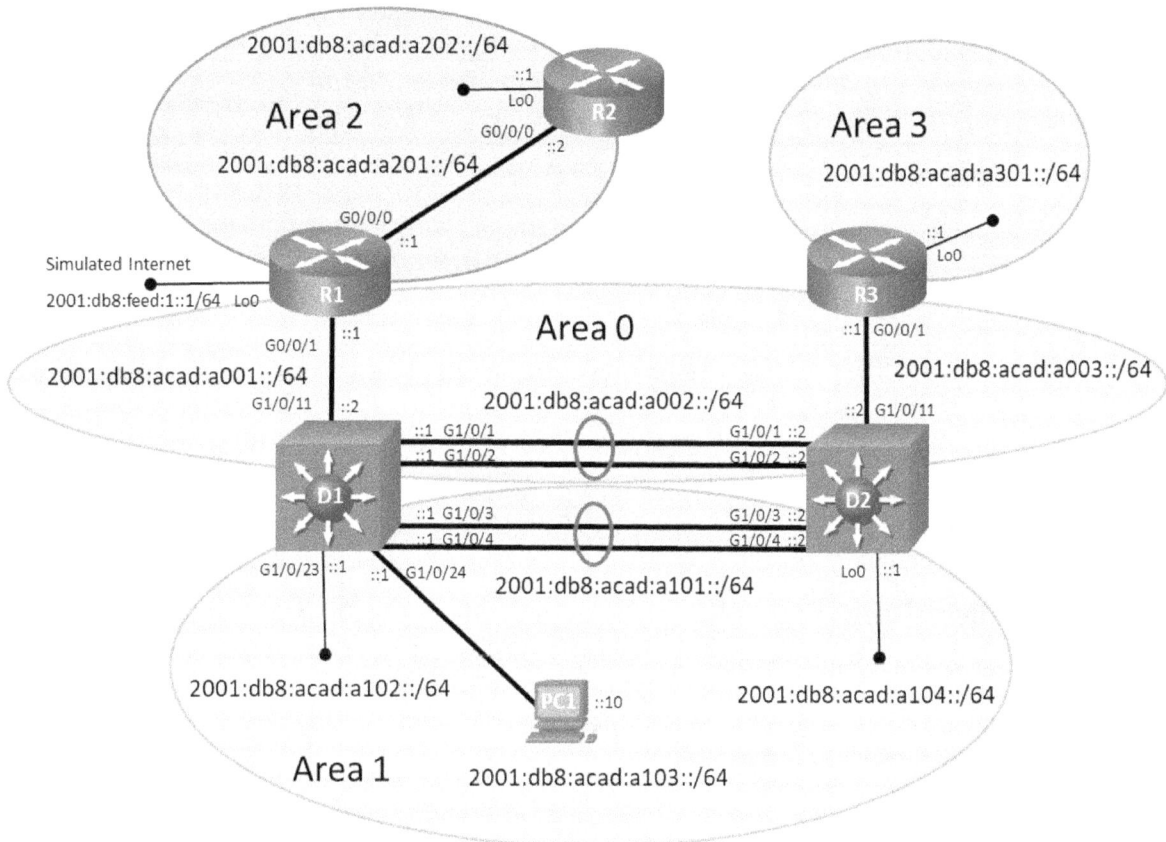

Addressing Table

Device	Interface	IPv6 Address	Link-Local
R1	G0/0/0	2001:db8:acad:a201::1/64	fe80::a201:1
	G0/0/1	2001:db8:acad:a001::1/64	fe80::a001:1
	Lo1	2001:db8:feed:1::1/64	fe80::1:1
R2	G0/0/0	2001:db8:acad:a201::2/64	fe80::a201:2
	Lo1	2001:db8:acad:a202::1/64	fe80::a202:1
R3	G0/0/1	2001:db8:acad:a003::1/64	fe80::a003:1
	Lo1	2001:db8:acad:a301::1/64	fe80::a301:1
D1	G1/0/1 (Po1)	2001:db8:acad:a002::1/64	fe80::a002:1
	G1/0/2 (Po1)	2001:db8:acad:a002::1/64	fe80::a002:1
	G1/0/3 (Po2)	2001:db8:acad:a101::1/64	fe80::a101:1

Device	Interface	IPv6 Address	Link-Local
	G1/0/4 (Po2)	2001:db8:acad:a101::1/64	fe80::a101:1
	G1/0/5	2001:db8:acad:a103::1/64	fe80::a103:1
	G1/0/11	2001:db8:acad:a001::2/64	fe80::a001:2
	G1/0/23	2001:db8:acad:a102::1/64	fe80::a102:1
D2	G1/0/1 (Po1)	2001:db8:acad:a002::2/64	fe80::a002:2
	G1/0/2 (Po1)	2001:db8:acad:a002::2/64	fe80::a002:2
	G1/0/3 (Po2)	2001:db8:acad:a101::2/64	fe80::a101:2
	G1/0/4 (Po2)	2001:db8:acad:a101::2/64	fe80::a101:2
	G1/0/11	2001:db8:acad:a003::2/64	fe80::a003:2
	Lo0	2001:db8:acad:a104::1/64	fe80::a104:1
PC1	N/A	SLAAC	SLAAC

Note: To make it easier to recognize IPv6 prefixes, familiarize yourself with the IPv6 GUA and LLA address formats.

- **GUA:** The GUA has a 16-bit subnet-ID, a<area-id>xx ("a" for area). For example, subnet-ID a201 is area 2, network 01.

- **LLA:** Following best practice the LLA address is unique on each interface. The LLA interface-ID uses the GUA subnet-ID:interface-ID for the last 64 bits. For example, fe80::a201:1 has an LLA interface-ID a201 (the subnet-ID of the GUA) and :1 (the Interface ID of the GUA).

Objectives

Troubleshoot network issues related to the configuration and operation of OSPFv3.

Background/Scenario

Although the topology has a limited number of routers, you should use the appropriate troubleshooting commands to help find and solve the problems in the three trouble tickets as if this were a much more complex topology with many more routers and networks.

You will be loading configurations with intentional errors onto the network. Your tasks are to FIND the error(s), document your findings and the command(s) or method(s) used to fix them, FIX the issue(s) presented here and then test the network to ensure both of the following conditions are met:

1. the complaint received in the ticket is resolved

2. full reachability is restored

Note: The routers used with CCNP hands-on labs are Cisco 4221 with Cisco IOS XE Release 16.9.4 (universalk9 image). The switches used in the labs are Cisco Catalyst 3650 with Cisco IOS XE Release 16.9.4 (universalk9 image). Other routers, switches, and Cisco IOS versions can be used. Depending on the model and Cisco IOS version, the commands available and output produced might vary from what is shown in the labs. Refer to the Router Interface Summary Table at the end of the lab for the correct interface identifiers.

Note: Make sure that the devices have been erased and have no startup configurations. If you are unsure, contact your instructor.

Required Resources

- 3 Routers (Cisco 4221 with Cisco IOS XE Release 16.9.4 universal image or comparable)
- 2 Switches (Cisco 3560 with Cisco IOS XE Release 16.9.4 universal image or comparable)
- 1 PC (Choice of operating system with terminal emulation program installed)
- Console cables to configure the Cisco IOS devices via the console ports
- Ethernet cables as shown in the topology

Instructions

Part 1: Trouble Ticket 10.1.2.1

Scenario:

Your University network has migrated to IPv6-only internally and is using multiarea OSPFv3 address families. Recently your team configured all the necessary devices for this migration, with different people responsible for configuring different parts of the network.

During testing and validation, the network team noticed that routers are not showing the 2001:db8:acad:a202::/64 prefix in their routing tables.

Use the commands listed below to load the configuration files for this trouble ticket:

Device	Command
R1	`copy flash:/enarsi/10.1.2.1-r1-config.txt run`
R2	`copy flash:/enarsi/10.1.2.1-r2-config.txt run`
R3	`copy flash:/enarsi/10.1.2.1-r3-config.txt run`
D1	`copy flash:/enarsi/10.1.2.1-d1-config.txt run`
D2	`copy flash:/enarsi/10.1.2.1-d2-config.txt run`

- All routers should have the 2001:db8:acad:a202::/64 network in their routing tables.
- All devices should be able to ping 2001:db8:acad:a202::2/64.
- When you have fixed the ticket, change the MOTD on EACH DEVICE using the following command:

    banner motd # This is $(hostname) FIXED from ticket <ticket number> #

- Save the configuration by issuing the **wri** command (on each device).
- Inform your instructor that you are ready for the next ticket.
- After the instructor approves your solution for this ticket, issue the **reset.now** privileged EXEC command. This script will clear your configurations and reload the devices.

Part 2: Trouble Ticket 10.1.2.2

Scenario:

After a regularly scheduled downtime for maintenance and IOS upgrades, users started reporting to the helpdesk that there is no access to the IPv6 prefix in area 3. Although there are backups for all device configurations it is suspected that some of the backups might not have been correct.

You have been tasked to find and resolve the issue with reaching the IPv6 prefix in area 3.

Use the commands listed below to load the configuration files for this trouble ticket:

Device	Command
R1	copy flash:/enarsi/10.1.2.2-r1-config.txt run
R2	copy flash:/enarsi/10.1.2.2-r2-config.txt run
R3	copy flash:/enarsi/10.1.2.2-r3-config.txt run
D1	copy flash:/enarsi/10.1.2.2-d1-config.txt run
D2	copy flash:/enarsi/10.1.2.2-d2-config.txt run

- All routers should have the 2001:db8:acad:a301::/64 network in their routing tables.
- All devices should be able to ping 2001:db8:acad:a301::1/64.
- When you have fixed the ticket, change the MOTD on EACH DEVICE using the following command:

  banner motd # This is $(hostname) FIXED from ticket <ticket number> #

- Then save the configuration by issuing the **wri** command (on each device).
- Inform your instructor that you are ready for the next ticket.
- After the instructor approves your solution for this ticket, issue the **reset.now** privileged EXEC command. This script will clear your configurations and reload the devices.

Part 3: Trouble Ticket 10.1.2.3

Scenario:

Network technicians recently installed a new router, R2 to area 2. However, during the verification phase the technicians are reporting that the other routers are not including the 2001:db8:acad:a202::/64 prefix in their routing tables. They call you for assistance.

Use the commands listed below to load the configuration files for this trouble ticket:

Device	Command
R1	copy flash:/enarsi/10.1.2.3-r1-config.txt run
R2	copy flash:/enarsi/10.1.2.3-r2-config.txt run
R3	copy flash:/enarsi/10.1.2.3-r3-config.txt run
D1	copy flash:/enarsi/10.1.2.3-d1-config.txt run
D2	copy flash:/enarsi/10.1.2.3-d2-config.txt run

Note: To simulate the link being cut, shutdown the G0/0/1 interface on R3.

- PC1 should be able to ping 2001:db8:acad:a301::1.

- Network 2001:db8:acad:a301::/64 must be in its own area.

- When you have fixed the ticket, change the MOTD on EACH DEVICE using the following command:

   banner motd # This is $(hostname) FIXED from ticket <ticket number> #

- Save the configuration by issuing the **wri** command (on each device).

- Inform your instructor that you are ready for the next ticket.

- After the instructor approves your solution for this ticket, issue the **reset.now** privileged EXEC command. This script will clear your configurations and reload the devices.

Device Config Final – Notes

_____

_____

_____

_____

_____

_____

_____

_____

_____

_____

_____

_____

_____

_____

_____

_____

_____

_____

_____

_____

_____

_____

_____

_____

_____

_____

_____

11.1.2 Lab - Implement eBGP for IPv4

Topology

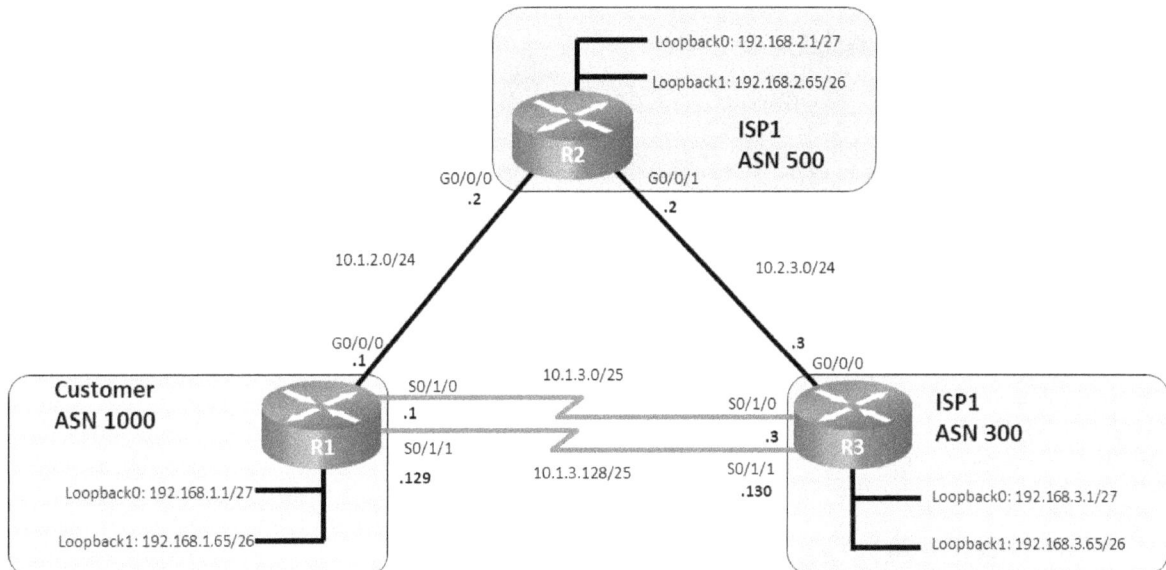

Addressing Table

Device	Interface	IPv4 Address
R1	G0/0/0	10.1.2.1/24
	S0/1/0	10.1.3.1/25
	S0/1/1	10.1.3.129/25
	Loopback0	192.168.1.1/27
	Loopback1	192.168.1.65/26
R2	G0/0/0	10.1.2.2/24
	G0/0/1	10.2.3.2/24
	Loopback0	192.168.2.1/27
	Loopback1	192.168.2.65/26
R3	G0/0/0	10.2.3.3/24
	S0/1/0	10.1.3.3/25
	S0/1/1	10.1.3.130/25
	Loopback0	192.168.3.1/27
	Loopback1	192.168.3.65/26

Objectives

Part 1: Build the Network and Configure Basic Device Settings and Interface Addressing

Part 2: Configure and Verify eBGP for IPv4 on all Routers

Part 3: Configure and Verify Route Summarization and Atomic Aggregate

Part 4: Configure and Verify Route Summarization with Atomic Aggregate and AS-Set

Part 5: Configure and Verify the Advertising of a Default Route

Background/Scenario

In this lab you will configure eBGP for IPv4.

Note: This lab is an exercise in developing, deploying, and verifying various path manipulation tools for BGP, and does not reflect networking best practices.

Note: The routers used with CCNP hands-on labs are Cisco 4221 with Cisco IOS XE Release 16.9.4 (universalk9 image). Other routers and Cisco IOS versions can be used. Depending on the model and Cisco IOS version, the commands available and the output produced might vary from what is shown in the labs.

Note: Ensure that the routers and switches have been erased and have no startup configurations. If you are unsure contact your instructor.

Required Resources

- 3 Routers (Cisco 4221 with Cisco IOS XE Release 16.9.4 universal image or comparable)
- 1 PC (Windows with a terminal emulation program, such as Tera Term)
- Console cables to configure the Cisco IOS devices via the console ports
- Ethernet and serial cables as shown in the topology

Instructions

Part 1: Build the Network and Configure Basic Device Settings and Interface Addressing

In Part 1, you will set up the network topology and configure basic settings and interface addressing on routers.

Step 1. Cable the network as shown in the topology.

Attach the devices as shown in the topology diagram, and cable as necessary.

Step 2. Configure basic settings for each router.

a. Console into each router, enter global configuration mode, and apply the basic settings and interface addressing. A command list for each router is provided below.

Router R1

```
hostname R1
no ip domain lookup
line con 0
logging sync
exec-time 0 0
exit
interface Loopback0
 ip address 192.168.1.1 255.255.255.224
 no shut
 exit
interface Loopback1
 ip address 192.168.1.65 255.255.255.192
 no shut
 exit
interface GigabitEthernet0/0/0
 ip address 10.1.2.1 255.255.255.0
 no shut
 exit
interface Serial0/1/0
 ip address 10.1.3.1 255.255.255.128
 no shut
 exit
interface Serial0/1/1
 ip address 10.1.3.129 255.255.255.128
 no shut
 exit
```

Router R2

```
hostname R2
no ip domain lookup
line con 0
logging sync
exec-time 0 0
exit
interface Loopback0
 ip address 192.168.2.1 255.255.255.224
 no shut
 exit
interface Loopback1
 ip address 192.168.2.65 255.255.255.192
 no shut
 exit
interface GigabitEthernet0/0/0
 ip address 10.1.2.2 255.255.255.0
 no shut
 exit
interface GigabitEthernet0/0/1
 ip address 10.2.3.2 255.255.255.0
 no shut
 exit
```

Router R3

```
hostname R3
no ip domain lookup
line con 0
 logging sync
 exec-time 0 0
 exit
interface Loopback0
 ip address 192.168.3.1 255.255.255.224
 no shut
 exit
interface Loopback1
 ip address 192.168.3.65 255.255.255.192
 no shut
 exit
interface GigabitEthernet0/0/0
 ip address 10.2.3.3 255.255.255.0
 negotiation auto
 no shut
 exit
interface Serial0/1/0
 ip address 10.1.3.3 255.255.255.128
 no shut
 exit
interface Serial0/1/1
 ip address 10.1.3.130 255.255.255.128
 no shut
 exit
```

b. Save the running configuration to startup-config.

Part 2: Configure and Verify eBGP for IPv4 on all Routers

Step 1.   Implement BGP and neighbor relationships on R1.

a. Enter BGP configuration mode from global configuration mode, specifying AS 1000.

```
R1(config)# router bgp 1000
```

b. Configure the BGP router-id for R1.

```
R1(config-router)# bgp router-id 1.1.1.1
```

c. Based on the topology diagram, configure all the designated neighbors for R1.

```
R1(config-router)# neighbor 10.1.2.2 remote-as 500
R1(config-router)# neighbor 10.1.3.3 remote-as 300
R1(config-router)# neighbor 10.1.3.130 remote-as 300
```

d. Configure R1 to advertise the IPv4 prefixes local to ASN 1000.

```
R1(config-router)# network 192.168.1.0 mask 255.255.255.224
R1(config-router)# network 192.168.1.64 mask 255.255.255.192
```

Step 2.   Implement BGP and neighbor relationships on R2.

a. Enter BGP configuration mode from global configuration mode, specifying AS 500.

```
R2(config)# router bgp 500
```

b. Configure the BGP router-id for R2.

```
R2(config-router)# bgp router-id 2.2.2.2
```

c. Based on the topology diagram, configure all the designated neighbors for R2.

```
R2(config-router)# neighbor 10.1.2.1 remote-as 1000
R2(config-router)# neighbor 10.2.3.3 remote-as 300
```

d. Configure R2 to advertise the IPv4 prefixes local to ASN 500.

```
R2(config-router)# network 192.168.2.0 mask 255.255.255.224
R2(config-router)# network 192.168.2.64 mask 255.255.255.192
```

Step 3. Implement BGP and neighbor relationships on R3.

a. Enter BGP configuration mode from global configuration mode, specifying AS 300.

```
R3(config)# router bgp 300
```

b. Configure the BGP router-id for R3.

```
R3(config-router)# bgp router-id 3.3.3.3
```

c. Unlike the configuration on R1 and R2, disable the default IPv4 unicast behavior.

```
R3(config-router)# no bgp default ipv4-unicast
```

The default behavior in IOS is **bgp default ipv4-unicast**. Routers R1 and R2 were configured using this default behavior. The **bgp default ipv4-unicast** command enables the automatic exchange of IPv4 address family prefixes. When this command is disabled using **no bgp default ipv4-unicast**, bgp neighbors must be activated within IPv4 address family (AF) configuration mode. BGP **network** commands must also be configured within IPv4 AF mode.

d. Based on the topology diagram, configure all the designated neighbors for R3.

```
R3(config-router)# neighbor 10.2.3.2 remote-as 500
R3(config-router)# neighbor 10.1.3.1 remote-as 1000
R3(config-router)# neighbor 10.1.3.129 remote-as 1000
```

Step 4. Verifying BGP neighbor relationships.

a. Examine the routing tables on each router. Notice that R1 and R2 are receiving BGP prefixes from each other but not receiving BGP prefixes from R3. And R3 is not receiving any prefixes from R1 or R2. This is because R3 was configured using **no bgp default ipv4-unicast** and the interfaces must be activated within IPv4 address configuration mode.

```
R1# show ip route bgp | begin Gateway
Gateway of last resort is not set


      192.168.2.0/24 is variably subnetted, 2 subnets, 2 masks
B        192.168.2.0/27 [20/0] via 10.1.2.2, 00:28:40
B        192.168.2.64/26 [20/0] via 10.1.2.2, 00:28:40


R2# show ip route bgp | begin Gateway
Gateway of last resort is not set


      192.168.1.0/24 is variably subnetted, 2 subnets, 2 masks
B        192.168.1.0/27 [20/0] via 10.1.2.1, 00:29:41
B        192.168.1.64/26 [20/0] via 10.1.2.1, 00:29:41
```

```
R3# show ip route bgp | begin Gateway
Gateway of last resort is not set
```

b. This can be further verified by examining the BGP neighbor adjacencies on R2. Notice the BGP state between R2 and R1 is established, while the BGP state between R2 and R3 is idle.

```
R2# show ip bgp neighbors
BGP neighbor is 10.1.2.1,  remote AS 1000, external link
  BGP version 4, remote router ID 1.1.1.1
  BGP state = Established, up for 00:35:34
  Last read 00:00:28, last write 00:00:35, hold time is 180, keepalive interval
is 60 seconds
  Neighbor sessions:
    1 active, is not multisession capable (disabled)
<output omitted>


BGP neighbor is 10.2.3.3,  remote AS 300, external link
  BGP version 4, remote router ID 0.0.0.0
  BGP state = Idle, down for never
  Neighbor sessions:
    0 active, is not multisession capable (disabled)
<output omitted>
```

c. The interfaces on R3 need to be activated in IPv4 AF configuration mode. The **neighbor activate** command in IPv4 AF configuration mode is required to enable the exchange of BGP information between neighbors. This will enable R3 to form an established neighbor adjacency with both R1 and R2. Additionally, because **bgp default ipv4-unicast** is disabled, **network** commands must be configured in IPv4 AF configuration mode.

```
R3(config-router)# address-family ipv4
R3(config-router-af)# neighbor 10.1.3.1 activate
R3(config-router-af)# neighbor 10.1.3.129 activate
R3(config-router-af)# neighbor 10.2.3.2 activate
R3(config-router-af)# network 192.168.3.0 mask 255.255.255.224
R3(config-router-af)# network 192.168.3.64 mask 255.255.255.192
```

d. Verify that all BGP speakers are receiving prefixes from their neighbors. The prefixes from R3 are highlighted in the routing tables of R1 and R2.

Note: The prefixes in the lab are for example purposes only. Most service providers do not accept prefixes larger than /24 for IPv4 (/25 through /32).

```
R1# show ip route bgp | begin Gateway
Gateway of last resort is not set


      192.168.2.0/24 is variably subnetted, 2 subnets, 2 masks
B        192.168.2.0/27 [20/0] via 10.1.2.2, 00:51:09
B        192.168.2.64/26 [20/0] via 10.1.2.2, 00:51:09
      192.168.3.0/24 is variably subnetted, 2 subnets, 2 masks
B        192.168.3.0/27 [20/0] via 10.1.3.3, 00:01:43
B        192.168.3.64/26 [20/0] via 10.1.3.3, 00:01:43
```

```
R2# show ip route bgp | begin Gateway
Gateway of last resort is not set

      192.168.1.0/24 is variably subnetted, 2 subnets, 2 masks
B        192.168.1.0/27 [20/0] via 10.1.2.1, 00:51:17
B        192.168.1.64/26 [20/0] via 10.1.2.1, 00:51:17
      192.168.3.0/24 is variably subnetted, 2 subnets, 2 masks
B        192.168.3.0/27 [20/0] via 10.2.3.3, 00:01:51
B        192.168.3.64/26 [20/0] via 10.2.3.3, 00:01:51


R3# show ip route bgp | begin Gateway
Gateway of last resort is not set

      192.168.1.0/24 is variably subnetted, 2 subnets, 2 masks
B        192.168.1.0/27 [20/0] via 10.1.3.1, 00:02:11
B        192.168.1.64/26 [20/0] via 10.1.3.1, 00:02:11
      192.168.2.0/24 is variably subnetted, 2 subnets, 2 masks
B        192.168.2.0/27 [20/0] via 10.2.3.2, 00:02:11
B        192.168.2.64/26 [20/0] via 10.2.3.2, 00:02:11
```

e. Verify that the BGP state between R2 and R3 has now been established.

```
R2# show ip bgp neighbors | begin BGP neighbor is 10.2.3.3
BGP neighbor is 10.2.3.3,  remote AS 300, external link
  BGP version 4, remote router ID 3.3.3.3
  BGP state = Established, up for 00:12:16
  Last read 00:00:37, last write 00:00:52, hold time is 180, keepalive interval
is 60 seconds
  Neighbor sessions:
    1 active, is not multisession capable (disabled)
<output omitted>
```

Step 5.   Examining the running-configs.

Examine the running-configs on all three routers. Because router R3 was configured using the **no bgp default ipv4-unicast** command, notice that the network commands were automatically entered under the IPv4 AF. This is the same configuration mode where the neighbors were activated to exchange BGP information.

```
R1# show running-config | section bgp
router bgp 1000
 bgp router-id 1.1.1.1
 bgp log-neighbor-changes
 network 192.168.1.0 mask 255.255.255.224
 network 192.168.1.64 mask 255.255.255.192
 neighbor 10.1.2.2 remote-as 500
 neighbor 10.1.3.3 remote-as 300
 neighbor 10.1.3.130 remote-as 300

R2# show running-config | section bgp
router bgp 500
 bgp router-id 2.2.2.2
 bgp log-neighbor-changes
```

```
        network 192.168.2.0 mask 255.255.255.224
        network 192.168.2.64 mask 255.255.255.192
        neighbor 10.1.2.1 remote-as 1000
        neighbor 10.2.3.3 remote-as 300


R3# show running-config | section bgp
router bgp 300
 bgp log-neighbor-changes
 no bgp default ipv4-unicast
 neighbor 10.1.3.1 remote-as 1000
 neighbor 10.1.3.129 remote-as 1000
 neighbor 10.2.3.2 remote-as 500
 !
 address-family ipv4
  network 192.168.3.0 mask 255.255.255.224
  network 192.168.3.64 mask 255.255.255.192
  neighbor 10.1.3.1 activate
  neighbor 10.1.3.129 activate
  neighbor 10.2.3.2 activate
 exit-address-family
```

Step 6.   Verifying BGP operations.

   a.   To verify the BGP operation on R2, issue the **show ip bgp** command.

```
R2# show ip bgp
BGP table version is 11, local router ID is 2.2.2.2
Status codes: s suppressed, d damped, h history, * valid, > best, i - internal,
              r RIB-failure, S Stale, m multipath, b backup-path, f RT-Filter,
              x best-external, a additional-path, c RIB-compressed,
              t secondary path, L long-lived-stale,
Origin codes: i - IGP, e - EGP, ? - incomplete
RPKI validation codes: V valid, I invalid, N Not found


     Network          Next Hop         Metric LocPrf Weight Path
 *   192.168.1.0/27   10.2.3.3                         0 300 1000 i
 *>                   10.1.2.1              0           0 1000 i
 *   192.168.1.64/26  10.2.3.3                         0 300 1000 i
 *>                   10.1.2.1              0           0 1000 i
 *>  192.168.2.0/27   0.0.0.0              0       32768 i
 *>  192.168.2.64/26  0.0.0.0              0       32768 i
 *>  192.168.3.0/27   10.2.3.3             0           0 300 i
 *                    10.1.2.1                         0 1000 300 i
 *>  192.168.3.64/26  10.2.3.3             0           0 300 i
 *                    10.1.2.1                         0 1000 300 i
```

Questions:

What does the * at the beginning of an entry indicate?

_____

What does the angle bracket (>) in an entry indicate?

_____

What is the address of the preferred next hop router to reach the 192.168.1.0/27 network? Explain.

_____

How can you verify that 10.1.2.1 is the next hop router used to reach 192.168.1.0/27?

_____

_____

What does a next hop of 0.0.0.0 indicate?

_____

b. Use the **show ip bgp** *ip-prefix* command to display all the paths for a specific route and the BGP path attributes for that route.

```
R2# show ip bgp 192.168.1.0
BGP routing table entry for 192.168.1.0/27, version 14
Paths: (2 available, best #2, table default)
  Advertised to update-groups:
    1
  Refresh Epoch 1
  300 1000
    10.2.3.3 from 10.2.3.3 (3.3.3.3)
      Origin IGP, localpref 100, valid, external
      rx pathid: 0, tx pathid: 0
  Refresh Epoch 2
  1000
    10.1.2.1 from 10.1.2.1 (1.1.1.1)
      Origin IGP, metric 0, localpref 100, valid, external, best
      rx pathid: 0, tx pathid: 0x0
```

Question:

What is the IPv4 address of the next hop router with the best path?

_____

c. Examine the BGP neighbor relationships on R2 using the **show ip bgp neighbors** command.

```
R2# show ip bgp neighbors
BGP neighbor is 10.1.2.1,  remote AS 1000, external link
  BGP version 4, remote router ID 1.1.1.1
  BGP state = Established, up for 00:00:51
  Last read 00:00:00, last write 00:00:51, hold time is 180, keepalive interval
is 60 seconds
  Neighbor sessions:
    1 active, is not multisession capable (disabled)
  Neighbor capabilities:
    Route refresh: advertised and received(new)
    Four-octets ASN Capability: advertised and received
    Address family IPv4 Unicast: advertised and received
    Enhanced Refresh Capability: advertised and received
    Multisession Capability:
    Stateful switchover support enabled: NO for session 1
```

```
    Message statistics:
      InQ depth is 0
      OutQ depth is 0


                            Sent         Rcvd
      Opens:                 1            1
      Notifications:         0            0
      Updates:               5            5
      Keepalives:            2            3
      Route Refresh:         0            0
      Total:                10           11
<output omitted>


BGP neighbor is 10.2.3.3,   remote AS 300, external link
  BGP version 4, remote router ID 3.3.3.3
  BGP state = Established, up for 16:23:45
  Last read 00:00:29, last write 00:00:51, hold time is 180, keepalive interval
is 60 seconds
  Neighbor sessions:
    1 active, is not multisession capable (disabled)
  Neighbor capabilities:
    Route refresh: advertised and received(new)
    Four-octets ASN Capability: advertised and received
    Address family IPv4 Unicast: advertised and received
    Enhanced Refresh Capability: advertised and received
    Multisession Capability:
    Stateful switchover support enabled: NO for session 1
  Message statistics:
    InQ depth is 0
    OutQ depth is 0


                            Sent         Rcvd
      Opens:                 1            1
      Notifications:         0            0
      Updates:               9            5
      Keepalives:         1082         1088
      Route Refresh:         0            0
      Total:              1096         1096
  Do log neighbor state changes (via global configuration)
  Default minimum time between advertisement runs is 30 seconds
<output omitted>
```

Questions:

How many neighbors does R2 have and what are their router IDs?

_____

What is the BGP state of both neighbors?

_____

What are the keepalive and hold time value for both neighbors?

_____

Part 3: Configure and Verify Route Summarization and Atomic Aggregate

Step 1.   Configure route summarization using atomic aggregate.

Summarizing prefixes conserves router resources and accelerates best-path calculation by reducing the size of the table. Summarization can be configured either for prefixes originated by the AS or prefixes received from downstream providers. Summarization also provides the benefits of stability by hiding flapping routes or having to install new prefixes when they are contained within a summary.

Although AS 1000 only has two prefixes 192.168.1.0/27 and 192.168.1.64/26, this customer has been allocated the entire 192.168.1.0/24 prefix. R3 in AS 300 has two prefixes 192.168.3.0/27 and 192.168.3.64/26 but has been allocated the entire 192.168.3.0/24 prefix.

Configure R1 and R3 to advertise a summary or aggregate route using the **aggregate-address** command. The **summary-only** option suppresses the specific prefixes that are summarized from also being advertised. Notice that this command is configured in **address-family ipv4** configuration mode on R3.

```
R1(config)# router bgp 1000
R1(config-router)# aggregate-address 192.168.1.0 255.255.255.0 summary-only


R3(config)# router bgp 300
R3(config-router)# address-family ipv4
R3(config-router-af)# aggregate-address 192.168.3.0 255.255.255.0 summary-only
```

Step 2.   Verify route summarization using atomic aggregate.

a.   Examine the routing tables on each router to verify the route summarization for the two prefixes. Verify that R1 and R3 are each receiving the summary route from the other router. Verify that R2 is receiving aggregate routes from both R1 and R3.

```
R1# show ip route bgp | begin Gateway
Gateway of last resort is not set


      192.168.1.0/24 is variably subnetted, 5 subnets, 4 masks
B        192.168.1.0/24 [200/0], 00:27:47, Null0
      192.168.2.0/24 is variably subnetted, 2 subnets, 2 masks
B        192.168.2.0/27 [20/0] via 10.1.2.2, 13:34:31
B        192.168.2.64/26 [20/0] via 10.1.2.2, 13:34:31
B     192.168.3.0/24 [20/0] via 10.1.3.3, 00:26:01


R2# show ip route bgp | begin Gateway
Gateway of last resort is not set


B     192.168.1.0/24 [20/0] via 10.1.2.1, 00:33:53
B     192.168.3.0/24 [20/0] via 10.2.3.3, 00:32:08


R3# show ip route bgp | begin Gateway
Gateway of last resort is not set


B     192.168.1.0/24 [20/0] via 10.1.3.1, 00:36:52
      192.168.2.0/24 is variably subnetted, 2 subnets, 2 masks
B        192.168.2.0/27 [20/0] via 10.2.3.2, 02:10:48
```

```
B         192.168.2.64/26 [20/0] via 10.2.3.2, 02:10:48
          192.168.3.0/24 is variably subnetted, 5 subnets, 4 masks
B         192.168.3.0/24 [200/0], 00:35:07, Null0
```

Question:

Why do R1 and R3 contain an entry with a next hop address of Null0? What is the result of having this Null0 route in the routing table?

_____

_____

_____

_____

b. Examine the BGP table on router R2 to verify the route summarization. When a prefix has the default classful mask, the subnet mask is not displayed. Both 192.168.1.0 and 192.168.3.0 prefixes have a /24 prefix length which would be the default mask for a Class C address.

```
R2# show ip bgp
BGP table version is 69, local router ID is 1.1.1.1
Status codes: s suppressed, d damped, h history, * valid, > best, i - internal,
<output omitted>
```

	Network	Next Hop	Metric	LocPrf	Weight	Path
*	192.168.1.0	10.2.3.3			0	300 1000 i
*>		10.1.2.1	0		0	1000 i
*>	192.168.2.0/27	0.0.0.0	0		32768	i
*>	192.168.2.64/26	0.0.0.0	0		32768	i
*	192.168.3.0	10.1.2.1			0	1000 300 i
*>		10.2.3.3	0		0	300 i

c. Examine the BGP table on routers R2 and R3 and verify that each router is receiving the summary route from the other router.

```
R1# show ip bgp
BGP table version is 69, local router ID is 1.1.1.1
Status codes: s suppressed, d damped, h history, * valid, > best, i - internal,
<output omitted>
```

	Network	Next Hop	Metric	LocPrf	Weight	Path
s>	192.168.1.0/27	0.0.0.0	0		32768	i
*>	192.168.1.0	0.0.0.0			32768	i
s>	192.168.1.64/26	0.0.0.0	0		32768	i
*	192.168.2.0/27	10.1.3.130			0	300 500 i
*		10.1.3.3			0	300 500 i
*>		10.1.2.2	0		0	500 i
*	192.168.2.64/26	10.1.3.130			0	300 500 i
*		10.1.3.3			0	300 500 i
*>		10.1.2.2	0		0	500 i
*	192.168.3.0	10.1.2.2			0	500 300 i
*		10.1.3.130	0		0	300 i
*>		10.1.3.3	0		0	300 i

```
R3# show ip bgp
BGP table version is 22, local router ID is 3.3.3.3
Status codes: s suppressed, d damped, h history, * valid, > best, i - internal,
              r RIB-failure, S Stale, m multipath, b backup-path, f RT-Filter,
              x best-external, a additional-path, c RIB-compressed,
              t secondary path, L long-lived-stale,
Origin codes: i - IGP, e - EGP, ? - incomplete
RPKI validation codes: V valid, I invalid, N Not found

     Network          Next Hop         Metric LocPrf Weight Path
 *   192.168.1.0      10.2.3.2                         0 500 1000 i
 *>                   10.1.3.1              0           0 1000 i
 *                    10.1.3.129           0           0 1000 i
 *   192.168.2.0/27   10.1.3.1                         0 1000 500 i
 *                    10.1.3.129                        0 1000 500 i
 *>                   10.2.3.2             0           0 500 i
 *   192.168.2.64/26  10.1.3.1                         0 1000 500 i
 *                    10.1.3.129                        0 1000 500 i
 *>                   10.2.3.2             0           0 500 i
 s>  192.168.3.0/27   0.0.0.0              0       32768 i
 *>  192.168.3.0      0.0.0.0                      32768 i
 s>  192.168.3.64/26  0.0.0.0              0       32768 i
```

Question:

Why do two of the entries have the status code of "s"? Specifically, this is the result of what command or option that was configured on these two routers?

_____

_____

_____

_____

d. Examine the explicit 192.168.1.0 prefix entry in R2's BGP table. The route's NLRI information indicates that the route was aggregated in AS 1000 by a router with the RID 1.1.1.1.

```
R2# show ip bgp 192.168.1.0
BGP routing table entry for 192.168.1.0/24, version 45
Paths: (2 available, best #2, table default)
  Advertised to update-groups:
     1
  Refresh Epoch 1
  300 1000, (aggregated by 1000 1.1.1.1)
    10.2.3.3 from 10.2.3.3 (3.3.3.3)
      Origin IGP, localpref 100, valid, external, atomic-aggregate
      rx pathid: 0, tx pathid: 0
  Refresh Epoch 2
  1000, (aggregated by 1000 1.1.1.1)
    10.1.2.1 from 10.1.2.1 (1.1.1.1)
      Origin IGP, metric 0, localpref 100, valid, external, atomic-aggregate,
best
      rx pathid: 0, tx pathid: 0x0
```

Part 4: Configure and Verify Route Summarization with Atomic Aggregate and AS-Set

Step 1.   Configure route summarization using atomic aggregate and AS-Set.

   a.   Shut down both serial interfaces on R1. This will create a single path from R1 (AS 1000) to R2 (AS 500) to R3 (AS 300).

```
R1(config)# interface s0/1/0
R1(config-if)# shutdown
R1(config-if)# exit
R1(config)# interface s0/1/1
R1(config-if)# shutdown
```

   b.   Remove route aggregation previously configured on R1.

```
R1(config)# router bgp 1000
R1(config-router)# no aggregate-address 192.168.1.0 255.255.255.0 summary-only
```

   c.   Verify that R3 is now receiving the non-summarized prefixes 192.168.1.0/27 and 192.168.1.64/26.

```
R3# show ip route 192.168.1.0
Routing entry for 192.168.1.0/24, 2 known subnets
  Variably subnetted with 2 masks
B        192.168.1.0/27 [20/0] via 10.2.3.2, 00:01:26
B        192.168.1.64/26 [20/0] via 10.2.3.2, 00:01:26
```

   d.   On R2, summarize the prefixes 192.168.1.0/27 and 192.168.1.64/26 received from R1 as 192.168.1.0/24.

```
R2(config)# router bgp 500
R2(config-router)# aggregate-address 192.168.1.0 255.255.255.0 summary-only
```

Step 2.   Verify route summarization using atomic aggregate and AS-Set.

   a.   Verify that R3 is receiving the aggregated prefix 192.168.1.0/24.

```
R3# show ip route bgp | begin Gateway
Gateway of last resort is not set

B     192.168.1.0/24 [20/0] via 10.2.3.2, 00:00:51
         192.168.2.0/24 is variably subnetted, 2 subnets, 2 masks
B        192.168.2.0/27 [20/0] via 10.2.3.2, 08:46:37
B        192.168.2.64/26 [20/0] via 10.2.3.2, 08:46:37
         192.168.3.0/24 is variably subnetted, 5 subnets, 4 masks
B        192.168.3.0/24 [200/0], 08:46:07, Null0
```

   b.   Examine R3's BGP table. Notice that the AS path only includes the AS that summarized the route, AS 500, router R2.

```
R3# show ip bgp
<output omitted>

      Network          Next Hop        Metric LocPrf Weight Path
 *>   192.168.1.0      10.2.3.2             0             0 500 i
 *>   192.168.2.0/27   10.2.3.2             0             0 500 i
 *>   192.168.2.64/26  10.2.3.2             0             0 500 i
 s>   192.168.3.0/27   0.0.0.0              0         32768 i
 *>   192.168.3.0      0.0.0.0                        32768 i
 s>   192.168.3.64/26  0.0.0.0              U         32768 i
```

c. On R2, remove the current route aggregation for the 192.168.1.0/24 prefix and configure it again, this time using the **as-set** option.

```
R2(config)# router bgp 500
R2(config-router)# no aggregate-address 192.168.1.0 255.255.255.0 summary-only
R2(config-router)# aggregate-address 192.168.1.0 255.255.255.0 as-set summary-
only
```

d. Verify that R3 is receiving the aggregated prefix 192.168.1.0/24.

```
R3# show ip route bgp | begin Gateway
Gateway of last resort is not set


B       192.168.1.0/24 [20/0] via 10.2.3.2, 00:01:35
        192.168.2.0/24 is variably subnetted, 2 subnets, 2 masks
B          192.168.2.0/27 [20/0] via 10.2.3.2, 08:50:02
B          192.168.2.64/26 [20/0] via 10.2.3.2, 08:50:02
        192.168.3.0/24 is variably subnetted, 5 subnets, 4 masks
B          192.168.3.0/24 [200/0], 08:49:32, Null0
```

e. Examine R3's BGP table again. Notice that the entry for 192.168.1.0 this time includes the entire AS path. The output from the **show ip bgp 192.168.1.0** command displays both AS numbers and identifies that R2 (2.2.2.2) aggregated the route.

```
R3# show ip bgp
<output omitted>
     Network          Next Hop        Metric LocPrf Weight Path
 *>  192.168.1.0      10.2.3.2             0             0 500 1000 i
 *>  192.168.2.0/27   10.2.3.2             0             0 500 i
 *>  192.168.2.64/26  10.2.3.2             0             0 500 i
 s>  192.168.3.0/27   0.0.0.0              0         32768 i
 *>  192.168.3.0      0.0.0.0                        32768 i
 s>  192.168.3.64/26  0.0.0.0              0         32768 i


R3# show ip bgp 192.168.1.0 | begin Refresh
  Refresh Epoch 7
  500 1000, (aggregated by 500 2.2.2.2)
    10.2.3.2 from 10.2.3.2 (2.2.2.2)
      Origin IGP, metric 0, localpref 100, valid, external, best
      rx pathid: 0, tx pathid: 0x0
```

Part 5: Configure and Verify the Advertising of a Default Route

Step 1. Configure default route advertisement on R2.

Configure R2 to advertise a default router to R1. R2 does not necessarily have to have a default route of its own. Core internet routers that have full internet routing tables and do not require a default route are referred to as being in a default-free zone (DFZ).

```
R2(config)# router bgp 500
R2(config-router)# neighbor 10.1.2.1 default-originate
```

Step 2.   Verify default route advertisement on R1.

    a.   Examine R1's routing table to verify that it has received a default route.

```
R1# show ip route bgp | begin Gateway
Gateway of last resort is 10.1.2.2 to network 0.0.0.0

B*    0.0.0.0/0 [20/0] via 10.1.2.2, 00:00:37
      192.168.2.0/24 is variably subnetted, 2 subnets, 2 masks
B        192.168.2.0/27 [20/0] via 10.1.2.2, 21:24:43
B        192.168.2.64/26 [20/0] via 10.1.2.2, 21:24:43
B     192.168.3.0/24 [20/0] via 10.1.2.2, 12:41:58
```

    b.   Examine R1's BGP table to verify that it has received a default route.

```
R1# show ip bgp
<output omitted>

       Network          Next Hop        Metric LocPrf Weight Path
 *>    0.0.0.0          10.1.2.2                          0 500 i
 *>    192.168.1.0/27   0.0.0.0              0        32768 i
 *>    192.168.1.64/26  0.0.0.0              0        32768 i
 *>    192.168.2.0/27   10.1.2.2            0          0 500 i
 *>    192.168.2.64/26  10.1.2.2            0          0 500 i
 *>    192.168.3.0      10.1.2.2                       0 500 300 i
```

Router Interface Summary Table

Router Model	Ethernet Interface #1	Ethernet Interface #2	Serial Interface #1	Serial Interface #2
1800	Fast Ethernet 0/0 (F0/0)	Fast Ethernet 0/1 (F0/1)	Serial 0/0/0 (S0/0/0)	Serial 0/0/1 (S0/0/1)
1900	Gigabit Ethernet 0/0 (G0/0)	Gigabit Ethernet 0/1 (G0/1)	Serial 0/0/0 (S0/0/0)	Serial 0/0/1 (S0/0/1)
2801	Fast Ethernet 0/0 (F0/0)	Fast Ethernet 0/1 (F0/1)	Serial 0/1/0 (S0/1/0)	Serial 0/1/1 (S0/1/1)
2811	Fast Ethernet 0/0 (F0/0)	Fast Ethernet 0/1 (F0/1)	Serial 0/0/0 (S0/0/0)	Serial 0/0/1 (S0/0/1)
2900	Gigabit Ethernet 0/0 (G0/0)	Gigabit Ethernet 0/1 (G0/1)	Serial 0/0/0 (S0/0/0)	Serial 0/0/1 (S0/0/1)
4221	Gigabit Ethernet 0/0/0 (G0/0/0)	Gigabit Ethernet 0/0/1 (G0/0/1)	Serial 0/1/0 (S0/1/0)	Serial 0/1/1 (S0/1/1)
4300	Gigabit Ethernet 0/0/0 (G0/0/0)	Gigabit Ethernet 0/0/1 (G0/0/1)	Serial 0/1/0 (S0/1/0)	Serial 0/1/1 (S0/1/1)

Note: To find out how the router is configured, look at the interfaces to identify the type of router and how many interfaces the router has. There is no way to effectively list all the combinations of configurations for each router class. This table includes identifiers for the possible combinations of Ethernet and Serial interfaces in the device. The table does not include any other type of interface, even though a specific router may contain one. An example of this might be an ISDN BRI interface. The string in parenthesis is the legal abbreviation that can be used in Cisco IOS commands to represent the interface.

Device Config Final – Notes

11.1.3 Lab - Implement MP-BGP

Topology

Addressing Table

Device	Interface	IPv4 Address	IPv6 Address	IPv6 Link-Local
R1	G0/0/0	10.1.2.1/24	2001:db8:acad:1012::1/64	fe80::1:1
	S0/1/0	10.1.3.1/25	2001:db8:acad:1013::1/64	fe80::1:2
	S0/1/1	10.1.3.129/25	2001:db8:acad:1014::1/64	fe80::1:3
	Loopback0	192.168.1.1/27	2001:db8:acad:1000::1/64	fe80::1:4
	Loopback1	192.168.1.65/26	2001:db8:acad:1001::1/64	fe80::1:5
R2	G0/0/0	10.1.2.2/24	2001:db8:acad:1012::2/64	fe80::2:1
	G0/0/1	10.2.3.2/24	2001:db8:acad:1023::2/64	fe80::2:2
	Loopback0	192.168.2.1/27	2001:db8:acad:2000::1/64	fe80::2:3
	Loopback1	192.168.2.65/26	2001:db8:acad:2001::1/64	fe80::2:4
R3	G0/0/0	10.2.3.3/24	2001:db8:acad:1023::3/64	fe80::3:1
	S0/1/0	10.1.3.3/25	2001:db8:acad:1013::3/64	fe80::3:2
	S0/1/1	10.1.3.130/25	2001:db8:acad:1014::3/64	fe80::3:3
	Loopback0	192.168.3.1/27	2001:db8:acad:3000::1/64	fe80::3:4
	Loopback1	192.168.3.65/26	2001:db8:acad:3001::1/64	fe80::3:5

Objectives

Part 1: Build the Network and Configure Basic Device Settings and Interface Addressing

Part 2: Configure MP-BGP on all Routers

Part 3: Verify MP-BGP

Part 4: Configure and Verify IPv6 Summarization

Background/Scenario

In this lab, you will configure MP-BGP, BGP for IPv4 and IPv6 using address families.

Note: This lab is an exercise in developing, deploying, and verifying various path manipulation tools for BGP, and does not reflect networking best practices.

Note: The routers used with CCNP hands-on labs are Cisco 4221 with Cisco IOS XE Release 16.9.4 (universalk9 image). Other routers and Cisco IOS versions can be used. Depending on the model and Cisco IOS version, the commands available and the output produced might vary from what is shown in the labs.

Note: Make sure that the routers and switches have been erased and have no startup configurations. If you are unsure contact your instructor.

Required Resources

- 3 Routers (Cisco 4221 with Cisco IOS XE Release 16.9.4 universal image or comparable)
- 1 PC (Choice of operating system with a terminal emulation program installed)
- Console cables to configure the Cisco IOS devices via the console ports
- Ethernet and serial cables as shown in the topology

Instructions

Part 1: Build the Network and Configure Basic Device Settings and Interface Addressing

In Part 1, you will set up the network topology and configure basic settings and interface addressing on routers.

Step 1. Cable the network as shown in the topology.

Attach the devices as shown in the topology diagram, and cable as necessary.

Step 2. Configure basic settings for each router.

a. Console into each router, enter global configuration mode, and apply the basic settings and interface addressing. A command list for each router is listed below to perform initial configuration.

Router R1

```
hostname R1
no ip domain lookup
line con 0
 logging sync
 exec-time 0 0
 exit
```

```
interface Loopback0
 ip address 192.168.1.1 255.255.255.224
 ipv6 address FE80::1:4 link-local
 ipv6 address 2001:DB8:ACAD:1000::1/64
 no shut
interface Loopback1
 ip address 192.168.1.65 255.255.255.192
 ipv6 address FE80::1:5 link-local
 ipv6 address 2001:DB8:ACAD:1001::1/64
 no shut
interface GigabitEthernet0/0/0
 ip address 10.1.2.1 255.255.255.0
 ipv6 address FE80::1:1 link-local
 ipv6 address 2001:DB8:ACAD:1012::1/64
 no shut
interface Serial0/1/0
 ip address 10.1.3.1 255.255.255.128
 ipv6 address FE80::1:2 link-local
 ipv6 address 2001:DB8:ACAD:1013::1/64
 no shut
interface Serial0/1/1
 ip address 10.1.3.129 255.255.255.128
 ipv6 address FE80::1:3 link-local
 ipv6 address 2001:DB8:ACAD:1014::1/64
 no shut
```

Router R2

```
hostname R2
no ip domain lookup
line con 0
 logging sync
 exec-time 0 0
 exit
interface Loopback0
 ip address 192.168.2.1 255.255.255.224
 ipv6 address FE80::2:3 link-local
 ipv6 address 2001:DB8:ACAD:2000::1/64
 no shut
interface Loopback1
 ip address 192.168.2.65 255.255.255.192
 ipv6 address FE80::2:4 link-local
 ipv6 address 2001:DB8:ACAD:2001::1/64
 no shut
interface GigabitEthernet0/0/0
 ip address 10.1.2.2 255.255.255.0
 ipv6 address FE80::2:1 link-local
 ipv6 address 2001:DB8:ACAD:1012::2/64
 no shut
interface GigabitEthernet0/0/1
 ip address 10.2.3.2 255.255.255.0
 ipv6 address FE80::2:2 link-local
```

```
  ipv6 address 2001:DB8:ACAD:1023::2/64
  no shut
```

Router R3

```
hostname R3
no ip domain lookup
line con 0
 logging sync
 exec-time 0 0
 exit
interface Loopback0
 ip address 192.168.3.1 255.255.255.224
 ipv6 address FE80::3:4 link-local
 ipv6 address 2001:DB8:ACAD:3000::1/64
 no shut
interface Loopback1
 ip address 192.168.3.65 255.255.255.192
 ipv6 address FE80::3:5 link-local
 ipv6 address 2001:DB8:ACAD:3001::1/64
 no shut
interface GigabitEthernet0/0/0
 ip address 10.2.3.3 255.255.255.0
 negotiation auto
 ipv6 address FE80::3:1 link-local
 ipv6 address 2001:DB8:ACAD:1023::3/64
 no shut
interface Serial0/1/0
 ip address 10.1.3.3 255.255.255.128
 ipv6 address FE80::3:2 link-local
 ipv6 address 2001:DB8:ACAD:1013::3/64
 no shut
interface Serial0/1/1
 ip address 10.1.3.130 255.255.255.128
 ipv6 address FE80::3:3 link-local
 ipv6 address 2001:DB8:ACAD:1014::3/64
 no shut
```

b. Save the running configuration to startup-config.

Part 2: Configure MP-BGP on all Routers

Step 1. Implement eBGP and neighbor relationships on R1 for IPv4 and IPv6.

a. Enable IPv6 routing.

```
R1(config)# ipv6 unicast-routing
```

b. Enter BGP configuration mode from global configuration mode, specifying AS 1000 and configure the router ID.

```
R1(config)# router bgp 1000
R1(config-router)# bgp router-id 1.1.1.1
```

   c. Based on the topology diagram, configure all the designated IPv4 neighbors for R1.

```
R1(config-router)# neighbor 10.1.2.2 remote-as 500
R1(config-router)# neighbor 10.1.3.3 remote-as 300
R1(config-router)# neighbor 10.1.3.130 remote-as 300
```

   d. Based on the topology diagram, configure all the designated IPv6 neighbors for R1.

```
R1(config-router)# neighbor 2001:db8:acad:1012::2 remote-as 500
R1(config-router)# neighbor 2001:db8:acad:1013::3 remote-as 300
R1(config-router)# neighbor 2001:db8:acad:1014::3 remote-as 300
```

   e. Enter address family configuration mode for IPv4 and activate each of the IPv4 neighbors.

```
R1(config-router)# address-family ipv4 unicast
R1(config-router-af)# neighbor 10.1.2.2 activate
R1(config-router-af)# neighbor 10.1.3.3 activate
R1(config-router-af)# neighbor 10.1.3.130 activate
R1(config-router-af)# exit
```

   f. Enter address family configuration mode for IPv6 and activate each of the IPv6 neighbors.

```
R1(config-router)# address-family ipv6 unicast
R1(config-router-af)# neighbor 2001:db8:acad:1012::2 activate
R1(config-router-af)# neighbor 2001:db8:acad:1013::3 activate
R1(config-router-af)# neighbor 2001:db8:acad:1014::3 activate
R1(config-router-af)# exit
```

Step 2. Implement eBGP and neighbor relationships on R2 for IPv4 and IPv6.

   a. Enable IPv6 routing.

```
R2(config)# ipv6 unicast-routing
```

   b. Enter BGP configuration mode from global configuration mode, specifying AS 500 and configure the router ID.

```
R2(config)# router bgp 500
R2(config-router)# bgp router-id 2.2.2.2
```

   c. Based on the topology diagram, configure all the designated IPv4 neighbors for R1.

```
R2(config-router)# neighbor 10.1.2.1 remote-as 1000
R2(config-router)# neighbor 10.2.3.3 remote-as 300
```

   d. Based on the topology diagram, configure all the designated IPv6 neighbors for R1.

```
R2(config-router)# neighbor 2001:db8:acad:1012::1 remote-as 1000
R2(config-router)# neighbor 2001:db8:acad:1023::3 remote-as 300
```

   e. Enter address family configuration mode for IPv4 and activate each of the IPv4 neighbors.

```
R2(config-router)# address-family ipv4 unicast
R2(config-router-af)# neighbor 10.1.2.1 activate
R2(config-router-af)# neighbor 10.2.3.3 activate
R2(config-router-af)# exit
```

   f. Enter address family configuration mode for IPv6 and activate each of the IPv6 neighbors.

```
R2(config-router)# address-family ipv6 unicast
R2(config-router-af)# neighbor 2001:db8:acad:1012::1 activate
```

```
R2(config-router-af)# neighbor 2001:db8:acad:1023::3 activate
R2(config-router-af)# exit
```

Step 3. Implement eBGP and neighbor relationships on R3 for IPv4 and IPv6.

a. Enable IPv6 routing.

```
R3(config)# ipv6 unicast-routing
```

b. Enter BGP configuration mode from global configuration mode, specifying AS 300 and configure the router ID.

```
R3(config)# router bgp 300
R3(config-router)# bgp router-id 3.3.3.3
```

c. Based on the topology diagram, configure all the designated IPv4 neighbors for R1.

```
R3(config-router)# neighbor 10.2.3.2 remote-as 500
R3(config-router)# neighbor 10.1.3.1 remote-as 1000
R3(config-router)# neighbor 10.1.3.129 remote-as 1000
```

d. Based on the topology diagram, configure all the designated IPv6 neighbors for R1.

```
R3(config-router)# neighbor 2001:db8:acad:1023::2 remote-as 500
R3(config-router)# neighbor 2001:db8:acad:1013::1 remote-as 1000
R3(config-router)# neighbor 2001:db8:acad:1014::1 remote-as 1000
```

e. Enter address family configuration mode for IPv4 and activate each of the IPv4 neighbors.

```
R3(config-router)# address-family ipv4 unicast
R3(config-router-af)# neighbor 10.1.3.1 activate
R3(config-router-af)# neighbor 10.1.3.129 activate
R3(config-router-af)# neighbor 10.2.3.2 activate
R3(config-router-af)# exit
```

f. Enter address family configuration mode for IPv6 and activate each of the IPv6 neighbors.

```
R3(config-router)# address-family ipv6 unicast
R3(config-router-af)# neighbor 2001:db8:acad:1023::2 activate
R3(config-router-af)# neighbor 2001:db8:acad:1013::1 activate
R3(config-router-af)# neighbor 2001:db8:acad:1014::1 activate
R3(config-router-af)# exit
```

Step 4. Advertise IPv4 and IPv6 prefixes on R1.

a. Enter address family configuration mode for IPv4 and advertise the IPv4 prefixes.

```
R1(config-router)# address-family ipv4 unicast
R1(config-router-af)# network 192.168.1.0 mask 255.255.255.224
R1(config-router-af)# network 192.168.1.64 mask 255.255.255.192
R1(config-router-af)# exit
```

b. Enter address family configuration mode for IPv6 and advertise the IPv6 prefixes.

```
R1(config-router)# address-family ipv6 unicast
R1(config-router-af)# network 2001:db8:acad:1000::/64
R1(config-router-af)# network 2001:db8:acad:1001::/64
R1(config-router-af)# exit
```

Step 5. Advertise IPv4 and IPv6 prefixes on R2.

a. Enter address family configuration mode for IPv4 and advertise the IPv4 prefixes.

```
R2(config-router)# address-family ipv4 unicast
R2(config-router-af)# network 192.168.2.0 mask 255.255.255.224
```

```
R2(config-router-af)# network 192.168.2.64 mask 255.255.255.192
R2(config-router-af)# exit
```

b. Enter address family configuration mode for IPv6 and advertise the IPv6 prefixes.

```
R2(config-router)# address-family ipv6 unicast
R2(config-router-af)# network 2001:db8:acad:2000::/64
R2(config-router-af)# network 2001:db8:acad:2001::/64
R2(config-router-af)# exit
```

Step 6. Advertise IPv4 and IPv6 prefixes on R3.

a. Enter address family configuration mode for IPv4 and advertise the IPv4 prefixes.

```
R3(config-router)# address-family ipv4 unicast
R3(config-router-af)# network 192.168.3.0 mask 255.255.255.224
R3(config-router-af)# network 192.168.3.64 mask 255.255.255.192
R3(config-router-af)# exit
```

b. Enter address family configuration mode for IPv6 and advertise the IPv6 prefixes.

```
R3(config-router)# address-family ipv6 unicast
R3(config-router-af)# network 2001:db8:acad:3000::/64
R3(config-router-af)# network 2001:db8:acad:3001::/64
R3(config-router-af)# exit
```

Note: Notice that the networks between the routers are not being advertised in eBGP. Typically, only the prefixes of the AS need to be advertised in eBGP. eBGP neighbors are typically directly connected and therefore will be able to form an adjacency. There is typically no need to advertise and inject the directly connected prefixes into the BGP routing table.

Part 3: Verify MP-BGP

Step 1. Display detailed neighbor adjacency information.

Use the **show bgp all neighbors** command on R2 to display detailed information about BGP connections to neighbors for all (IPv4 and IPv6) address families. Each neighbor shows that it is in the "Established" state. This indicates that the router can send and receive BGP messages. R2 has two neighbor addresses, R1 and R3, for each address family, IPv4 and IPv6.

```
R2# show bgp all neighbors
For address family: IPv4 Unicast
BGP neighbor is 10.1.2.1,  remote AS 1000, external link
  BGP version 4, remote router ID 1.1.1.1
  BGP state = Established, up for 01:56:25
  Last read 00:00:48, last write 00:00:50, hold time is 180, keepalive interval is
60 seconds
<output omitted>


BGP neighbor is 10.2.3.3,  remote AS 300, external link
  BGP version 4, remote router ID 3.3.3.3
  BGP state = Established, up for 01:55:47
  Last read 00:00:04, last write 00:00:41, hold time is 180, keepalive interval is
60 seconds
<output omitted>


For address family: IPv6 Unicast
BGP neighbor is 2001:DB8:ACAD:1012::1,  remote AS 1000, external link
```

```
    BGP version 4, remote router ID 1.1.1.1
    BGP state = Established, up for 01:56:39
    Last read 00:00:07, last write 00:00:04, hold time is 180, keepalive interval is
60 seconds
<output omitted>


BGP neighbor is 2001:DB8:ACAD:1023::3,   remote AS 300, external link
    BGP version 4, remote router ID 3.3.3.3
    BGP state = Established, up for 01:56:09
    Last read 00:00:32, last write 00:00:48, hold time is 180, keepalive interval is
60 seconds
<output omitted>
```

Note: Most information displayed using the **show bgp all neighbors** command has been omitted for brevity. The command **show bgp neighbors** is used to display only BGP for IPv4 adjacencies. To display the same information for only IPv6 neighbors, use the command **show bgp ipv6 neighbors**.

Questions:

What is the BGP state for each neighbor adjacency?

_____

How often are BGP keepalives sent?

_____

How many seconds will a BGP session remain open if no further keepalive messages are received?

_____

Step 2.   Display summary neighbor adjacency information.

Use the **show bgp ipv4 unicast summary** and **show bgp ipv6 unicast summary** commands on R2 to display a summary of IPv4/IPv6 peering information with R1 and R3. The information displayed using the **show bgp ipv4 unicast summary** is a subset of the **show ip all bgp** command.

```
R2# show bgp ipv4 unicast summary
BGP router identifier 2.2.2.2, local AS number 500
BGP table version is 11, main routing table version 11
6 network entries using 1488 bytes of memory
10 path entries using 1360 bytes of memory
5/3 BGP path/bestpath attribute entries using 1400 bytes of memory
4 BGP AS-PATH entries using 128 bytes of memory
0 BGP route-map cache entries using 0 bytes of memory
0 BGP filter-list cache entries using 0 bytes of memory
BGP using 4376 total bytes of memory
BGP activity 12/0 prefixes, 20/0 paths, scan interval 60 secs

Neighbor        V          AS MsgRcvd MsgSent   TblVer  InQ OutQ Up/Down  State/PfxRcd
10.1.2.1        4        1000     152     151       11    0    0 02:12:36  4
10.2.3.3        4         300     150     150       11    0    0 02:11:51  4


R2# show bgp ipv6 unicast summary
BGP router identifier 2.0.0.0, local AS number 500
```

```
BGP table version is 9, main routing table version 9
6 network entries using 1632 bytes of memory
10 path entries using 1520 bytes of memory
5/3 BGP path/bestpath attribute entries using 1400 bytes of memory
4 BGP AS-PATH entries using 128 bytes of memory
0 BGP route-map cache entries using 0 bytes of memory
0 BGP filter-list cache entries using 0 bytes of memory
BGP using 4680 total bytes of memory
BGP activity 12/0 prefixes, 20/0 paths, scan interval 60 secs


Neighbor          V          AS MsgRcvd MsgSent   TblVer  InQ OutQ Up/Down  State/PfxRcd
2001:DB8:ACAD:1012::1
                  4        1000     150     150        9    0    0 02:12:39  4
2001:DB8:ACAD:1023::3
                  4         300     151     150        9    0    0 02:11:54  4
```

Question:

What is the difference between the "local AS number" and the "AS" number displayed in the list of BGP neighbors?

_____

_____

Step 3.   Verify BGP tables for IPv4 and IPv6.

    a.   Use the **show bgp ipv4 unicast** command on R2 to display its IPv4 BGP table. This command is equivalent to the **show ip bgp** command and either command can be used. Notice that R1 shows six IPv4 networks in its IPv4 BGP table. Each network is valid "*" and has one path which is the best path ">". Amongst other information, the next hop IPv4 address and the AS path are included.

```
R2# show bgp ipv4 unicast
BGP table version is 11, local router ID is 2.2.2.2
Status codes: s suppressed, d damped, h history, * valid, > best, i - internal,
              r RIB-failure, S Stale, m multipath, b backup-path, f RT-Filter,
              x best-external, a additional-path, c RIB-compressed,
              t secondary path, L long-lived-stale,
Origin codes: i - IGP, e - EGP, ? - incomplete
RPKI validation codes: V valid, I invalid, N Not found

     Network          Next Hop         Metric LocPrf Weight Path
 *   192.168.1.0/27   10.2.3.3                         0 300 1000 i
 *>                   10.1.2.1              0           0 1000 i
 *   192.168.1.64/26  10.2.3.3                         0 300 1000 i
 *>                   10.1.2.1              0           0 1000 i
 *>  192.168.2.0/27   0.0.0.0              0       32768 i
 *>  192.168.2.64/26  0.0.0.0              0       32768 i
 *   192.168.3.0/27   10.1.2.1                         0 1000 300 i
 *>                   10.2.3.3             0           0 300 i
 *   192.168.3.64/26  10.1.2.1                         0 1000 300 i
 *>                   10.2.3.3             0           0 300 i
```

b. Use the **show bgp ipv6 unicast** command on R2 to display similar information for its IPv6 BGP table.

```
R2# show bgp ipv6 unicast
BGP table version is 9, local router ID is 2.2.2.2
Status codes: s suppressed, d damped, h history, * valid, > best, i - internal,
              r RIB-failure, S Stale, m multipath, b backup-path, f RT-Filter,
              x best-external, a additional-path, c RIB-compressed,
              t secondary path, L long-lived-stale,
Origin codes: i - IGP, e - EGP, ? - incomplete
RPKI validation codes: V valid, I invalid, N Not found

     Network          Next Hop            Metric LocPrf Weight Path
 *   2001:DB8:ACAD:1000::/64
                      2001:DB8:ACAD:1023::3
                                                            0 300 1000 i
 *>                   2001:DB8:ACAD:1012::1
                                              0             0 1000 i
 *   2001:DB8:ACAD:1001::/64
                      2001:DB8:ACAD:1023::3
                                                            0 300 1000 i
 *>                   2001:DB8:ACAD:1012::1
                                              0             0 1000 i
 *>  2001:DB8:ACAD:2000::/64
                      ::                      0         32768 i
 *>  2001:DB8:ACAD:2001::/64
                      ::                      0         32768 i
 *   2001:DB8:ACAD:3000::/64
                      2001:DB8:ACAD:1012::1
                                                            0 1000 300 i
 *>                   2001:DB8:ACAD:1023::3
                                              0             0 300 i
 *   2001:DB8:ACAD:3001::/64
                      2001:DB8:ACAD:1012::1
                                                            0 1000 300 i
 *>                   2001:DB8:ACAD:1023::3
                                              0             0 300 i
```

Questions:

In the first output **show bgp ipv4 unicast,** why is 10.1.2.1 the preferred next hop address for 192.168.1.0 instead of 10.2.3.3?

_____

_____

Why do some entries in the **show bgp ipv6 unicast** output include a next hop address of "::"?

_____

_____

Step 4.   Viewing explicit routes and path attributes.

    a.   Use the **show bgp ipv4 unicast** *ipv4-prefix subnet-mask* command on R2 to display all the paths for a specific route and BGP path attributes for that route.

```
R2# show bgp ipv4 unicast 192.168.1.0 255.255.255.224
BGP routing table entry for 192.168.1.0/27, version 2
Paths: (2 available, best #2, table default)
  Advertised to update-groups:
     1
  Refresh Epoch 1
  300 1000
    10.2.3.3 from 10.2.3.3 (3.3.3.3)
      Origin IGP, localpref 100, valid, external
      rx pathid: 0, tx pathid: 0
  Refresh Epoch 1
  1000
    10.1.2.1 from 10.1.2.1 (1.1.1.1)
      Origin IGP, metric 0, localpref 100, valid, external, best
      rx pathid: 0, tx pathid: 0x0
```

The **show bgp ipv6 unicast** *ipv6-prefix prefix-length* command displays similar information for IPv6 prefixes.

```
R2# show bgp ipv6 unicast 2001:db8:acad:1000::/64
BGP routing table entry for 2001:DB8:ACAD:1000::/64, version 2
Paths: (2 available, best #2, table default)
  Flag: 0x100
  Advertised to update-groups:
     1
  Refresh Epoch 1
  300 1000
    2001:DB8:ACAD:1023::3 (FE80::3:1) from 2001:DB8:ACAD:1023::3 (3.3.3.3)
      Origin IGP, localpref 100, valid, external
      rx pathid: 0, tx pathid: 0
  Refresh Epoch 1
  1000
    2001:DB8:ACAD:1012::1 (FE80::1:1) from 2001:DB8:ACAD:1012::1 (1.1.1.1)
      Origin IGP, metric 0, localpref 100, valid, external, best
      rx pathid: 0, tx pathid: 0x0
```

Question:

Why does the output for the **show bgp ipv6 unicast** command include the link-local address following the global unicast address?

---

    b.   Use the **show bgp ipv4 unicast neighbors** *ipv4-prefix* **advertised-routes** command on R2 to display IPv4 routes advertised to a specific neighbor.

```
R2# show bgp ipv4 unicast neighbors 10.1.2.1 advertised-routes
BGP table version is 11, local router ID is 2.2.2.2
Status codes: s suppressed, d damped, h history, * valid, > best, i - internal,
              r RIB-failure, S Stale, m multipath, b backup-path, f RT-Filter,
```

```
               x best-external, a additional-path, c RIB-compressed,
               t secondary path, L long-lived-stale,
Origin codes: i - IGP, e - EGP, ? - incomplete
RPKI validation codes: V valid, I invalid, N Not found


     Network          Next Hop          Metric LocPrf Weight Path
 *>  192.168.1.0/27   10.1.2.1               0            0 1000 i
 *>  192.168.1.64/26  10.1.2.1               0            0 1000 i
 *>  192.168.2.0/27   0.0.0.0                0        32768 i
 *>  192.168.2.64/26  0.0.0.0                0        32768 i
 *>  192.168.3.0/27   10.2.3.3               0            0 300 i
 *>  192.168.3.64/26  10.2.3.3               0            0 300 i


Total number of prefixes 6
```

c. Use the **show bgp ipv6 unicast** *ipv5-prefix prefix-length* command to display similar information for IPv6 advertised routes.

```
R2# show bgp ipv6 unicast neighbors 2001:db8:acad:1012::1 advertised-routes
BGP table version is 9, local router ID is 2.2.2.2
Status codes: s suppressed, d damped, h history, * valid, > best, i - internal,
              r RIB-failure, S Stale, m multipath, b backup-path, f RT-Filter,
              x best-external, a additional-path, c RIB-compressed,
              t secondary path, L long-lived-stale,
Origin codes: i - IGP, e - EGP, ? - incomplete
RPKI validation codes: V valid, I invalid, N Not found


     Network          Next Hop          Metric LocPrf Weight Path
 *>  2001:DB8:ACAD:1000::/64
                      2001:DB8:ACAD:1012::1
                                             0            0 1000 i
 *>  2001:DB8:ACAD:1001::/64
                      2001:DB8:ACAD:1012::1
                                             0            0 1000 i
 *>  2001:DB8:ACAD:2000::/64
                      ::                     0        32768 i
 *>  2001:DB8:ACAD:2001::/64
                      ::                     0        32768 i
 *>  2001:DB8:ACAD:3000::/64
                      2001:DB8:ACAD:1023::3
                                             0            0 300 i
 *>  2001:DB8:ACAD:3001::/64
                      2001:DB8:ACAD:1023::3
                                             0            0 300 i


Total number of prefixes 6
```

Question:

Why do some entries in the **show bgp ipv4 unicast neighbors** output include a next hop address of 0.0.0.0 and the **show bgp ipv6 unicast neighbors** output includes a next hop address of "::"?

Step 5.   Verifying the IP routing tables for IPv4 and IPv6.

a.   By examining the IPv4 and IPv6 routing tables on R2, you can verify that BGP is receiving the IPv4 and IPv6 prefixes from R1 and R3.

```
R2# show ip route bgp | begin Gateway
Gateway of last resort is not set

      192.168.1.0/24 is variably subnetted, 2 subnets, 2 masks
B        192.168.1.0/27 [20/0] via 10.1.2.1, 04:29:03
B        192.168.1.64/26 [20/0] via 10.1.2.1, 04:28:32
      192.168.3.0/24 is variably subnetted, 2 subnets, 2 masks
B        192.168.3.0/27 [20/0] via 10.2.3.3, 04:17:14
B        192.168.3.64/26 [20/0] via 10.2.3.3, 04:16:44


R2# show ipv6 route bgp | section 2001
B   2001:DB8:ACAD:1000::/64 [20/0]
     via FE80::1:1, GigabitEthernet0/0/0
B   2001:DB8:ACAD:1001::/64 [20/0]
     via FE80::1:1, GigabitEthernet0/0/0
B   2001:DB8:ACAD:3000::/64 [20/0]
     via FE80::3:1, GigabitEthernet0/0/1
B   2001:DB8:ACAD:3001::/64 [20/0]
     via FE80::3:1, GigabitEthernet0/0/1
```

Part 4: Configure and Verify IPv6 Route Summarization

Summarizing prefixes conserves router resources and accelerates best-path calculation by reducing the size of the table. Summarization can be configured either for prefixes originated by the AS or prefixes received from downstream providers. Summarization also provides the benefits of stability by hiding flapping routes or having to install new prefixes when they are contained within a summary.

a.   Verify R2 and R3 are receiving 2001:db8:acad:1000::/64 and 2001:db8:acad:1001::/64 from R1.

```
R2# show ipv6 route bgp | section 2001
B   2001:DB8:ACAD:1000::/64 [20/0]
     via FE80::1:1, GigabitEthernet0/0/0
B   2001:DB8:ACAD:1001::/64 [20/0]
     via FE80::1:1, GigabitEthernet0/0/0
B   2001:DB8:ACAD:3000::/64 [20/0]
     via FE80::3:1, GigabitEthernet0/0/1
B   2001:DB8:ACAD:3001::/64 [20/0]
     via FE80::3:1, GigabitEthernet0/0/1


R3# show ipv6 route bgp | section 2001
B   2001:DB8:ACAD:1000::/64 [20/0]
     via FE80::1:2, Serial0/1/0
B   2001:DB8:ACAD:1001::/64 [20/0]
     via FE80::1:2, Serial0/1/0
B   2001:DB8:ACAD:2000::/64 [20/0]
     via FE80::2:2, GigabitEthernet0/0/0
```

```
B    2001:DB8:ACAD:2001::/64 [20/0]
        via FE80::2:2, GigabitEthernet0/0/0
```

b. Although AS 1000 only has two IPv6 prefixes - 2001:db8:acad:1000::/64 and 2001:db8:acad:1001::/64, this customer has been allocated the entire 2001:db8:acad:1000::/52 prefix (2001:db8:acad:1xxx).

R1 is configured using the **aggregate-address** command in IPv6 AF mode to summarize its IPv6 prefixes. This is known as a summary route or aggregate route. The **summary-only** option suppresses the more specific prefixes from also being advertised.

```
R1(config)# router bgp 1000
R1(config-router)# address-family ipv6 unicast
R1(config-router-af)# aggregate-address 2001:db8:acad:1000::/52 summary-only
```

c. Verify that R2 and R3 are now receiving the aggregate route and installing it in the IPv6 BGP table.

```
R2# show bgp ipv6 unicast | begin Network
     Network          Next Hop          Metric LocPrf Weight Path
 *   2001:DB8:ACAD:1000::/52
                      2001:DB8:ACAD:1023::3
                                                       0 300 1000 i
 *>                   2001:DB8:ACAD:1012::1
                                          0            0 1000 i
<output omitted>


R3# show bgp ipv6 unicast | begin Network
     Network          Next Hop          Metric LocPrf Weight Path
 *   2001:DB8:ACAD:1000::/52
                      2001:DB8:ACAD:1023::2
                                                       0 500 1000 i
 *                    2001:DB8:ACAD:1014::1
                                          0            0 1000 i
<output omitted>
```

d. Verify that R2 and R3 are now receiving the aggregate route and it is installed in the IPv6 routing table.

```
R2# show ipv6 route bgp | section 2001
B    2001:DB8:ACAD:1000::/52 [20/0]
        via FE80::1:1, GigabitEthernet0/0/0
B    2001:DB8:ACAD:3000::/64 [20/0]
        via FE80::3:1, GigabitEthernet0/0/1
B    2001:DB8:ACAD:3001::/64 [20/0]
        via FE80::3:1, GigabitEthernet0/0/1


R3# show ipv6 route bgp | section 2001
B    2001:DB8:ACAD:1000::/52 [20/0]
        via FE80::1:2, Serial0/1/0
B    2001:DB8:ACAD:2000::/64 [20/0]
        via FE80::2:2, GigabitEthernet0/0/0
B    2001:DB8:ACAD:2001::/64 [20/0]
        via FE80::2:2, GigabitEthernet0/0/0
```

Question:

If R1's 2001:db8:acad:1000::/64 network went down, what would be the effect, if any, on the routing tables of R2 and R3? Explain.

_____

_____

_____

Router Interface Summary Table

Router Model	Ethernet Interface #1	Ethernet Interface #2	Serial Interface #1	Serial Interface #2
1800	Fast Ethernet 0/0 (F0/0)	Fast Ethernet 0/1 (F0/1)	Serial 0/0/0 (S0/0/0)	Serial 0/0/1 (S0/0/1)
1900	Gigabit Ethernet 0/0 (G0/0)	Gigabit Ethernet 0/1 (G0/1)	Serial 0/0/0 (S0/0/0)	Serial 0/0/1 (S0/0/1)
2801	Fast Ethernet 0/0 (F0/0)	Fast Ethernet 0/1 (F0/1)	Serial 0/1/0 (S0/1/0)	Serial 0/1/1 (S0/1/1)
2811	Fast Ethernet 0/0 (F0/0)	Fast Ethernet 0/1 (F0/1)	Serial 0/0/0 (S0/0/0)	Serial 0/0/1 (S0/0/1)
2900	Gigabit Ethernet 0/0 (G0/0)	Gigabit Ethernet 0/1 (G0/1)	Serial 0/0/0 (S0/0/0)	Serial 0/0/1 (S0/0/1)
4221	Gigabit Ethernet 0/0/0 (G0/0/0)	Gigabit Ethernet 0/0/1 (G0/0/1)	Serial 0/1/0 (S0/1/0)	Serial 0/1/1 (S0/1/1)
4300	Gigabit Ethernet 0/0/0 (G0/0/0)	Gigabit Ethernet 0/0/1 (G0/0/1)	Serial 0/1/0 (S0/1/0)	Serial 0/1/1 (S0/1/1)

Note: To find out how the router is configured, look at the interfaces to identify the type of router and how many interfaces the router has. There is no way to effectively list all the combinations of configurations for each router class. This table includes identifiers for the possible combinations of Ethernet and Serial interfaces in the device. The table does not include any other type of interface, even though a specific router may contain one. An example of this might be an ISDN BRI interface. The string in parenthesis is the legal abbreviation that can be used in Cisco IOS commands to represent the interface.

Device Config Final – Notes

_____

_____

_____

_____

_____

_____

_____

_____

_____

CHAPTER 12

Advanced BGP

12.1.2 Lab - Implement BGP Path Manipulation

Topology

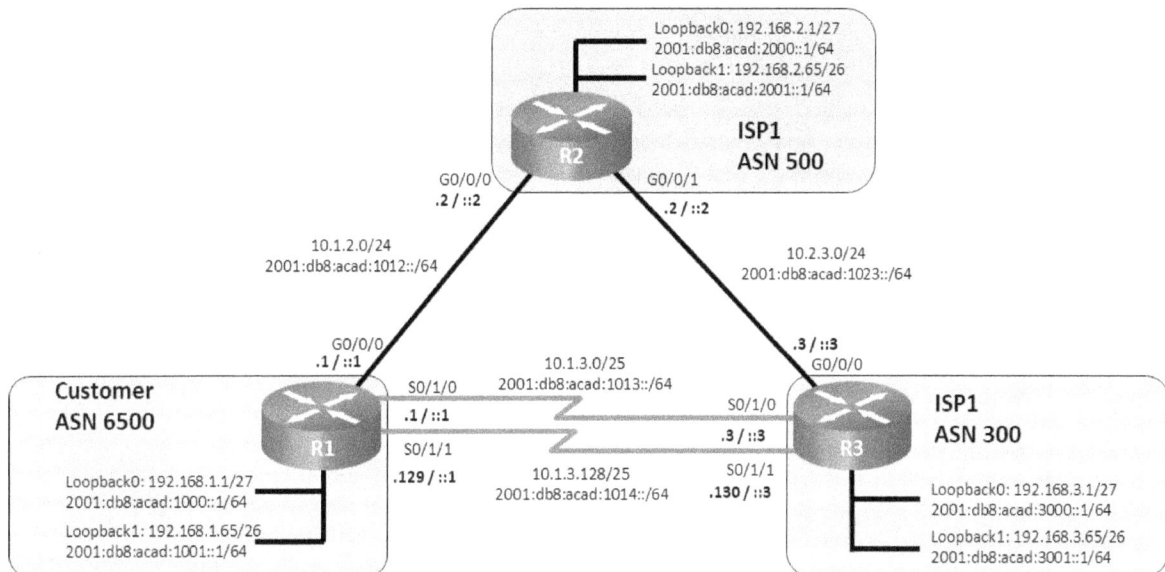

Addressing Table

Device	Interface	IPv4 Address	IPv6 Address	IPv6 Link-Local
R1	G0/0/0	10.1.2.1/24	2001:db8:acad:1012::1/64	fe80::1:1
	S0/1/0	10.1.3.1/25	2001:db8:acad:1013::1/64	fe80::1:2
	S0/1/1	10.1.3.129/25	2001:db8:acad:1014::1/64	fe80::1:3
	Loopback0	192.168.1.1/27	2001:db8:acad:1000::1/64	fe80::1:4
	Loopback1	192.168.1.65/26	2001:db8:acad:1001::1/64	fe80::1:5
R2	G0/0/0	10.1.2.2/24	2001:db8:acad:1012::2/64	fe80::2:1
	G0/0/1	10.2.3.2/24	2001:db8:acad:1023::2/64	fe80::2:2
	Loopback0	192.168.2.1/27	2001:db8:acad:2000::1/64	fe80::2:4
	Loopback1	192.168.2.65/26	2001:db8:acad:2001::1/64	fe80::2:4
R3	G0/0/0	10.2.3.3/24	2001:db8:acad:1023::3/64	fe80::3:1
	S0/1/0	10.1.3.3/25	2001:db8:acad:1013::3/64	fe80::3:2
	S0/1/1	10.1.3.130/25	2001:db8:acad:1014::3/64	fe80::3:3
	Loopback0	192.168.3.1/27	2001:db8:acad:3000::1/64	fe80::3:4
	Loopback1	192.168.3.65/26	2001:db8:acad:3001::1/64	fe80::3:5

Objectives

Part 1: Build the Network and Configure Basic Device Settings and Interface Addressing

Part 2: Configure and Verify Multi-Protocol BGP on all Routers

Part 3: Configure and Verify BGP Path Manipulation Settings on all Routers

Background/Scenario

The default settings in BGP allow for a great deal of undesired route information to pass between autonomous systems. In this lab you will configure Multi-Protocol BGP and implement various path manipulation options for both IPv4 and IPv6.

Note: This lab is an exercise in developing, deploying, and verifying various path manipulation tools for BGP, and does not reflect networking best practices.

Note: The routers used with CCNP hands-on labs are Cisco 4221 with Cisco IOS XE Release 16.9.4 (universalk9 image). Other routers and Cisco IOS versions can be used. Depending on the model and Cisco IOS version, the commands available and the output produced might vary from what is shown in the labs.

Note: Ensure that the routers have been erased and have no startup configurations. If you are unsure contact your instructor.

Required Resources

- 3 Routers (Cisco 4221 with Cisco IOS XE Release 16.9.4 universal image or comparable)
- 1 PC (Choice of operating system with a terminal emulation program installed)
- Console cables to configure the Cisco IOS devices via the console ports
- Ethernet and serial cables as shown in the topology

Instructions

Part 1: Build the Network and Configure Basic Device Settings and Interface Addressing

In Part 1, you will set up the network topology and configure basic settings and interface addressing on routers.

Step 1.   Cable the network as shown in the topology.

Attach the devices as shown in the topology diagram, and cable as necessary.

Step 2.   Configure basic settings for each router.

    a.   Console into each router, enter global configuration mode, and apply the basic settings and interface addressing. A command list for each router is listed below to perform initial configuration.

Router R1

```
no ip domain lookup
hostname R1
line con 0
 exec-timeout 0 0
 logging synchronous
banner motd # This is R1, BGP Path Manipulation Lab #
ipv6 unicast-routing
interface g0/0/0
 ip address 10.1.2.1 255.255.255.0
 ipv6 address fe80::1:1 link-local
 ipv6 address 2001:db8:acad:1012::1/64
 no shutdown
interface s0/1/0
 ip address 10.1.3.1 255.255.255.128
 ipv6 address fe80::1:2 link-local
 ipv6 address 2001:db8:acad:1013::1/64
 no shutdown
interface s0/1/1
 ip address 10.1.3.129 255.255.255.128
 ipv6 address fe80::1:3 link-local
 ipv6 address 2001:db8:acad:1014::1/64
 no shutdown
interface loopback 0
 ip address 192.168.1.1 255.255.255.224
 ipv6 address fe80::1:4 link-local
 ipv6 address 2001:db8:acad:1000::1/64
 no shutdown
interface loopback 1
 ip address 192.168.1.65 255.255.255.192
 ipv6 address fe80::1:5 link-local
 ipv6 address 2001:db8:acad:1001::1/64
 no shutdown
```

Router R2

```
no ip domain lookup
hostname R2
line con 0
 exec-timeout 0 0
 logging synchronous
banner motd # This is R2, BGP Path Manipulation Lab #
ipv6 unicast-routing
interface g0/0/0
 ip address 10.1.2.2 255.255.255.0
 ipv6 address fe80::2:1 link-local
 ipv6 address 2001:db8:acad:1012::2/64
 no shutdown
interface g0/0/1
 ip address 10.2.3.2 255.255.255.0
 ipv6 address fe80::2:2 link-local
 ipv6 address 2001:db8:acad:1023::2/64
```

```
 no shutdown
interface loopback 0
 ip address 192.168.2.1 255.255.255.224
 ipv6 address fe80::2:3 link-local
 ipv6 address 2001:db8:acad:2000::1/64
 no shutdown
interface loopback 1
 ip address 192.168.2.65 255.255.255.192
 ipv6 address fe80::2:4 link-local
 ipv6 address 2001:db8:acad:2001::1/64
 no shutdown
```

Router R3

```
no ip domain lookup
hostname R3
line con 0
 exec-timeout 0 0
 logging synchronous
banner motd # This is R3, BGP Path Manipulation Lab #
ipv6 unicast-routing
interface g0/0/0
 ip address 10.2.3.3 255.255.255.0
 ipv6 address fe80::3:1 link-local
 ipv6 address 2001:db8:acad:1023::3/64
 no shutdown
interface s0/1/0
 ip address 10.1.3.3 255.255.255.128
 ipv6 address fe80::3:2 link-local
 ipv6 address 2001:db8:acad:1013::3/64
 no shutdown
interface s0/1/1
 ip address 10.1.3.130 255.255.255.128
 ipv6 address fe80::3:3 link-local
 ipv6 address 2001:db8:acad:1014::3/64
 no shutdown
interface loopback 0
 ip address 192.168.3.1 255.255.255.224
 ipv6 address fe80::3:4 link-local
 ipv6 address 2001:db8:acad:3000::1/64
 no shutdown
interface loopback 1
 ip address 192.168.3.65 255.255.255.192
 ipv6 address fe80::3:5 link-local
 ipv6 address 2001:db8:acad:3001::1/64
 no shutdown
```

b. Set the clock on each router to UTC time.

c. Save the running configuration to startup-config.

Part 2: Configure and Verify Multi-Protocol BGP on all Routers

In Part 2, you will configure and verify Multi-Protocol BGP on all routers to achieve full connectivity between the routers. The text below provides you with the complete configuration for R1. You will use this to inform your configuration of R2 and R3. The configuration being used here is not meant to represent best practice, but to assess your ability to complete the required configurations.

Step 1.   On R1, create the core BGP configuration.

    a.   Enter BGP configuration mode from global configuration mode, specifying AS 6500.

```
R1(config)# router bgp 6500
```

    b.   Configure the BGP router-id for R1.

```
R1(config-router)# bgp router-id 1.1.1.1
```

    c.   Disable the default IPv4 unicast address family behavior.

```
R1(config-router)# no bgp default ipv4-unicast
```

    d.   Based on the topology diagram, configure all the designated neighbors for R1.

```
R1(config-router)# neighbor 10.1.2.2 remote-as 500
R1(config-router)# neighbor 10.1.3.3 remote-as 300
R1(config-router)# neighbor 10.1.3.130 remote-as 300
R1(config-router)# neighbor 2001:db8:acad:1012::2 remote-as 500
R1(config-router)# neighbor 2001:db8:acad:1013::3 remote-as 300
R1(config-router)# neighbor 2001:db8:acad:1014::3 remote-as 300
```

Step 2.   On R1, configure the IPv4 unicast address family.

    a.   Enter the IPv4 unicast address family configuration mode.

```
R1(config-router)# address-family ipv4 unicast
```

    b.   Configure network statements for the IPv4 networks attached to interfaces loopback0 and loopback1. Remember that BGP does not work the same way that an IGP does, and that the network statement has no impact on neighbor adjacency; it is used solely for advertising purposes.

```
R1(config-router-af)# network 192.168.1.0 mask 255.255.255.224
R1(config-router-af)# network 192.168.1.64 mask 255.255.255.192
```

    c.   Deactivate the IPv6 neighbors and activate the IPv4 neighbors.

```
R1(config-router-af)# no neighbor 2001:db8:acad:1012::2 activate
R1(config-router-af)# no neighbor 2001:db8:acad:1013::3 activate
R1(config-router-af)# no neighbor 2001:db8:acad:1014::3 activate
R1(config-router-af)# neighbor 10.1.2.2 activate
R1(config-router-af)# neighbor 10.1.3.3 activate
R1(config-router-af)# neighbor 10.1.3.130 activate
```

Step 3.   On R1, configure the IPv6 unicast address family.

    a.   Enter the IPv6 unicast address family configuration mode.

```
R1(config-router)# address-family ipv6 unicast
```

    b.   Configure network statements for the IPv6 networks that are attached to interfaces loopback0 and loopback1. Remember that BGP does not work the same way that an IGP does; therefore, the network statement has no impact on neighbor adjacency; it is used solely for advertising purposes.

```
R1(config-router-af)# network 2001:db8:acad:1000::/64
R1(config-router-af)# network 2001:db8:acad:1001::/64
```

c. Activate the IPv6 neighbors that are configured for BGP.

```
R1(config-router-af)# neighbor 2001:db8:acad:1012::2 activate
R1(config-router-af)# neighbor 2001:db8:acad:1013::3 activate
R1(config-router-af)# neighbor 2001:db8:acad:1014::3 activate
```

Step 4. Configure MP-BGP on R2 and R3 as you did in the previous step.

_____

_____

_____

_____

Step 5. Verify that MP-BGP is operational.

a. Use the **show bgp ipv4 unicast summary** and **show bgp ipv6 unicast summary** commands to verify that BGP has established three IPv4 and three IPv6 adjacencies and received four prefixes from each neighbor.

```
R1# show bgp ipv4 unicast summary
BGP router identifier 1.1.1.1, local AS number 6500
BGP table version is 9, main routing table version 9
6 network entries using 1488 bytes of memory
14 path entries using 1904 bytes of memory
5/3 BGP path/bestpath attribute entries using 1400 bytes of memory
4 BGP AS-PATH entries using 128 bytes of memory
0 BGP route-map cache entries using 0 bytes of memory
0 BGP filter-list cache entries using 0 bytes of memory
BGP using 4920 total bytes of memory
BGP activity 12/0 prefixes, 28/0 paths, scan interval 60 secs

Neighbor        V    AS MsgRcvd MsgSent   TblVer  InQ OutQ Up/Down  State/PfxRcd
10.1.2.2        4   500       8       8        9    0    0 00:02:42  4
10.1.3.3        4   300       8       8        9    0    0 00:02:12  4
10.1.3.130      4   300       8       8        9    0    0 00:02:11  4

R1# show bgp ipv6 unicast summary
BGP router identifier 1.1.1.1, local AS number 6500
BGP table version is 9, main routing table version 9
6 network entries using 1632 bytes of memory
14 path entries using 2128 bytes of memory
5/3 BGP path/bestpath attribute entries using 1400 bytes of memory
4 BGP AS-PATH entries using 128 bytes of memory
0 BGP route-map cache entries using 0 bytes of memory
0 BGP filter-list cache entries using 0 bytes of memory
BGP using 5288 total bytes of memory
BGP activity 12/0 prefixes, 28/0 paths, scan interval 60 secs

Neighbor  V         AS MsgRcvd MsgSent   TblVer  InQ OutQ Up/Down  State/PfxRcd
2001:DB8:ACAD:1012::2
          4        500       8       8        9    0    0 00:02:50  4
```

```
2001:DB8:ACAD:1013::3
          4        300       8        8        9     0     0 00:02:14   4
2001:DB8:ACAD:1014::3
          4        300       8        8        9     0     0 00:02:13   4
```

b. Use the **show bgp ipv4 unicast** and **show bgp ipv6 unicast** commands to view the specified BGP tables. Note that R1 has multiple paths to each destination network. Take note of the next hop address for the destination networks marked with the ">" symbol.

```
R1# show bgp ipv4 unicast | begin Network
     Network          Next Hop          Metric LocPrf Weight Path
 *>  192.168.1.0/27   0.0.0.0                0          32768 i
 *>  192.168.1.64/26  0.0.0.0                0          32768 i
 *   192.168.2.0/27   10.1.3.130                           0 300 500 i
 *>                   10.1.2.2               0              0 500 i
 *                    10.1.3.3                              0 300 500 i
 *   192.168.2.64/26  10.1.3.130                            0 300 500 i
 *>                   10.1.2.2               0              0 500 i
 *                    10.1.3.3                              0 300 500 i
 *   192.168.3.0/27   10.1.3.130             0              0 300 i
 *                    10.1.2.2                              0 500 300 i
 *>                   10.1.3.3               0              0 300 i
 *   192.168.3.64/26  10.1.3.130             0              0 300 i
 *                    10.1.2.2                              0 500 300 i
 *>                   10.1.3.3               0              0 300 i


R1# show bgp ipv6 unicast | begin Network
     Network          Next Hop          Metric LocPrf Weight Path
 *>  2001:DB8:ACAD:1000::/64
                      ::                     0          32768 i
 *>  2001:DB8:ACAD:1001::/64
                      ::                     0          32768 i
 *   2001:DB8:ACAD:2000::/64
                      2001:DB8:ACAD:1013::3
                                                            0 300 500 i
 *>                   2001:DB8:ACAD:1012::2
                                             0              0 500 i
 *                    2001:DB8:ACAD:1014::3
                                                            0 300 500 i
 *   2001:DB8:ACAD:2001::/64
                      2001:DB8:ACAD:1013::3
                                                            0 300 500 i
 *>                   2001:DB8:ACAD:1012::2
                                             0              0 500 i
 *                    2001:DB8:ACAD:1014::3
                                                            0 300 500 i
 *>  2001:DB8:ACAD:3000::/64
                      2001:DB8:ACAD:1013::3
                                             0              0 300 i
 *                    2001:DB8:ACAD:1012::2
                                                            0 500 300 i
```

```
*                         2001:DB8:ACAD:1014::3
                                                    0                 0 300 i
*>    2001:DB8:ACAD:3001::/64
                          2001:DB8:ACAD:1013::3
                                                    0                 0 300 i
*                         2001:DB8:ACAD:1012::2
                                                                0 500 300 i
*                         2001:DB8:ACAD:1014::3
                                                    0                 0 300 i
```

c. Use the **show ip route bgp** and **show ipv6 route bgp** commands to view the routing tables. Note that there is only one route to each destination, and that the routes included in the routing table have the same next hop as those with the ">" symbol in the BGP tables.

```
R1# show ip route bgp | begin Gateway
Gateway of last resort is not set

      192.168.2.0/24 is variably subnetted, 2 subnets, 2 masks
B        192.168.2.0/27 [20/0] via 10.1.2.2, 00:04:10
B        192.168.2.64/26 [20/0] via 10.1.2.2, 00:04:10
      192.168.3.0/24 is variably subnetted, 2 subnets, 2 masks
B        192.168.3.0/27 [20/0] via 10.1.3.3, 00:04:09
B        192.168.3.64/26 [20/0] via 10.1.3.3, 00:04:09


R1# show ipv6 route bgp
IPv6 Routing Table - default - 15 entries
Codes: C - Connected, L - Local, S - Static, U - Per-user Static route
       B - BGP, R - RIP, H - NHRP, I1 - ISIS L1
       I2 - ISIS L2, IA - ISIS interarea, IS - ISIS summary, D - EIGRP
       EX - EIGRP external, ND - ND Default, NDp - ND Prefix, DCE - Destination
       NDr - Redirect, RL - RPL, O - OSPF Intra, OI - OSPF Inter
       OE1 - OSPF ext 1, OE2 - OSPF ext 2, ON1 - OSPF NSSA ext 1
       ON2 - OSPF NSSA ext 2, a - Application
B    2001:DB8:ACAD:2000::/64 [20/0]
     via FE80::2:1, GigabitEthernet0/0/0
B    2001:DB8:ACAD:2001::/64 [20/0]
     via FE80::2:1, GigabitEthernet0/0/0
B    2001:DB8:ACAD:3000::/64 [20/0]
     via FE80::3:2, Serial0/1/0
B    2001:DB8:ACAD:3001::/64 [20/0]
     via FE80::3:2, Serial0/1/0
```

Part 3: Configure and Verify BGP Path Manipulation Settings on all Routers

In Part 3, you will configure path manipulation tools for BGP. The way these tools are being used here is not meant to represent best practice, but to assess your ability to complete the required configurations.

Step 1.   Configure ACL-based route filtering.

In this step, you will configure R3 so that it only sends ASN300 networks to R1; it will not tell R1 that it knows about the networks in ASN200.

a. On R1, issue the command **show bgp ipv4 unicast | i 300** to see what prefixes ASN300 is sharing via BGP. Take note of those prefixes that do not originate in ASN300.

```
R1# show bgp ipv4 unicast | i 300
 *     192.168.2.0/27    10.1.3.3                          0 300 500 i
 *                       10.1.3.130                        0 300 500 i
 *     192.168.2.64/26   10.1.3.3                          0 300 500 i
 *                       10.1.3.130                        0 300 500 i
 *     192.168.3.0/27    10.1.2.2                          0 500 300 i
 *>                      10.1.3.3              0           0 300 i
 *                       10.1.3.130           0           0 300 i
 *     192.168.3.64/26   10.1.2.2                          0 500 300 i
 *>                      10.1.3.3             0            0 300 i
 *                       10.1.3.130           0            0 300 i
```

b. On R3, configure an access list designed to match the source address and mask of the networks belonging to ASN300:

```
R3(config)# ip access-list extended ALLOWED_TO_R1
R3(config-ext-nacl)# permit ip 192.168.3.0 0.0.0.0 255.255.255.224 0.0.0.0
R3(config-ext-nacl)# permit ip 192.168.3.64 0.0.0.0 255.255.255.192 0.0.0.0
R3(config-ext-nacl)# exit
```

c. On R3, apply the ALLOWED_TO_R1 ACL as a distribute list to the IPv4 neighbor adjacencies with R1.

```
R3(config)# router bgp 300
R3(config-router)# address-family ipv4 unicast
R3(config-router-af)# neighbor 10.1.3.1 distribute-list ALLOWED_TO_R1 out
R3(config-router-af)# neighbor 10.1.3.129 distribute-list ALLOWED_TO_R1 out
R3(config-router-af)# end
```

d. Perform a reset of the IPv4 adjacency with R1 for the outbound traffic without tearing down the session.

```
R3# clear bgp ipv4 unicast 6500 out
```

e. On R1, issue the command **show bgp ipv4 unicast | i 300** to see what prefixes route ASN300 is now sharing via BGP. All of the prefixes should now originate in ASN300:

```
R1# show bgp ipv4 unicast | i 300
 *     192.168.3.0/27    10.1.2.2                          0 500 300 i
 *>                      10.1.3.3             0            0 300 i
 *                       10.1.3.130           0            0 300 i
 *     192.168.3.64/26   10.1.2.2                          0 500 300 i
 *>                      10.1.3.3             0            0 300 i
 *                       10.1.3.130           0            0 300 i
```

Step 2. Configure prefix-list-based route filtering.

In this step, you will configure R1 so that it only accepts ASN500 networks from R2; it will not accept information about ASN300 networks from R2.

a. On R1, issue the command **show bgp ipv4 unicast | begin 192.168.3** to see what prefixes ASN500 is sharing via BGP. Take note of those prefixes that do not originate in ASN500.

```
R1# show bgp ipv4 unicast | begin 192.168.3
 *     192.168.3.0/27    10.1.3.130           0            0 300 i
 *                       10.1.2.2                          0 500 300 i
```

```
    *>                      10.1.3.3                0            0 300 i
    *     192.168.3.64/26   10.1.3.130              0            0 300 i
    *                       10.1.2.2                             0 500 300 i
    *>                      10.1.3.3                0            0 300 i
```

b. On R1, configure a prefix list designed to match the source address and mask of networks belonging to ASN500.

```
R1(config)# ip prefix-list ALLOWED_FROM_R2 seq 5 permit 192.168.2.0/24 le 27
```

c. Apply the ALLOWED_FROM_R2 prefix list to the IPv4 neighbor adjacencies for R2.

```
R1(config)# router bgp 6500
R1(config-router)# address-family ipv4 unicast
R1(config-router-af)# neighbor 10.1.2.2 prefix-list ALLOWED_FROM_R2 in
R1(config-router-af)# end
```

d. Perform a reset of the IPv4 adjacency with R2 for the inbound traffic without tearing down the session.

```
R1# clear bgp ipv4 unicast 500 in
```

e. On R1, issue the command **show bgp ipv4 unicast | i 500** to see what prefixes route ASN500 is now sharing via BGP. All of the prefixes should now originate in ASN500.

```
R1# show bgp ipv4 unicast | i 500
    *>   192.168.2.0/27    10.1.2.2                0            0 500 i
    *>   192.168.2.64/26   10.1.2.2                0            0 500 i
```

Step 3. Configure an AS-PATH ACL to filter routes being advertised.

In this step, you will configure R1 so that it only sends ASN100 networks to R2; it will not forward information about prefixes from any other ASN to ASN500.

a. On R2, issue the command **show bgp ipv4 unicast | begin Network** to see what prefixes ASN6500 is sharing via BGP. Take note of those prefixes that do not originate in ASN6500. Advertising these routes could set ASN6500 up as a transit AS, and that is not a desirable scenario.

```
R2# show bgp ipv4 unicast | begin Network
     Network          Next Hop        Metric LocPrf Weight Path
  *   192.168.1.0/27   10.2.3.3                         0 300 6500 i
  *>                   10.1.2.1            0             0 6500 i
  *   192.168.1.64/26  10.2.3.3                         0 300 6500 i
  *>                   10.1.2.1            0             0 6500 i
  *>  192.168.2.0/27   0.0.0.0            0         32768 i
  *>  192.168.2.64/26  0.0.0.0            0         32768 i
  *   192.168.3.0/27   10.1.2.1                         0 6500 300 i
  *>                   10.2.3.3            0             0 300 i
  *   192.168.3.64/26  10.1.2.1                         0 6500 300 i
  *>                   10.2.3.3            0             0 300 i
```

b. On R1, configure AS-PATH ACL to match the routes from the local ASN.

```
R1(config)# ip as-path access-list 1 permit ^$
```

c. On R1, apply the AS-PATH ACL as a filter-list on the adjacency configured with R2.

```
R1(config)# router bgp 6500
R1(config-router)# address-family ipv4 unicast
R1(config-router-af)# neighbor 10.1.2.2 filter-list 1 out
R1(config-router-af)# end
```

d. On R1, perform a reset of the IPv4 adjacency with R2 for the outbound traffic without tearing down the session.

```
R1# clear bgp ipv4 unicast 500 out
```

e. On R2, issue the **command show bgp ipv4 unicast | i 6500** to see what prefixes route ASN6500 is now sharing via BGP. All of the prefixes should now originate in ASN6500.

```
R2# show bgp ipv4 unicast | i 6500
 *     192.168.1.0/27   10.2.3.3                          0 300 6500 i
 *>                     10.1.2.1          0               0 6500 i
 *     192.168.1.64/26  10.2.3.3                          0 300 6500 i
 *>                     10.1.2.1          0               0 6500 i
```

Step 4. Configure IPv6 prefix-list-based route filtering.

In this step, you will configure R1 so that it only accepts ASN500 IPv6 networks from R2. It will not accept information about ASN300 IPv6 networks from R2.

a. On R1, issue the command **show bgp ipv6 unicast neighbors 2001:db8:acad:1012::2 routes** to see what IPv6 prefixes ASN500 is sharing via BGP. Take note of those IPv6 prefixes that do not originate in ASN500.

```
R1# show bgp ipv6 unicast neighbors 2001:db8:acad:1012::2 routes
BGP table version is 9, local router ID is 1.1.1.1
Status code001s: s suppressed, d damped, h history, * valid, > best, i - inter-
nal,
              r RIB-failure, S Stale, m multipath, b backup-path, f RT-Filter,
              x best-external, a additional-path, c RIB-compressed,
              t secondary path, L long-lived-stale,
Origin codes: i - IGP, e - EGP, ? - incomplete
RPKI validation codes: V valid, I invalid, N Not found

     Network          Next Hop          Metric LocPrf Weight Path
 *>  2001:DB8:ACAD:2000::/64
                      2001:DB8:ACAD:1012::2
                                             0             0 500 i
 *>  2001:DB8:ACAD:2001::/64
                      2001:DB8:ACAD:1012::2
                                             0             0 500 i
 *   2001:DB8:ACAD:3000::/64
                      2001:DB8:ACAD:1012::2
                                                           0 500 300 i
 *   2001:DB8:ACAD:3001::/64
                      2001:DB8:ACAD:1012::2
                                                           0 500 300 i

Total number of prefixes 4
```

b. On R1, configure an IPv6 prefix list designed to match the source address and mask of networks belonging to ASN500.

```
R1(config)# ipv6 prefix-list IPV6_ALLOWED_FROM_R2 seq 5 permit
2001:db8:acad:2000::/64

R1(config)# ipv6 prefix-list IPV6_ALLOWED_FROM_R2 seq 10 permit
2001:db8:acad:2001::/64
```

c. Apply the IPV6_ALLOWED_FROM_R2 prefix list to the IPv6 neighbor adjacencies for R2.

```
R1(config)# router bgp 6500
R1(config-router)# address-family ipv6 unicast
R1(config-router-af)# neighbor 2001:db8:acad:1012::2 prefix-list IPV6_ALLOWED_
FROM_R2 in
R1(config-router-af)# end
```

d. Perform a reset of the IPv6 adjacency with R2 for the inbound traffic without tearing down the session.

```
R1# clear bgp ipv6 unicast 500 in
```

e. On R1, issue the command **show bgp ipv6 unicast neighbors 2001:db8:acad:1012::2 routes** to see what IPv6 prefixes route ASN500 is now sharing via BGP. All of the IPv6 prefixes should now originate in ASN500.

```
R1# show bgp ipv6 unicast neighbors 2001:db8:acad:1012::2 routes
BGP table version is 9, local router ID is 1.1.1.1
Status codes: s suppressed, d damped, h history, * valid, > best, i - internal,
              r RIB-failure, S Stale, m multipath, b backup-path, f RT-Filter,
              x best-external, a additional-path, c RIB-compressed,
              t secondary path, L long-lived-stale,
Origin codes: i - IGP, e - EGP, ? - incomplete
RPKI validation codes: V valid, I invalid, N Not found

     Network          Next Hop            Metric LocPrf Weight Path
 *>  2001:DB8:ACAD:2000::/64
                      2001:DB8:ACAD:1012::2
                                              0              0 500 i
 *>  2001:DB8:ACAD:2001::/64
                      2001:DB8:ACAD:1012::2
                                              0              0 500 i

Total number of prefixes 2
```

f. Configure and apply an IPv6 filter to do the same thing on the adjacency with ASN300.

Step 5. Configure BGP path attribute manipulation to affect routing.

In this step, you will configure R1 so that it prefers the next-hop address of 192.168.3.130 over 192.168.3.3, which would normally be the preferred path to ASN300 networks. You will do this by using a prefix list to identify the destination networks and then use a route map to match the prefix list and set the matched networks to have a local preference of 250.

a. On R1, issue the command **show ip route bgp** and take note of the next hop addresses for the 192.168.3.0/27 and 192.168.3.64/26 networks. Then issue the command **show bpg ipv4 unicast** and note that the 10.1.3.130 is a valid next hop. (It's just not the best next hop, according to the BGP path selection algorithm.) Lastly, issue the command **show bgp ipv4 unicast 192.168.3.0** to see details about all the paths available and which one was selected.

```
R1# show bgp ipv4 unicast 192.168.3.0
BGP routing table entry for 192.168.3.0/27, version 8
Paths: (2 available, best #1, table default)
  Advertised to update-groups:
     1
```

```
Refresh Epoch 1
300
  10.1.3.3 from 10.1.3.3 (3.3.3.3)
    Origin IGP, metric 0, localpref 100, valid, external, best
    rx pathid: 0, tx pathid: 0x0
Refresh Epoch 1
300
  10.1.3.130 from 10.1.3.130 (3.3.3.3)
    Origin IGP, metric 0, localpref 100, valid, external
    rx pathid: 0, tx pathid: 0
```

b. On R1, configure a prefix list designed to match the source address and mask of networks belonging to ASN300.

```
R1(config)# ip prefix-list PREFERRED_IPV4_PATH seq 5 permit 192.168.3.0/24 le 27
```

c. Create a route-map named USE_THIS_PATH_FOR_IPV4 that matches on the prefix list you just created and sets the local preference to 250.

```
R1(config)# route-map USE_THIS_PATH_FOR_IPV4 permit 10
R1(config)# match ip address prefix-list PERFERRED_IPV4_PATH
R1(config)# set local-preference 250
```

d. Next, apply this route map to the BGP neighbor 10.1.3.130.

```
R1(config)# router bgp 6500
R1(config-router)# address-family ipv4 unicast
R1(config-router-af)# neighbor 10.1.3.130 route-map USE_THIS_PATH_FOR_IPV4 in
R1(config-router-af)# end
```

e. Perform a reset of the IPv4 adjacency with R3 for the inbound traffic without tearing down the session.

```
R1# clear bgp ipv4 unicast 300 in
```

f. On R1, issue the command **show ip route bgp** and take note of the next hop addresses for the 192.168.3.0/27 and 192.168.3.64/26 networks; it should be 10.1.3.130 for both. Issue the command **show bgp ipv4 unicast** and you should see the local preference value in the appropriate column.

```
R1# show ip route bgp | begin Gateway
Gateway of last resort is not set

      192.168.2.0/24 is variably subnetted, 2 subnets, 2 masks
B        192.168.2.0/27 [20/0] via 10.1.2.2, 00:35:17
B        192.168.2.64/26 [20/0] via 10.1.2.2, 00:35:17
      192.168.3.0/24 is variably subnetted, 2 subnets, 2 masks
B        192.168.3.0/27 [20/0] via 10.1.3.130, 00:00:08
B        192.168.3.64/26 [20/0] via 10.1.3.130, 00:00:08


R1# show bgp ipv4 unicast | begin Network
     Network          Next Hop         Metric LocPrf Weight Path
 *>  192.168.1.0/27   0.0.0.0               0        32768 i
 *>  192.168.1.64/26  0.0.0.0               0        32768 i
 *>  192.168.2.0/27   10.1.2.2              0            0 500 i
 *>  192.168.2.64/26  10.1.2.2              0            0 500 i
 *   192.168.3.0/27   10.1.3.3             0            0 300 i
 *>                   10.1.3.130           0    250      0 300 i
```

```
*     192.168.3.64/26  10.1.3.3             0           0 300 i
*>                     10.1.3.130           0     250    0 300 i
```

Router Interface Summary Table

Router Model	Ethernet Interface #1	Ethernet Interface #2	Serial Interface #1	Serial Interface #2
1800	Fast Ethernet 0/0 (F0/0)	Fast Ethernet 0/1 (F0/1)	Serial 0/0/0 (S0/0/0)	Serial 0/0/1 (S0/0/1)
1900	Gigabit Ethernet 0/0 (G0/0)	Gigabit Ethernet 0/1 (G0/1)	Serial 0/0/0 (S0/0/0)	Serial 0/0/1 (S0/0/1)
2801	Fast Ethernet 0/0 (F0/0)	Fast Ethernet 0/1 (F0/1)	Serial 0/1/0 (S0/1/0)	Serial 0/1/1 (S0/1/1)
2811	Fast Ethernet 0/0 (F0/0)	Fast Ethernet 0/1 (F0/1)	Serial 0/0/0 (S0/0/0)	Serial 0/0/1 (S0/0/1)
2900	Gigabit Ethernet 0/0 (G0/0)	Gigabit Ethernet 0/1 (G0/1)	Serial 0/0/0 (S0/0/0)	Serial 0/0/1 (S0/0/1)
4221	Gigabit Ethernet 0/0/0 (G0/0/0)	Gigabit Ethernet 0/0/1 (G0/0/1)	Serial 0/1/0 (S0/1/0)	Serial 0/1/1 (S0/1/1)
4300	Gigabit Ethernet 0/0/0 (G0/0/0)	Gigabit Ethernet 0/0/1 (G0/0/1)	Serial 0/1/0 (S0/1/0)	Serial 0/1/1 (S0/1/1)

Note: To find out how the router is configured, look at the interfaces to identify the type of router and how many interfaces the router has. There is no way to effectively list all the combinations of configurations for each router class. This table includes identifiers for the possible combinations of Ethernet and Serial interfaces in the device. The table does not include any other type of interface, even though a specific router may contain one. An example of this might be an ISDN BRI interface. The string in parenthesis is the legal abbreviation that can be used in Cisco IOS commands to represent the interface.

Device Config Final – Notes

_____

_____

_____

_____

_____

_____

_____

_____

_____

_____

_____

_____

_____

_____

13.1.2 Lab - Implement BGP Communities

Topology

Addressing Table

Device	Interface	IPv4 Address	IPv6 Address	IPv6 Link-Local
R1	G0/0/0	10.1.2.1/24	2001:db8:acad:1012::1/64	fe80::1:1
	S0/1/0	10.1.3.1/25	2001:db8:acad:1013::1/64	fe80::1:2
	S0/1/1	10.1.3.129/25	2001:db8:acad:1014::1/64	fe80::1:3
	Loopback0	192.168.1.1/27	2001:db8:acad:1000::1/64	fe80::1:4
	Loopback1	192.168.1.65/26	2001:db8:acad:1001::1/64	fe80::1:5
R2	G0/0/0	10.1.2.2/24	2001:db8:acad:1012::2/64	fe80::2:1
	G0/0/1	10.2.3.2/24	2001:db8:acad:1023::2/64	fe80::2:2
	Loopback0	192.168.2.1/27	2001:db8:acad:2000::1/64	fe80::2:3
	Loopback1	192.168.2.65/26	2001:db8:acad:2001::1/64	fe80::2:4
R3	G0/0/0	10.2.3.3/24	2001:db8:acad:1023::3/64	fe80::3:1
	S0/1/0	10.1.3.3/25	2001:db8:acad:1013::3/64	fe80::3:2
	S0/1/1	10.1.3.130/25	2001:db8:acad:1014::3/64	fe80::3:3
	Loopback0	192.168.3.1/27	2001:db8:acad:3000::1/64	fe80::3:4
	Loopback1	192.168.3.65/26	2001:db8:acad:3001::1/64	fe80::3:5

Objectives

Part 1: Build the Network and Configure Basic Device Settings and Interface Addressing

Part 2: Configure and Verify Multi-Protocol BGP on all Routers

Part 3: Configure and Verify BGP Communities on all Routers

Background/Scenario

As you saw in the previous lab, standard path manipulation tools generally require a lot of configuration. Imagine if the last lab was implemented in a large and complex routing environment. Using BGP communities gives you an option for tagging and controlling routing information in a less labor-intensive manner. In this lab you will configure Multi-Protocol BGP and implement BGP community configurations for both IPv4 and IPv6.

Note: This lab is an exercise in developing, deploying, and verifying BGP communities, and does not reflect networking best practices.

Note: The routers used with CCNP hands-on labs are Cisco 4221 with Cisco IOS XE Release 16.9.4 (universalk9 image). Other routers and Cisco IOS versions can be used. Depending on the model and Cisco IOS version, the commands available and the output produced might vary from what is shown in the labs.

Note: Ensure that the routers have been erased and have no startup configurations. If you are unsure contact your instructor.

Required Resources

- 3 Routers (Cisco 4221 with Cisco IOS XE Release 16.9.4 universal image or comparable)
- 1 PC (Choice of operating system with a terminal emulation program installed)
- Console cables to configure the Cisco IOS devices via the console ports
- Ethernet and serial cables as shown in the topology

Instructions

Part 1: Build the Network and Configure Basic Device Settings and Interface Addressing

In Part 1, you will set up the network topology and configure basic settings and interface addressing on routers.

Step 1. Cable the network as shown in the topology.

Attach the devices as shown in the topology diagram, and cable as necessary.

Step 2. Configure basic settings for each router.

  a. Console into each router, enter global configuration mode, and apply the basic settings and interface addressing. A command list for each router is listed below for initial configuration.

Router R1

```
no ip domain lookup
hostname R1
line con 0
 exec-timeout 0 0
 logging synchronous
 exit
banner motd # This is R1, BGP Path Manipulation Lab #
ipv6 unicast-routing
interface g0/0/0
 ip address 10.1.2.1 255.255.255.0
 ipv6 address fe80::1:1 link-local
 ipv6 address 2001:db8:acad:1012::1/64
 no shutdown
interface s0/1/0
 ip address 10.1.3.1 255.255.255.128
 ipv6 address fe80::1:2 link-local
 ipv6 address 2001:db8:acad:1013::1/80
 no shutdown
interface s0/1/1
 ip address 10.1.3.129 255.255.255.128
 ipv6 address fe80::1:3 link-local
 ipv6 address 2001:db8:acad:1014::1/80
 no shutdown
interface loopback 0
 ip address 192.168.1.1 255.255.255.224
 ipv6 address fe80::1:4 link-local
 ipv6 address 2001:db8:acad:1000::1/64
 no shutdown
interface loopback 1
 ip address 192.168.1.65 255.255.255.192
 ipv6 address fe80::1:5 link-local
 ipv6 address 2001:db8:acad:1001::1/64
 no shutdown
 exit
```

Router R2

```
no ip domain lookup
hostname R2
line con 0
 exec-timeout 0 0
 logging synchronous
 exit
banner motd # This is R2, BGP Path Manipulation Lab #
ipv6 unicast-routing
interface g0/0/0
 ip address 10.1.2.2 255.255.255.0
 ipv6 address fe80::2:1 link-local
 ipv6 address 2001:db8:acad:1012::2/64
 no shutdown
```

```
interface g0/0/1
 ip address 10.2.3.2 255.255.255.0
 ipv6 address fe80::2:2 link-local
 ipv6 address 2001:db8:acad:1023::2/64
 no shutdown
interface loopback 0
 ip address 192.168.2.1 255.255.255.224
 ipv6 address fe80::2:3 link-local
 ipv6 address 2001:db8:acad:2000::1/64
 no shutdown
interface loopback 1
 ip address 192.168.2.65 255.255.255.192
 ipv6 address fe80::2:4 link-local
 ipv6 address 2001:db8:acad:2001::1/64
 no shutdown
```

Router R3

```
no ip domain lookup
hostname R3
line con 0
 exec-timeout 0 0
 logging synchronous
 exit
banner motd # This is R3, BGP Path Manipulation Lab #
ipv6 unicast-routing
interface g0/0/0
 ip address 10.2.3.3 255.255.255.0
 ipv6 address fe80::3:1 link-local
 ipv6 address 2001:db8:acad:1023::3/64
 no shutdown
interface s0/1/0
 ip address 10.1.3.3 255.255.255.128
 ipv6 address fe80::3:2 link-local
 ipv6 address 2001:db8:acad:1013::3/80
 no shutdown
interface s0/1/1
 ip address 10.1.3.130 255.255.255.128
 ipv6 address fe80::3:3 link-local
 ipv6 address 2001:db8:acad:1014::3/80
 no shutdown
interface loopback 0
 ip address 192.168.3.1 255.255.255.224
 ipv6 address fe80::3:4 link-local
 ipv6 address 2001:db8:acad:3000::1/64
 no shutdown
interface loopback 1
 ip address 192.168.3.65 255.255.255.192
 ipv6 address fe80::3:5 link-local
 ipv6 address 2001:db8:acad:3001::1/64
 no shutdown
```

b. Set the clock on each router to UTC time.

c. Save the running configuration to startup-config.

Part 2: Configure and Verify Multi-Protocol BGP on all Routers

In Part 2, you will configure and verify Multi-Protocol BGP on all routers to achieve full connectivity between the routers. The text below provides you with the complete configuration for R1. You will use this to inform your configuration of R2 and R3. The configuration being used here is not meant to represent best practice, but to assess your ability to complete the required configurations.

Step 1. On R1, create the core BGP configuration.

a. Enter BGP configuration mode from global configuration mode, specifying AS 6500.

```
R1(config)# router bgp 6500
```

b. Configure the BGP router-id for R1.

```
R1(config-router)# bgp router-id 1.1.1.1
```

c. Disable the default IPv4 unicast behavior

```
R1(config-router)# no bgp default ipv4-unicast
```

d. Based on the topology diagram, configure all the designated neighbors for R1.

```
R1(config-router)# neighbor 10.1.2.2 remote-as 500
R1(config-router)# neighbor 10.1.3.3 remote-as 300
R1(config-router)# neighbor 10.1.3.130 remote-as 300
R1(config-router)# neighbor 2001:db8:acad:1012::2 remote-as 500
R1(config-router)# neighbor 2001:db8:acad:1013::3 remote-as 300
R1(config-router)# neighbor 2001:db8:acad:1014::3 remote-as 300
```

Step 2. On R1, configure the IPv4 unicast address family.

a. Enter the IPv4 unicast address family configuration mode.

```
R1(config-router)# address-family ipv4 unicast
```

b. Configure network statements for the IPv4 networks that are attached to interfaces loopback0 and loopback1. Remember that BGP does not work the same way that an IGP does, and that the network statement has no impact on neighbor adjacency; it is used solely for advertising purposes.

```
R1(config-router-af)# network 192.168.1.0 mask 255.255.255.224
R1(config-router-af)# network 192.168.1.64 mask 255.255.255.192
```

c. Deactivate the IPv6 neighbors and activate the IPv4 neighbors.

```
R1(config-router-af)# no neighbor 2001:db8:acad:1012::2 activate
R1(config-router-af)# no neighbor 2001:db8:acad:1013::3 activate
R1(config-router-af)# no neighbor 2001:db8:acad:1014::3 activate
R1(config-router-af)# neighbor 10.1.2.2 activate
R1(config-router-af)# neighbor 10.1.3.3 activate
R1(config-router-af)# neighbor 10.1.3.130 activate
```

Step 3. On R1, configure the IPv4 unicast address family.

a. Enter the IPv6 unicast address family configuration mode.

```
R1(config-router)# address-family ipv6 unicast
```

b. Configure network statements for the IPv6 networks that are attached to interfaces loopback0 and loopback1. Remember that BGP does not work the same way that an

IGP does, and the network statement has no impact on neighbor adjacency; it is used solely for advertising purposes.

```
R1(config-router-af)# network 2001:db8:acad:1000::/64
R1(config-router-af)# network 2001:db8:acad:1001::/64
```

c. Activate the IPv6 neighbors that are configured for BGP.

```
R1(config-router-af)# neighbor 2001:db8:acad:1012::2 activate
R1(config-router-af)# neighbor 2001:db8:acad:1013::3 activate
R1(config-router-af)# neighbor 2001:db8:acad:1014::3 activate
```

d. Configure MP-BGP on R2 and R3 as in previous steps.

_____

_____

_____

_____

_____

Step 4.   Verify that MP-BGP is operational.

a. Use the **show bgp ipv4 unicast summary** and **show bgp ipv6 unicast summary** commands to verify that BGP has established adjacencies and received prefixes.

```
R1# show bgp ipv4 unicast summary
BGP router identifier 1.1.1.1, local AS number 6500
BGP table version is 9, main routing table version 9
6 network entries using 1488 bytes of memory
14 path entries using 1904 bytes of memory
5/3 BGP path/bestpath attribute entries using 1400 bytes of memory
4 BGP AS-PATH entries using 128 bytes of memory
0 BGP route-map cache entries using 0 bytes of memory
0 BGP filter-list cache entries using 0 bytes of memory
BGP using 4920 total bytes of memory
BGP activity 12/0 prefixes, 28/0 paths, scan interval 60 secs

Neighbor        V    AS MsgRcvd MsgSent    TblVer  InQ OutQ Up/Down  State/PfxRcd
10.1.2.2        4   500       7       7         9    0    0 00:01:46 4
10.1.3.3        4   300       7       7         9    0    0 00:00:45 4
10.1.3.130      4   300       7       7         9    0    0 00:00:44 4


R1# show bgp ipv6 unicast summary
BGP router identifier 1.1.1.1, local AS number 6500
BGP table version is 9, main routing table version 9
6 network entries using 1632 bytes of memory
14 path entries using 2128 bytes of memory
5/3 BGP path/bestpath attribute entries using 1400 bytes of memory
4 BGP AS-PATH entries using 128 bytes of memory
0 BGP route-map cache entries using 0 bytes of memory
0 BGP filter-list cache entries using 0 bytes of memory
BGP using 5288 total bytes of memory
BGP activity 12/0 prefixes, 28/0 paths, scan interval 60 secs
```

```
Neighbor          V   AS MsgRcvd MsgSent   TblVer  InQ OutQ Up/Down  State/PfxRcd
2001:DB8:ACAD:1012::2
                  4  500      7       7        9    0    0 00:02:08  4
2001:DB8:ACAD:1013::3
                  4  300      8       7        9    0    0 00:01:09  4
2001:DB8:ACAD:1014::3
                  4  300      8       7        9    0    0 00:01:09  4
```

b. Use the **show bgp ipv4 unicast** and **show bgp ipv6 unicast** commands to view the specified BGP tables. Note that R1 has multiple paths to each destination network. Take note of the next hop address for the destination networks marked with the ">" symbol.

```
R1# show bgp ipv4 unicast
BGP table version is 9, local router ID is 1.1.1.1
Status codes: s suppressed, d damped, h history, * valid, > best, i - internal,
              r RIB-failure, S Stale, m multipath, b backup-path, f RT-Filter,
              x best-external, a additional-path, c RIB-compressed,
              t secondary path, L long-lived-stale,
Origin codes: i - IGP, e - EGP, ? - incomplete
RPKI validation codes: V valid, I invalid, N Not found

     Network          Next Hop          Metric LocPrf Weight Path
 *>  192.168.1.0/27   0.0.0.0                0         32768 i
 *>  192.168.1.64/26  0.0.0.0                0         32768 i
 *   192.168.2.0/27   10.1.3.3                             0 300 500 i
 *                    10.1.3.130                           0 300 500 i
 *>                   10.1.2.2               0             0 500 i
 *   192.168.2.64/26  10.1.3.3                             0 300 500 i
 *                    10.1.3.130                           0 300 500 i
 *>                   10.1.2.2               0             0 500 i
 *   192.168.3.0/27   10.1.2.2                             0 500 300 i
 *>                   10.1.3.3               0             0 300 i
 *                    10.1.3.130             0             0 300 i
 *   192.168.3.64/26  10.1.2.2                             0 500 300 i
 *>                   10.1.3.3               0             0 300 i
 *                    10.1.3.130             0             0 300 i


R1# show bgp ipv6 unicast
BGP table version is 9, local router ID is 1.1.1.1
Status codes: s suppressed, d damped, h history, * valid, > best, i - internal,
              r RIB-failure, S Stale, m multipath, b backup-path, f RT-Filter,
              x best-external, a additional-path, c RIB-compressed,
              t secondary path, L long-lived-stale,
Origin codes: i - IGP, e - EGP, ? - incomplete
RPKI validation codes: V valid, I invalid, N Not found

     Network          Next Hop          Metric LocPrf Weight Path
 *>  2001:DB8:ACAD:1000::/64
                      ::                     0         32768 i
 *>  2001:DB8:ACAD:1001::/64
                      ::                     0         32768 i
```

```
*     2001:DB8:ACAD:2000::/64
                      2001:DB8:ACAD:1014::3
                                                              0 300 500 i
*                     2001:DB8:ACAD:1013::3
                                                              0 300 500 i
*>                    2001:DB8:ACAD:1012::2
                                               0              0 500 i
*     2001:DB8:ACAD:2001::/64
                      2001:DB8:ACAD:1014::3
                                                              0 300 500 i
*                     2001:DB8:ACAD:1013::3
                                                              0 300 500 i
*>                    2001:DB8:ACAD:1012::2
                                               0              0 500 i
*     2001:DB8:ACAD:3000::/64
                      2001:DB8:ACAD:1012::2
                                                              0 500 300 i
*                     2001:DB8:ACAD:1014::3
                                               0              0 300 i
*>                    2001:DB8:ACAD:1013::3
                                               0              0 300 i
*     2001:DB8:ACAD:3001::/64
                      2001:DB8:ACAD:1012::2
                                                              0 500 300 i
*                     2001:DB8:ACAD:1014::3
                                               0              0 300 i
*>                    2001:DB8:ACAD:1013::3
                                               0              0 300 i
```

c. Use the **show ip route bgp** and **show ipv6 route bgp** commands to view the routing tables. Note that there is only one route to each destination, and that the routes included in the routing table have the same next hop as those with the ">" symbol in the BGP tables.

```
R1# show ip route bgp | begin Gateway
Gateway of last resort is not set

      192.168.2.0/24 is variably subnetted, 2 subnets, 2 masks
B        192.168.2.0/27 [20/0] via 10.1.2.2, 00:06:20
B        192.168.2.64/26 [20/0] via 10.1.2.2, 00:06:20
      192.168.3.0/24 is variably subnetted, 2 subnets, 2 masks
B        192.168.3.0/27 [20/0] via 10.1.3.3, 00:06:19
B        192.168.3.64/26 [20/0] via 10.1.3.3, 00:06:19


R1# show ipv6 route bgp
IPv6 Routing Table - default - 15 entries
Codes: C - Connected, L - Local, S - Static, U - Per-user Static route
       B - BGP, R - RIP, H - NHRP, I1 - ISIS L1
       I2 - ISIS L2, IA - ISIS interarea, IS - ISIS summary, D - EIGRP
       EX - EIGRP external, ND - ND Default, NDp - ND Prefix, DCE - Destination
       NDr - Redirect, RL - RPL, O - OSPF Intra, OI - OSPF Inter
```

```
            OE1 - OSPF ext 1, OE2 - OSPF ext 2, ON1 - OSPF NSSA ext 1
            ON2 - OSPF NSSA ext 2, a - Application
B    2001:DB8:ACAD:2000::/64 [20/0]
      via FE80::2, GigabitEthernet0/0/0
B    2001:DB8:ACAD:2001::/64 [20/0]
      via FE80::2, GigabitEthernet0/0/0
B    2001:DB8:ACAD:3000::/64 [20/0]
      via FE80::3:2, Serial0/1/0
B    2001:DB8:ACAD:3001::/64 [20/0]
      via FE80::3:2, Serial0/1/0
```

Part 3: Configure and Verify BGP Communities on all Routers

In Part 3, you will configure BGP communities and various community attributes to see their effect on routing decisions. The way these tools are being used is not meant to represent best practice, but to assess your ability to complete the required configurations.

Step 1.   Configure all routers to send community information.

In this step, you will configure all of the routers to support the **new-format** for exchanging community information and enable sending community information to all neighbors on all routers. A BGP community is a 32-bit number that can be included as a flag or tag in a route. The BGP community can be configured and displayed as two 16 bit numbers **AA:NN** commonly referred to as **new-format**. To configure and display using the **AA:NN**, issue the **ip bgp-community new-format** command. The first part of the **AA:NN** represents the AS number and the second part represents a 2-byte number.

The configuration for R1 is shown below. Use this as an example and complete the configuration on R2 and R3 on your own.

a.   Issue the global configuration command that enables configuration and display of community information using the AA:NN format.

```
R1(config)# ip bgp-community new-format
```

b.   Add a neighbor statement for each neighbor with the send community parameter.

```
R1(config)# router bgp 6500
R1(config-router)# address-family ipv4 unicast
R1(config-router-af)# neighbor 10.1.2.2 send-community
R1(config-router-af)# neighbor 10.1.3.3 send-community
R1(config-router-af)# neighbor 10.1.3.130 send-community
R1(config-router-af)# address-family ipv6 unicast
R1(config-router-af)# neighbor 2001:db8:acad:1012::2 send-community
R1(config-router-af)# neighbor 2001:db8:acad:1013::3 send-community
R1(config-router-af)# neighbor 2001:db8:acad:1014::3 send-community
R1(config-router-af)# exit
```

_____

_____

_____

_____

_____

c. At this point, the routers are ready to send community information, but there is no community information available. On R2, issue the command **show bgp ipv4 unicast 192.168.2.0/27**, and you will see there is no community information listed.

_____

_____

_____

_____

_____

Step 2. Configure and verify the effect of the no-export community.

In this step, you will configure R3 so that it sets the well-known **no-export** community value on the updates describing its local networks that are being sent to R1. The effect of this is that R1 will not pass along information about these paths to other eBGP neighbors.

a. On R2, issue the command **show bgp ipv4 unicast 192.168.3.0/27** to see the available BGP paths to 192.168.3.0/27 from R2.

```
R2# show bgp ipv4 unicast 192.168.3.0/27
BGP routing table entry for 192.168.3.0/27, version 6
Paths: (2 available, best #2, table default)
  Advertised to update-groups:
     2
  Refresh Epoch 1
  6500 300
    10.1.2.1 from 10.1.2.1 (1.1.1.1)
      Origin IGP, localpref 100, valid, external
      rx pathid: 0, tx pathid: 0
  Refresh Epoch 1
  300
    10.2.3.3 from 10.2.3.3 (3.3.3.3)
      Origin IGP, metric 0, localpref 100, valid, external, best
      rx pathid: 0, tx pathid: 0x0
```

b. In this case, note that there are two paths, one directly from R3/ASN300 and the other from R1/ASN6500. This might not be desirable, because it sets ASN6500 up as a transit network. To fix this issue, a prefix list on R3 can be created to match the source address and mask of networks belonging to ASN300.

```
R3(config)# ip prefix-list LOCAL_NETWORK_COMMSET seq 5 permit 192.168.3.0/24 le
27
R3(config)# ipv6 prefix-list LOCAL_6_NETWORK_COMMSET seq 5 permit
2001:db8:acad:3000::/64
R3(config)# ipv6 prefix-list LOCAL_6_NETWORK_COMMSET seq 10 permit
2001:db8:acad:3001::/64
```

c. Next, build a route map for IPv4 and IPv6 on R3 that uses the prefix list to set the **no-export additive** community on networks matching the prefix list, and the **internet additive** community on networks that do not match the prefix list. By default, when setting a community, any existing communities are over-written, but can be preserved using the optional **additive** keyword.

```
R3(config)# route-map COMMSET permit 10
R3(config-route-map)# match ip address prefix-list LOCAL_NETWORK_COMMSET
```

```
R3(config-route-map)# set community no-export additive
R3(config-route-map)# exit
R3(config)# route-map COMMSET permit 20
R3(config-route-map)# set community internet additive
R3(config-route-map)# exit
R3(config)# route-map COMMSET_6 permit 10
R3(config-route-map)# match ipv6 address prefix-list LOCAL_6_NETWORK_COMMSET
R3(config-route-map)# set community no-export additive
R3(config-route-map)# exit
R3(config)# route-map COMMSET_6 permit 20
R3(config-route-map)# set community internet additive
```

d. Next, apply these route maps to the neighbor statements associated with R1.

```
R3(config)# router bgp 300
R3(config-router)# address-family ipv4 unicast
R3(config-router-af)# neighbor 10.1.3.1 route-map COMMSET out
R3(config-router-af)# neighbor 10.1.3.129 route-map COMMSET out
R3(config-router-af)# address-family ipv6 unicast
R3(config-router-af)# neighbor 2001:db8:acad:1013::1 route-map COMMSET_6 out
R3(config-router-af)# neighbor 2001:db8:acad:1014::1 route-map COMMSET_6 out
```

e. Perform a reset of the adjacencies with the outbound traffic to R1 without tearing down the session.

```
R3# clear bgp ipv4 unicast 6500 out
R3# clear bgp ipv6 unicast 6500 out
```

f. On R2, issue the command **show bgp ipv4 unicast 192.168.3.0/27** to see the available BGP paths to 192.168.3.0/27 from R2. This time, you should not see a path to 192.168.3.0/27 via the next-hop 10.1.2.1. If you use the command **show bgp ipv6 unicast 2001:db8:acad:3000::/64**, you will see only one next-hop address, and that is 2001:db8:acad:1023::3.

```
R2# show bgp ipv4 unicast 192.168.3.0/27
BGP routing table entry for 192.168.3.0/27, version 6
Paths: (1 available, best #1, table default)
  Advertised to update-groups:
     2
  Refresh Epoch 1
  300
    10.2.3.3 from 10.2.3.3 (3.3.3.3)
      Origin IGP, metric 0, localpref 100, valid, external, best
      rx pathid: 0, tx pathid: 0x0


R2# show bgp ipv6 unicast 2001:db8:acad:3000::/64
BGP routing table entry for 2001:DB8:ACAD:3000::/64, version 8
Paths: (1 available, best #1, table default)
  Advertised to update-groups:
     2
  Refresh Epoch 2
  300
    2001:DB8:ACAD:1023::3 (FE80::3:1) from 2001:DB8:ACAD:1023::3 (3.3.3.3)
      Origin IGP, metric 0, localpref 100, valid, external, best
      rx pathid: 0, tx pathid: 0x0
```

Step 3.   Add private community information to routes advertised by R1.

In this step, you will configure R1 so that it adds custom community strings to IPv4 and IPv6 routes that it advertises to R2/ASN500.

    a.   On R1, create two route maps. One route map will add the community 650:400 to all IPv4 routes advertised to R2/ASN500, and the second route map adds the community 650:600 to all IPv6 routes advertised to R2/ASN500.

```
R1(config)# route-map ADDCOMM permit 10
R1(config-route-map)# set community 650:400 additive
R1(config-route-map)# exit
R1(config)# route-map ADDCOMM_6 permit 10
R1(config-route-map)# set community 650:600 additive
R1(config-route-map)# exit
```

    b.   On R1, apply the appropriate route map to the appropriate R2 neighbor statement.

```
R1(config)# router bgp 6500
R1(config-router)# address-family ipv4 unicast
R1(config-router-af)# neighbor 10.1.2.2 route-map ADDCOMM out
R1(config-router-af)# address-family ipv6 unicast
R1(config-router-af)# neighbor 2001:db8:acad:1012::2 route-map ADDCOMM_6 out
R1(config-router-af)# end
```

    c.   On R1, perform a reset of the adjacencies with the outbound traffic to R2 without tearing down the session.

```
R1# clear bgp ipv4 unicast 500 out
R1# clear bgp ipv6 unicast 500 out
```

    d.   On R2, verify the community tags are present by issuing the commands **show bgp ipv4 unicast 192.168.1.0/27 | i Community** and **show bgp ipv6 unicast 2001:db8:acad:1000::/64 | i Community**.

```
R2# show bgp ipv4 unicast 192.168.1.0/27 | i Community
    Community: 650:400
R2# show bgp ipv6 unicast 2001:db8:acad:1000::/64 | i Community
    Community: 650:600
```

If you run those same commands on R3, you will see that the community tags are present there as well. Because Community is an optional transitive attribute, it is passed on to eBGP neighbors by default.

Step 4.   Configure community-based route filtering and manipulation.

In this step, you will configure R3 so that it drops all routes coming from R2 with the 650:400 community attribute. Then configure R3 so that it sets a higher local preference for all routes coming from R2 with the 650:600 community attribute.

    a.   On R3, create two community lists; one that matches the 650:400 attribute and another that matches the 650:600 attribute.

```
R3(config)# ip community-list 100 permit 650:400
R3(config)# ip community-list 101 permit 650:600
```

    b.   On R3, create a pair of route maps that use the newly created community lists. The first route map will drop routes with the 650:400 community set and permit all others. The second route map will match the community 650:600 and set the local preference value to 250. Routes not matching the community 650:600 will not be modified.

```
R3(config)# route-map COMMCHECK_4 deny 10
R3(config-route-map)# match community 100
R3(config-route-map)# route-map COMMCHECK_4 permit 20
R3(config-route-map)# exit
R3(config)# route-map COMMCHECK_6 permit 10
R3(config-route-map)# match community 101
R3(config-route-map)# set local-preference 250
R3(config-route-map)# route-map COMMCHECK_6 permit 20
R3(config-route-map)# exit
```

c. On R3, apply the appropriate route map to the appropriate R2 neighbor statement.

```
R3(config)# router bgp 300
R3(config-router)# address-family ipv4 unicast
R3(config-router-af)# neighbor 10.2.3.2 route-map COMMCHECK_4 in
R3(config-router-af)# address-family ipv6 unicast
R3(config-router-af)# neighbor 2001:db8:acad:1023::2 route-map COMMCHECK_6 in
R3(config-router-af)# end
```

d. Perform a reset of the adjacencies with the inbound traffic to R2 without tearing down the session.

```
R3# clear bgp ipv4 unicast 500 in
R3# clear bgp ipv6 unicast 500 in
```

e. On R3, verify the IPv4 policy is working. Issue the command **show bgp ipv4 unicast 192.168.1.0/27 | i Community** and you will see that there is no output. Follow this with the **show bgp ipv4 unicast** command and you will see that there are no paths to the ASN6500 networks via R2. They have all been filtered.

```
R3# show bgp ipv4 unicast 192.168.1.0/27 | i Community

R3# show bgp ipv4 unicast
BGP table version is 9, local router ID is 3.3.3.3
Status codes: s suppressed, d damped, h history, * valid, > best, i - internal,
              r RIB-failure, S Stale, m multipath, b backup-path, f RT-Filter,
              x best-external, a additional-path, c RIB-compressed,
              t secondary path, L long-lived-stale,
Origin codes: i - IGP, e - EGP, ? - incomplete
RPKI validation codes: V valid, I invalid, N Not found

     Network          Next Hop            Metric LocPrf Weight Path
 *>  192.168.1.0/27   10.1.3.1                 0             0 6500 i
 *                    10.1.3.129               0             0 6500 i
 *>  192.168.1.64/26  10.1.3.1                 0             0 6500 i
 *                    10.1.3.129               0             0 6500 i
 *   192.168.2.0/27   10.1.3.1                               0 6500 500 i
 *                    10.1.3.129                              0 6500 500 i
 *>                   10.2.3.2                 0             0 500 i
 *   192.168.2.64/26  10.1.3.1                               0 6500 500 i
 *                    10.1.3.129                              0 6500 500 i
 *>                   10.2.3.2                 0             0 500 i
 *>  192.168.3.0/27   0.0.0.0                  0         32768 i
 *>  192.168.3.64/26  0.0.0.0                  0         32768 i
```

f. On R3, verify the IPv6 policy is working. Issue the command **show bgp ipv6 unicast** and note the local preference has been assigned to the ASN500 routes advertised from R2.

```
R3# show bgp ipv6 unicast
BGP table version is 11, local router ID is 3.3.3.3
Status codes: s suppressed, d damped, h history, * valid, > best, i - internal,
              r RIB-failure, S Stale, m multipath, b backup-path, f RT-Filter,
              x best-external, a additional-path, c RIB-compressed,
              t secondary path, L long-lived-stale,
Origin codes: i - IGP, e - EGP, ? - incomplete
RPKI validation codes: V valid, I invalid, N Not found
```

	Network	Next Hop	Metric	LocPrf	Weight	Path
*>	2001:DB8:ACAD:1000::/64					
		2001:DB8:ACAD:1023::2				
				250	0	500 6500 i
*		2001:DB8:ACAD:1014::1				
			0		0	6500 i
*		2001:DB8:ACAD:1013::1				
			0		0	6500 i
*>	2001:DB8:ACAD:1001::/64					
		2001:DB8:ACAD:1023::2				
				250	0	500 6500 i
*		2001:DB8:ACAD:1014::1				
			0		0	6500 i
*		2001:DB8:ACAD:1013::1				

	Network	Next Hop	Metric	LocPrf	Weight	Path
			0		0	6500 i
*	2001:DB8:ACAD:2000::/64					
		2001:DB8:ACAD:1014::1				
					0	6500 500 i
*		2001:DB8:ACAD:1013::1				
					0	6500 500 i
*>		2001:DB8:ACAD:1023::2				
			0		0	500 i
*	2001:DB8:ACAD:2001::/64					
		2001:DB8:ACAD:1014::1				
					0	6500 500 i
*		2001:DB8:ACAD:1013::1				
					0	6500 500 i
*>		2001:DB8:ACAD:1023::2				
			0		0	500 i
*>	2001:DB8:ACAD:3000::/64					
		::	0		32768	i
*>	2001:DB8:ACAD:3001::/64					
		::	0		32768	i

Reflection Questions

1. Name the three common well-known communities.

   _____

   _____

2. When setting a community inside a route map using the **set community** command, what is the function of the optional "**additive**" keyword?

   _____

   _____

Router Interface Summary Table

Router Model	Ethernet Interface #1	Ethernet Interface #2	Serial Interface #1	Serial Interface #2
1800	Fast Ethernet 0/0 (F0/0)	Fast Ethernet 0/1 (F0/1)	Serial 0/0/0 (S0/0/0)	Serial 0/0/1 (S0/0/1)
1900	Gigabit Ethernet 0/0 (G0/0)	Gigabit Ethernet 0/1 (G0/1)	Serial 0/0/0 (S0/0/0)	Serial 0/0/1 (S0/0/1)
2801	Fast Ethernet 0/0 (F0/0)	Fast Ethernet 0/1 (F0/1)	Serial 0/1/0 (S0/1/0)	Serial 0/1/1 (S0/1/1)
2811	Fast Ethernet 0/0 (F0/0)	Fast Ethernet 0/1 (F0/1)	Serial 0/0/0 (S0/0/0)	Serial 0/0/1 (S0/0/1)
2900	Gigabit Ethernet 0/0 (G0/0)	Gigabit Ethernet 0/1 (G0/1)	Serial 0/0/0 (S0/0/0)	Serial 0/0/1 (S0/0/1)
4221	Gigabit Ethernet 0/0/0 (G0/0/0)	Gigabit Ethernet 0/0/1 (G0/0/1)	Serial 0/1/0 (S0/1/0)	Serial 0/1/1 (S0/1/1)
4300	Gigabit Ethernet 0/0/0 (G0/0/0)	Gigabit Ethernet 0/0/1 (G0/0/1)	Serial 0/1/0 (S0/1/0)	Serial 0/1/1 (S0/1/1)

Note: To find out how the router is configured, look at the interfaces to identify the type of router and how many interfaces the router has. There is no way to effectively list all the combinations of configurations for each router class. This table includes identifiers for the possible combinations of Ethernet and Serial interfaces in the device. The table does not include any other type of interface, even though a specific router may contain one. An example of this might be an ISDN BRI interface. The string in parenthesis is the legal abbreviation that can be used in Cisco IOS commands to represent the interface.

Device Config Final – Notes

   _____

   _____

   _____

   _____

   _____

   _____

   _____

CHAPTER 14

Troubleshooting BGP

14.1.2 Lab - Troubleshoot BGP

Topology

Note: The advertised prefixes in AS 65100, 65200 and in the "the cloud" are associated with the interface addresses shown (.1 and ::1) for verification purpose. These prefixes are advertised by D1 and D2 to their respective BGP peers.

Note: AS 65100 advertises a shorter AS path to 2001::db8:cafe::/48.

Addressing Table

Device	Interface	IPv6 Address IPv4 Address	IPv6 Link-Local
R1	G0/0/0	2001:db8:c0c0:a001::1/64	fe80::a001:1
		192.168.2.1/24	N/A
	G0/0/1	2001:db8:cab:f001::2/64	fe80::f001:2
		172.16.1.2/30	N/A

Device	Interface	IPv6 Address IPv4 Address	IPv6 Link-Local
	Lo0 (iBGP Peering)	2001:db8:c0c0:99::1/128	fe80::99:1
		192.168.99.1	N/A
	Lo1	2001:db8:c0c0:a011::1/64	fe80::a011:1
		192.168.1.1/24	N/A
R2	G0/0/0	2001:db8:c0c0:a001::2/64	fe80::a001:2
		192.168.2.2/24	N/A
	G0/0/1	2001:db8:c0c0:a002::2/64	fe80::a002:2
		192.168.5.2/24	N/A
	Lo0 (iBGP Peering)	2001:db8:c0c0:99::2/128	fe80::99:2
		192.168.99.2/24	N/A
	Lo1	2001:db8:c0c0:a021::1/64	fe80::a021:1
		192.168.3.1/24	N/A
	Lo2	2001:db8:c0c0:a022::1/64	fe80::a022:1
		192.168.4.1/24	N/A
R3	G0/0/0	2001:db8:c0c0:a002::1/64	fe80::a002:1
		192.168.5.1/24	N/A
	G0/0/1	2001:db8:b0b:f001::2/64	fe80::f001:2
		172.16.2.2/30	N/A
	Lo0 (iBGP Peering)	2001:db8:c0c0:99::3/128	fe80::99:3
		192.168.99.3/24	N/A
	Lo1	2001:db8:c0c0:a031::1/64	fe80::a031:1
		192.168.6.1/24	N/A
D1	G1/0/11	2001:db8:cab:f001::1/64	fe80::cab:f001:1
		172.16.1.1/30	N/A
	Lo1	2001:db8:cab:f002::1/64	fe80::cab:f002:1
		10.1.1.1/24	N/A
	Lo10	2001:db8:cab1::1/48	EUI-64
		10.1.10.1/24	N/A
	Lo20	2001:db8:cab2::1/48	EUI-64
		10.1.20.1/24	N/A
	Lo30	2001:db8:cab3::1/48	EUI-64
		10.1.30.1/24	N/A
	Lo48	2001:db8:cafe::1/48	EUI-64
D2	G1/0/11	2001:db8:b0b:f001::1/64	EUI-64
		172.16.2.1/30	N/A
	Lo1	2001:db8:b0b:f002::1/64	EUI-64
		10.2.1.1/24	N/A
	Lo10	2001:db8:b0b1::1/48	EUI-64
		10.2.10.1/24	N/A

Device	Interface	IPv6 Address IPv4 Address	IPv6 Link-Local
	Lo20	2001:db8:b0b2::1/48	EUI-64
		10.2.20.1/24	N/A
	Lo30	2001:db8:b0b3::1/48	EUI-64
		10.2.30.1/24	N/A
	Lo48	2001:db8:cafe::1/48	EUI-64

Note: To make it easier to recognize IPv6 prefixes, familiarize yourself with the IPv6 GUA and LLA address formats.

- GUA: The GUA has a 16-bit subnet-ID, a<area-id><router><network> ("a" for area). For example, subnet-ID a021 is area 0, router 2, network 1.

- LLA: Following best practice the LLA is unique on each interface. The LLA interface-ID uses the GUA subnet-ID:interface-ID for the last 64 bits. For example, fe80::a201:1 has an LLA interface-ID a201 (the subnet-ID of the GUA) and :1 (the interface ID of the GUA).

Objectives

Troubleshoot network issues related to the configuration and operation of MP-BGP with address families.

Background/Scenario

Lombardia Cocoa Company (AS 65000) peers with two ISPs (AS 65100 and AS 65200) and receives the following prefixes from each:

- From ISP1 Hopper Cable ISP (AS 65100):

    - 10.1.1.0/24, 10.1.10.0/24, 10.1.20.0/24, 10.1.30.0/24

    - 2001:db8:cab:f002::/64, 2001:db8:cab1::/48, 2001:db8:cab2::/48, 2001:db8:cab3::/48, 2001:db8:cafe::/48 (shorter AS path)

- From ISP2 Bob's ISP (AS 65200):

    - 10.2.1.0/24, 10.2.10.0/24, 10.2.20.0/24, 10.2.30.0/24

    - 2001:db8:b0b:f002::/64, 2001:db8:b0b2::/48, 2001:db8:b0b2::/48, 2001:db8:b0b3::/48, 2001:db8:cafe::/48 (longer AS path)

Both ISPs also receive specific prefixes from Lombardia.

- Lombardia advertises to both ISPs:

    - 192.168.1.0/24, 192.168.3.0/24, 192.168.4.0/24, 192.168.6.0/24

    - 2001:db8:c0c0:a011::/64, 2001:db8:c0c0:a021::/64, 2001:db8:c0c0:a022::/64, 2001:db8:c0c0:a031::/64

Note: Inter-router links are not advertised by BGP.

Note: Lombardia iBGP peering is done using loopback 0 addresses.

Although the topology has a limited number of routers, you should use the appropriate troubleshooting commands to help find and solve the problems in the three trouble tickets as if this were a much more complex topology with many more routers and networks.

You will be loading configurations with intentional errors onto the network. Your tasks are to FIND the error(s), document your findings and the command(s) or method(s) used to fix them, FIX the issue(s) presented here, and then test the network to ensure both of the following conditions are met:

1. the complaint received in the ticket is resolved

2. full reachability is restored

Note: The routers used with CCNP hands-on labs are Cisco 4221 with Cisco IOS XE Release 16.9.4 (universalk9 image). The switches used in the labs are Cisco Catalyst 3650 with Cisco IOS XE Release 16.9.4 (universalk9 image). Other routers, switches, and Cisco IOS versions can be used. Depending on the model and Cisco IOS version, the commands available and output produced might vary from what is shown in the labs. Refer to the Router Interface Summary Table at the end of the lab for the correct interface identifiers.

Note: Make sure that the devices have been erased and have no startup configurations. If you are unsure, contact your instructor.

Required Resources

- 3 Routers (Cisco 4221 with Cisco IOS XE Release 16.9.4 universal image or comparable)

- 2 Switches (Cisco 3560 with Cisco IOS XE Release 16.9.4 universal image or comparable)

- Console cables to configure the Cisco IOS devices via the console ports

- Ethernet cables as shown in the topology

Part 1: Trouble Ticket 14.1.2.1

Scenario:

Lombardia Cocoa Company recently added router R2 between routers R1 and R3. Previously, R1 and R3 were connected directly. Routers R1, R2 and R3 were reconfigured to implement these changes. During testing and validation, the network team noticed that routers R1 and R3 only have BGP routes from their directly connected eBGP peers. All three AS 65000 routers should be receiving routes indicated in the network documentation (see the previous Background/Scenario section).

You have been asked to help find the problem and implement any necessary changes.

Use the commands listed below to load the configuration files for this trouble ticket:

Device	Command
R1	`copy flash:/enarsi/14.1.2.1-r1-config.txt run`
R2	`copy flash:/enarsi/14.1.2.1-r2-config.txt run`
R3	`copy flash:/enarsi/14.1.2.1-r3-config.txt run`
D1	`copy flash:/enarsi/14.1.2.1-d1-config.txt run`
D2	`copy flash:/enarsi/14.1.2.1-d2-config.txt run`

- All routers should receive the prefixes shown in the previous Background/Scenario section.

- Verification: The following pings should be successful. If a ping does not work, the ticket is not resolved:

```
R1# ping 10.2.1.1 source lo1
R1# ping 2001:db8:b0b:f002::1 source lo1
R3# ping 10.1.1.1 source lo1
R3# ping 2001:db8:cab:f002::1 source lo1
```

- When you have fixed the ticket, change the MOTD on EACH DEVICE using the following command:

  banner motd # This is $(hostname) FIXED from ticket <ticket number> #

- Then save the configuration by issuing the **wri** command (on each device).

- Inform your instructor that you are ready for the next ticket.

- After the instructor approves your solution for this ticket, issue the **reset.now** privileged EXEC command. This script will clear your configurations and reload the devices.

Part 2: Trouble Ticket 14.1.2.2

Scenario:

The preferred default path to the 2001:db8:cafe::/48 prefix is via AS 65100 because it is a shorter AS path. Your peering agreements with both ISPs state that you have to pay for traffic transiting via AS 65100 but not through AS 65200. To decrease costs, management has tasked the network team to send traffic for 2001:db8:cafe::/48 to AS 65200.

Policy changes were made to forward traffic to the 2001:db8:cafe::/48 prefix to AS 65200. However, verification commands indicate that only router R3 is forwarding traffic to the 2001:db8:cafe::/48 prefix to AS 65200. Routers R1 and R2 are still forwarding traffic for the 2001:db8:cafe::/48 prefix to AS 65100.

You have been tasked to find and resolve the issue.

Use the commands listed below to load the configuration files for this trouble ticket:

Device	Command
R1	`copy flash:/enarsi/14.1.2.2-r1-config.txt run`
R2	`copy flash:/enarsi/14.1.2.2-r2-config.txt run`
R3	`copy flash:/enarsi/14.1.2.2-r3-config.txt run`
D1	`copy flash:/enarsi/14.1.2.2-d1-config.txt run`
D2	`copy flash:/enarsi/14.1.2.2-d2-config.txt run`

- All AS 65000 routers should forward traffic for 2001:db8:cafe::/48 via AS 65200.

- IPv6 BGP tables in all AS 65000 routers should verify that AS 65200 is the preferred (best) path.

- When you have fixed the ticket, change the MOTD on EACH DEVICE using the following command:

  banner motd # This is $(hostname) FIXED from ticket <ticket number> #

- Then save the configuration by issuing the **wri** command (on each device).

- Inform your instructor that you are ready for the next ticket.

- After the instructor approves your solution for this ticket, issue the **reset.now** privileged EXEC command. This script will clear your configurations and reload the devices.

Router Interface Summary Table

Router Model	Ethernet Interface #1	Ethernet Interface #2	Serial Interface #1	Serial Interface #2
1800	Fast Ethernet 0/0 (F0/0)	Fast Ethernet 0/1 (F0/1)	Serial 0/0/0 (S0/0/0)	Serial 0/0/1 (S0/0/1)
1900	Gigabit Ethernet 0/0 (G0/0)	Gigabit Ethernet 0/1 (G0/1)	Serial 0/0/0 (S0/0/0)	Serial 0/0/1 (S0/0/1)
2801	Fast Ethernet 0/0 (F0/0)	Fast Ethernet 0/1 (F0/1)	Serial 0/1/0 (S0/1/0)	Serial 0/1/1 (S0/1/1)
2811	Fast Ethernet 0/0 (F0/0)	Fast Ethernet 0/1 (F0/1)	Serial 0/0/0 (S0/0/0)	Serial 0/0/1 (S0/0/1)
2900	Gigabit Ethernet 0/0 (G0/0)	Gigabit Ethernet 0/1 (G0/1)	Serial 0/0/0 (S0/0/0)	Serial 0/0/1 (S0/0/1)
4221	Gigabit Ethernet 0/0/0 (G0/0/0)	Gigabit Ethernet 0/0/1 (G0/0/1)	Serial 0/1/0 (S0/1/0)	Serial 0/1/1 (S0/1/1)
4300	Gigabit Ethernet 0/0/0 (G0/0/0)	Gigabit Ethernet 0/0/1 (G0/0/1)	Serial 0/1/0 (S0/1/0)	Serial 0/1/1 (S0/1/1)

Note: To find out how the router is configured, look at the interfaces to identify the type of router and how many interfaces the router has. There is no way to effectively list all the combinations of configurations for each router class. This table includes identifiers for the possible combinations of Ethernet and Serial interfaces in the device. The table does not include any other type of interface, even though a specific router may contain one. An example of this might be an ISDN BRI interface. The string in parenthesis is the legal abbreviation that can be used in Cisco IOS commands to represent the interface.

Device Config Final – Notes

CHAPTER 15

Route Maps and Conditional Forwarding

15.1.2 Lab - Control Routing Updates

Topology

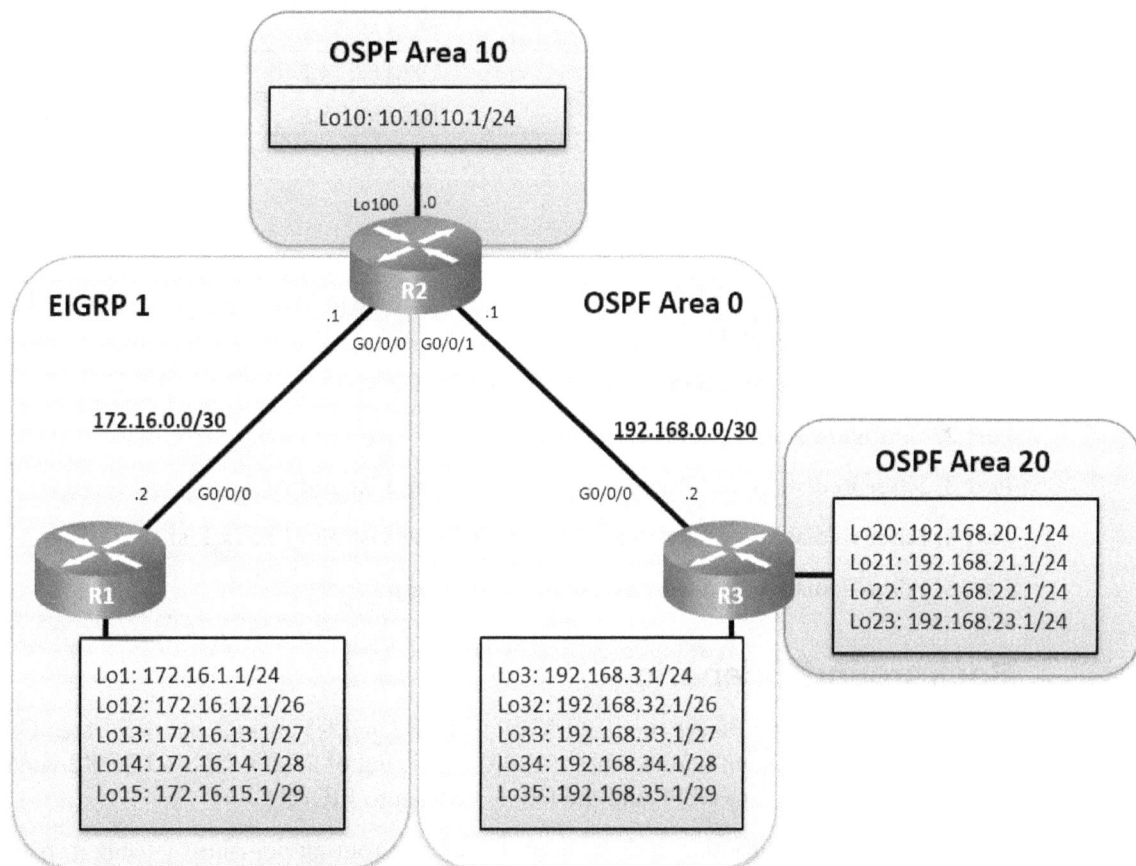

Addressing Table

Device	Interface	IPv4 Address	Subnet Mask
R1	G0/0/0	172.16.0.2	255.255.255.252
	Loopback 1	172.16.1.1	255.255.255.0
	Loopback 12	172.16.12.1	255.255.255.192
	Loopback 13	172.16.13.1	255.255.255.224
	Loopback 14	172.16.14.1	255.255.255.240
	Loopback 15	172.16.15.1	255.255.255.248
R2	G0/0/0	172.16.0.1	255.255.255.252
	G0/0/1	192.168.0.1	255.255.255.252
	Loopback 10	10.10.10.1	255.255.255.0

Device	Interface	IPv4 Address	Subnet Mask
R3	G0/0/0	192.168.0.2	255.255.255.252
	Loopback 3	192.168.3.1	255.255.255.0
	Loopback 20	192.168.20.1	255.255.255.0
	Loopback 21	192.168.21.1	255.255.255.0
	Loopback 22	192.168.22.1	255.255.255.0
	Loopback 23	192.168.23.1	255.255.255.0
	Loopback 32	192.168.32.1	255.255.255.192
	Loopback 33	192.168.33.1	255.255.255.224
	Loopback 34	192.168.34.1	255.255.255.240
	Loopback 35	192.168.35.1	255.255.255.248

Objectives

In this lab you will learn how to control routing updates in an EIGRP and OSPF redistributed routing network. Specifically, you will:

Part 1: Build the Network and Configure Basic Device Settings

Part 2: Configure Routing and Redistribution

Part 3: Filter Redistributed Routes using a Distribute List and ACL

Part 4: Filter Redistributed Routes using a Distribute List and Prefix List

Part 5: Filter Redistributed Routes using a Route Map

Background/Scenario

In this scenario, R1 and R2 are running EIGRP while R2 and R3 are running multi-area OSPF. R2 is the OSPF autonomous system border router (ASBR) consisting of areas 0, 10, and 20. R2 will also redistribute the EIGRP routes into OSPF and the OSPF routes into EIGRP.

Your task is to control routing updates between the two routing domains by using distribute lists, prefix lists, and route maps.

Note: The routers used with CCNP hands-on labs are Cisco 4221 with Cisco IOS XE Release 16.9.4 (universalk9 image). The switches used in the labs are Cisco Catalyst 3650 with Cisco IOS XE Release 16.9.4 (universalk9 image). Other routers, switches, and Cisco IOS versions can be used. Depending on the model and Cisco IOS version, the commands available and the output produced might vary from what is shown in the labs. Refer to the Router Interface Summary Table at the end of the lab for the correct interface identifiers.

Note: Make sure that the routers and switches have been erased and have no startup configurations. If you are unsure, contact your instructor.

Required Resources

- 3 Routers (Cisco 4221 with Cisco IOS XE Release 16.9.4 universal image or comparable)
- Console cables to configure the Cisco IOS devices via the console ports

- 1 PC (Choice of operating system with terminal emulation program installed)
- Ethernet cables as shown in the topology

Instructions

Part 1: Build the Network and Configure Basic Device Settings

In Part 1, you will set up the network topology and configure basic settings and interface addressing on routers.

Step 1.   Cable the network as shown in the topology.

Attach the devices as shown in the topology diagram, and cable as necessary.

Step 2.   Configure basic settings for each device.

    a.   Console into each device, enter global configuration mode, and apply the basic settings. The startup configurations for each device are provided below.

Router R1

```
hostname R1
no ip domain lookup
line con 0
 logging sync
 exec-time 0 0
 exit
banner motd # This is R1, Controlling Routing Updates #
interface g0/0/0
 description Connection to R2
 ip add 172.16.0.2 255.255.255.252
 no shut
 exit
interface Lo1
 ip address 172.16.1.1 255.255.255.0
exit
interface Lo12
 ip address 172.16.12.1 255.255.255.192
exit
interface Lo13
 ip address 172.16.13.1 255.255.255.224
exit
interface Lo14
 ip address 172.16.14.1 255.255.255.240
exit
interface Lo15
 ip address 172.16.15.1 255.255.255.248
end
```

Router R2

```
hostname R2
no ip domain lookup
line con 0
```

```
 logging sync
 exec-time 0 0
exit
banner motd # This is R2, Controlling Routing Updates #
interface g0/0/0
 description Connection to R1
 ip add 172.16.0.1 255.255.255.252
 no shut
exit
interface GigabitEthernet0/0/1
 description Connection to R3
 ip address 192.168.0.1 255.255.255.252
 no shut
exit
int lo10
 ip add 10.10.10.1 255.255.255.0
 ip ospf network point-to-point
end
```

Router R3

```
hostname R3
no ip domain lookup
line con 0
 logging sync
 exec-time 0 0
 exit
banner motd # This is R3, Controlling Routing Updates #
interface g0/0/0
 description Connection to R2
 ip add 192.168.0.2 255.255.255.252
 no shut
 exit
interface Lo3
 ip add 192.168.3.1 255.255.255.0
 ip ospf network point-to-point
exit
interface Lo32
 ip add 192.168.32.1 255.255.255.0
 ip ospf network point-to-point
exit
interface Lo33
 ip add 192.168.33.1 255.255.255.0
 ip ospf network point-to-point
exit
interface Lo34
 ip add 192.168.34.1 255.255.255.0
 ip ospf network point-to-point
exit
interface Lo35
 ip add 192.168.35.1 255.255.255.0
 ip ospf network point-to-point
```

```
  exit
  interface Lo20
   ip add 192.168.20.1 255.255.255.0
   ip ospf network point-to-point
  exit
  interface Lo21
   ip add 192.168.21.1 255.255.255.0
   ip ospf network point-to-point
  exit
  interface Lo22
   ip add 192.168.22.1 255.255.255.0
   ip ospf network point-to-point
  exit
  interface Lo23
   ip add 192.168.23.1 255.255.255.0
   ip ospf network point-to-point
  end
```

b. Save the running configuration to startup-config.

Part 2: Configure Routing and Redistribution

In Part 2, you will implement EIGRP and OSPF routing and redistribute the routes into each routing protocol. Although redistribution is not covered in this chapter, it is covered here for example purposes.

Step 1. Configure routing.

In this step, you will configure EIGRP on R1 and R2, and OSPF on R2 and R3.

a. On R1, advertise the connected networks using EIGRP in autonomous system 1. Assign R1 the router ID of 1.1.1.1.

```
R1(config)# router eigrp 1
R1(config-router)# eigrp router-id 1.1.1.1
R1(config-router)# network 172.16.0.0 0.0.0.3
R1(config-router)# network 172.16.1.0 0.0.0.255
R1(config-router)# network 172.16.12.0 0.0.0.63
R1(config-router)# network 172.16.13.0 0.0.0.31
R1(config-router)# network 172.16.14.0 0.0.0.15
R1(config-router)# network 172.16.15.0 0.0.0.7
R1(config-router)# end
```

b. On R3, advertise the connected networks using OSPF process ID 123 for area 0 and area 20.

```
R3(config)# router ospf 123
R3(config-router)# router-id 3.3.3.3
R3(config-router)# network 192.168.0.0 0.0.0.3 area 0
R3(config-router)# network 192.168.3.0 0.0.0.255 area 0
R3(config-router)# network 192.168.32.0 0.0.0.63 area 0
R3(config-router)# network 192.168.33.0 0.0.0.31 area 0
R3(config-router)# network 192.168.34.0 0.0.0.15 area 0
R3(config-router)# network 192.168.35.0 0.0.0.7 area 0
R3(config-router)# network 192.168.20.0 0.0.3.255 area 20
R3(config-router)# end
```

Note: The **network 192.168.20.0 0.0.3.255 area 20** command enables OSPF for loopbacks 20, 21, 22, and 23.

    c. On R2, configure EIGRP and redistribute the OSPF networks into EIGRP AS 1. Then configure OSPF and redistribute and summarize the EIGRP networks into OSPF.

```
R2(config)# router eigrp 1
R2(config-router)# eigrp router-id 2.2.2.2
R2(config-router)# network 172.16.0.0 0.0.0.3
R2(config-router)# exit
*Mar 18 13:08:33.786: %DUAL-5-NBRCHANGE: EIGRP-IPv4 1: Neighbor 172.16.0.2
(GigabitEthernet0/0/0) is up: new adjacency
R2(config)#
R2(config)# router ospf 123
R2(config-router)# router-id 2.2.2.2
R2(config-router)# network 192.168.0.0 0.0.0.3 area 0
R2(config-router)# network 10.10.10.0 0.0.0.255 area 10
R2(config-router)# end
*Mar 18 13:08:36.024: %OSPF-5-ADJCHG: Process 123, Nbr 3.3.3.3 on
GigabitEthernet0/0/1 from LOADING to FULL, Loading Done
```

Step 2.   Verify EIGRP and OSPF routing.

    a. Verify the EIGRP routing table entries on R1. No routes are displayed because R1 is directly connected to all of the EIGRP routes.

```
R1# show ip route eigrp | begin Gateway
Gateway of last resort is not set
```

    b. Verify the OSPF routing table entries on R3. R3 has an inter-area route entry for the OSPF Area 10 network.

```
R3# show ip route ospf | begin Gateway
Gateway of last resort is not set


      10.0.0.0/24 is subnetted, 1 subnets
O IA     10.10.10.0 [110/2] via 192.168.0.1, 00:04:29, GigabitEthernet0/0/0
```

    c. Verify the EIGRP and OSPF routing table entries on R2. R2 has entries for all of the EIGRP networks and the OSPF networks including the Area 20 networks.

```
R2# show ip route eigrp | begin Gateway
Gateway of last resort is not set


      172.16.0.0/16 is variably subnetted, 7 subnets, 7 masks
D        172.16.1.0/24
            [90/130816] via 172.16.0.2, 00:05:52, GigabitEthernet0/0/0
D        172.16.12.0/26
            [90/130816] via 172.16.0.2, 00:05:52, GigabitEthernet0/0/0
D        172.16.13.0/27
            [90/130816] via 172.16.0.2, 00:05:52, GigabitEthernet0/0/0
D        172.16.14.0/28
            [90/130816] via 172.16.0.2, 00:05:52, GigabitEthernet0/0/0
D        172.16.15.0/29
            [90/130816] via 172.16.0.2, 00:05:52, GigabitEthernet0/0/0
```

```
R2# show ip route ospf | begin Gateway
Gateway of last resort is not set


O     192.168.3.0/24 [110/2] via 192.168.0.2, 00:12:22, GigabitEthernet0/0/1
O IA  192.168.20.0/24 [110/2] via 192.168.0.2, 00:12:22, GigabitEthernet0/0/1
O IA  192.168.21.0/24 [110/2] via 192.168.0.2, 00:12:22, GigabitEthernet0/0/1
O IA  192.168.22.0/24 [110/2] via 192.168.0.2, 00:12:22, GigabitEthernet0/0/1
O IA  192.168.23.0/24 [110/2] via 192.168.0.2, 00:12:22, GigabitEthernet0/0/1
O     192.168.32.0/24 [110/2] via 192.168.0.2, 00:12:22, GigabitEthernet0/0/1
O     192.168.33.0/24 [110/2] via 192.168.0.2, 00:00:30, GigabitEthernet0/0/1
O     192.168.34.0/24 [110/2] via 192.168.0.2, 00:12:22, GigabitEthernet0/0/1
O     192.168.35.0/24 [110/2] via 192.168.0.2, 00:12:22, GigabitEthernet0/0/1
```

   d. Verify connectivity to an EIGRP and OSPF network using the **ping** command as shown. R2 has connectivity to the EIGRP and OSPF networks.

```
R2# ping 172.16.13.1
Type escape sequence to abort.
Sending 5, 100-byte ICMP Echos to 172.16.13.1, timeout is 2 seconds:
!!!!!
Success rate is 100 percent (5/5), round-trip min/avg/max = 1/1/1 ms
R2# ping 192.168.20.1
Type escape sequence to abort.
Sending 5, 100-byte ICMP Echos to 192.168.20.1, timeout is 2 seconds:
!!!!!
Success rate is 100 percent (5/5), round-trip min/avg/max = 1/1/1 ms
```

Step 3. Configure redistribution on R2.

In this step, you will configure redistribution between the EIGRP and OSPF routing domains. Redistribution is configured on the router that connects to both domains. In our topology that is R2.

It is important to note that route redistribution is always performed outbound. This means that the routing table of the router doing the redistribution does not change. Only the routing tables of the routers receiving the redistributed routes will be updated.

   a. On R2, redistribute the OSPF routes in EIGRP. Routes redistributed into EIGRP require that a metric be assigned.

```
R2(config)# router eigrp 1
R2(config-router)# redistribute ospf 123 metric 10000 100 255 1 1500
R2(config-router)# exit
```

   b. On R2, redistribute the EIGRP routes in OSPF. Routes redistributed into OSPF are automatically assigned a metric of 20. In our example, you are assigning a higher cost metric of 100 to redistributed routes.

```
R2(config)# router ospf 123
R2(config-router)# redistribute eigrp 1 subnets metric 100
R2(config-router)# end
```

Step 4. Verify redistribution.

   a. On R2, verify the EIGRP and OSPF routing table. Notice how the R2 routing table has not changed.

```
R2# show ip route eigrp | begin Gateway
Gateway of last resort is not set
```

```
        172.16.0.0/16 is variably subnetted, 7 subnets, 7 masks
D        172.16.1.0/24
           [90/130816] via 172.16.0.2, 00:14:22, GigabitEthernet0/0/0
D        172.16.12.0/26
           [90/130816] via 172.16.0.2, 00:14:22, GigabitEthernet0/0/0
D        172.16.13.0/27
           [90/130816] via 172.16.0.2, 00:14:22, GigabitEthernet0/0/0
D        172.16.14.0/28
           [90/130816] via 172.16.0.2, 00:14:22, GigabitEthernet0/0/0
D        172.16.15.0/29
           [90/130816] via 172.16.0.2, 00:14:22, GigabitEthernet0/0/0
```

R2# **show ip route ospf | begin Gateway**
Gateway of last resort is not set

```
O     192.168.3.0/24 [110/2] via 192.168.0.2, 00:14:32, GigabitEthernet0/0/1
O IA  192.168.20.0/24 [110/2] via 192.168.0.2, 00:14:32, GigabitEthernet0/0/1
O IA  192.168.21.0/24 [110/2] via 192.168.0.2, 00:14:32, GigabitEthernet0/0/1
O IA  192.168.22.0/24 [110/2] via 192.168.0.2, 00:14:32, GigabitEthernet0/0/1
O IA  192.168.23.0/24 [110/2] via 192.168.0.2, 00:14:32, GigabitEthernet0/0/1
O     192.168.32.0/24 [110/2] via 192.168.0.2, 00:14:32, GigabitEthernet0/0/1
O     192.168.33.0/24 [110/2] via 192.168.0.2, 00:14:32, GigabitEthernet0/0/1
O     192.168.34.0/24 [110/2] via 192.168.0.2, 00:14:32, GigabitEthernet0/0/1
O     192.168.35.0/24 [110/2] via 192.168.0.2, 00:14:32, GigabitEthernet0/0/1
```

b. On R1, verify the EIGRP routing table. Originally the R1 routing table displayed no entries as R1 was directly connected to all EIGRP networks. However, R1 now knows about the external routes redistributed from the R2 OSPF routing domain. The highlighted entries identify all of the OSPF routes.

R1# **show ip route eigrp | begin Gateway**
Gateway of last resort is not set

```
        10.0.0.0/24 is subnetted, 1 subnets
D EX     10.10.10.0
           [170/281856] via 172.16.0.1, 00:02:02, GigabitEthernet0/0/0
        192.168.0.0/30 is subnetted, 1 subnets
D EX     192.168.0.0
           [170/281856] via 172.16.0.1, 00:02:02, GigabitEthernet0/0/0
D EX  192.168.3.0/24
           [170/281856] via 172.16.0.1, 00:02:02, GigabitEthernet0/0/0
D EX  192.168.20.0/24
           [170/281856] via 172.16.0.1, 00:02:02, GigabitEthernet0/0/0
D EX  192.168.21.0/24
           [170/281856] via 172.16.0.1, 00:02:02, GigabitEthernet0/0/0
D EX  192.168.22.0/24
           [170/281856] via 172.16.0.1, 00:02:02, GigabitEthernet0/0/0
D EX  192.168.23.0/24
           [170/281856] via 172.16.0.1, 00:02:02, GigabitEthernet0/0/0
D EX  192.168.32.0/24
           [170/281856] via 172.16.0.1, 00:02:02, GigabitEthernet0/0/0
```

```
D EX  192.168.33.0/24
              [170/281856] via 172.16.0.1, 00:02:02, GigabitEthernet0/0/0
D EX  192.168.34.0/24
              [170/281856] via 172.16.0.1, 00:02:02, GigabitEthernet0/0/0
D EX  192.168.35.0/24
              [170/281856] via 172.16.0.1, 00:02:02, GigabitEthernet0/0/0
```

c. Verify the EIGRP routing table on R3. Previously, R3 only had the Area 10 network in
    its routing table. R3 now knows about the external EIGRP routes redistributed by R2.
    Also notice that the redistribution command assigned a metric of 100.

```
R3# show ip route ospf | begin Gateway
Gateway of last resort is not set


      10.0.0.0/24 is subnetted, 1 subnets
O IA    10.10.10.0 [110/2] via 192.168.0.1, 00:16:36, GigabitEthernet0/0/0
      172.16.0.0/16 is variably subnetted, 6 subnets, 6 masks
O E2    172.16.0.0/30
              [110/100] via 192.168.0.1, 00:02:42, GigabitEthernet0/0/0
O E2    172.16.1.0/24
              [110/100] via 192.168.0.1, 00:02:42, GigabitEthernet0/0/0
O E2    172.16.12.0/26
              [110/100] via 192.168.0.1, 00:02:42, GigabitEthernet0/0/0
O E2    172.16.13.0/27
              [110/100] via 192.168.0.1, 00:02:42, GigabitEthernet0/0/0
O E2    172.16.14.0/28
              [110/100] via 192.168.0.1, 00:02:42, GigabitEthernet0/0/0
O E2    172.16.15.0/29
              [110/100] via 192.168.0.1, 00:02:42, GigabitEthernet0/0/0
```

d. From all routers, verify connectivity to all configured destinations using the following
    TCL script. All pings should be successful. Troubleshoot if necessary.

```
tclsh

foreach address {
192.168.0.1
192.168.20.1
192.168.21.1
192.168.22.1
192.168.23.1
192.168.3.1
192.168.32.1
192.168.33.1
192.168.34.1
192.168.35.1
10.10.10.1
172.16.0.1
172.16.0.2
172.16.1.1
172.16.12.1
172.16.13.1
```

```
                 172.16.14.1
                 172.16.15.1
                 } { ping $address }
```

Part 3: Filter Redistributed Routes using a Distribute List and ACL

Routes can be filtered using a variety of techniques including:

- **Distribute list and ACL** - A distribute list allows access control lists (ACLs) to be applied to routing updates.

- **Distribute list and prefix list** - A distribute list with a prefix list is an alternative to ACLs designed to filter routes. Prefix lists are not exclusively used with distribute lists but can also be used with route maps and other commands.

- **Route maps** - Route maps are complex access lists that allow conditions to be tested against a packet or route, and then actions taken to modify attributes of the packet or route.

In this part, you will filter routes using the distribute list and ACL technique.

Step 1.   Configure an ACL and distribute list on R2.

In this step, you will use a distribute list and ACL to filter routes being advertised from R2 to R1. Specifically, you will filter the OSPF 20 routes (i.e., 192.168.20.0/22) from being advertised by R2 to R1.

a.   On R1, verify the routing table entry for the 192.168.20.0/22 route. R1 displays the entry for the 192.169.20.0 network.

```
R1# show ip route 192.168.20.0
Routing entry for 192.168.20.0/24
   Known via "eigrp 1", distance 170, metric 28416, type external
   Redistributing via eigrp 1
   Last update from 172.16.0.1 on GigabitEthernet0/0/0, 01:54:32 ago
   Routing Descriptor Blocks:
   * 172.16.0.1, from 172.16.0.1, 01:54:32 ago, via GigabitEthernet0/0/0
       Route metric is 28416, traffic share count is 1
       Total delay is 1010 microseconds, minimum bandwidth is 10000 Kbit
       Reliability 255/255, minimum MTU 1500 bytes
       Loading 1/255, Hops 1
```

b.   You will filter the Area 20 networks from being advertised into the EIGRP domain. Although a distribute list could be implemented on the receiving router (i.e., R1), it is usually best to filter routes from the redistributing router. Therefore, on R2, create a standard named ACL called **OSPF20-FILTER** that denies the 192.168.20.0/22 route. The ACL must also permit all other routes otherwise, no OSPF routes would be redistributed into EIGRP.

```
R2(config)# ip access-list standard OSPF20-FILTER
R2(config-std-nacl)# remark Used with DList to filter OSPF 20 routes
R2(config-std-nacl)# deny 192.168.20.0 0.0.3.255
R2(config-std-nacl)# permit any
R2(config-std-nacl)# exit
```

    c. Next configure a distribute list under the EIGRP process to filter routes propagated to R1 using the pre-configured ACL.

```
R2(config)# router eigrp 1
R2(config-router)# distribute-list OSPF20-FILTER out ospf 123
R2(config-router)# end
```

Step 2. Verify the configuration

    a. On R1, verify if the 192.168.20.0 route is now missing from the R1 routing table. The output confirms that the 192.168.20.0/24, 192.168.21.0/24, 192.168.22.0/24, 192.168.23.0/24 (192.168.20.0/22) routes are no longer in the routing table of R1.

```
R1# show ip route 192.168.20.0
% Network not in table

R1# show ip route eigrp | begin Gateway
Gateway of last resort is not set

      10.0.0.0/24 is subnetted, 1 subnets
D EX    10.10.10.0
           [170/281856] via 172.16.0.1, 01:56:57, GigabitEthernet0/0/0
        192.168.0.0/30 is subnetted, 1 subnets
D EX    192.168.0.0
           [170/281856] via 172.16.0.1, 01:56:57, GigabitEthernet0/0/0
D EX  192.168.3.0/24
           [170/281856] via 172.16.0.1, 01:56:57, GigabitEthernet0/0/0
D EX  192.168.32.0/24
           [170/281856] via 172.16.0.1, 01:56:57, GigabitEthernet0/0/0
D EX  192.168.33.0/24
           [170/281856] via 172.16.0.1, 01:56:57, GigabitEthernet0/0/0
D EX  192.168.34.0/24
           [170/281856] via 172.16.0.1, 01:56:57, GigabitEthernet0/0/0
D EX  192.168.35.0/24
           [170/281856] via 172.16.0.1, 01:56:57, GigabitEthernet0/0/0
```

Note: If additional router filtering was required, only the OSPF20-FILTER ACL on R2 would need to be altered.

Part 4: Filter Redistributed Routes using a Distribute List and Prefix List

In this part, you will filter routes using the distribute list and prefix list technique.

Using a distribute list with an ACL or with a prefix list basically achieves the same result of filtering routes. However, in large enterprise networks, route filtering can be quite complex. The ACLs can be very extensive and therefore taxing on router resources. For this reason, prefix lists should be used instead of ACLs since they are more efficient and less taxing on router resources than ACLs.

Note: Prefix lists are not exclusively used with distribute lists but can also be used with route maps and other commands.

Step 1.   Filter redistributed routes using a distribute list and prefix list.

In this step, a prefix list will be configured with a distribute list to filter R1 routes being advertised from R2 to R3.

a.   On R3, verify the routing table entry for the routes learned externally identified with the O E2 source entry. The output displays route entries for the EIGRP networks connected to R1.

```
R3# show ip route ospf | include O E2
O E2     172.16.0.0/30
O E2     172.16.1.0/24
O E2     172.16.12.0/26
O E2     172.16.13.0/27
O E2     172.16.14.0/28
O E2     172.16.15.0/29
```

b.   Configure R2 with a prefix list identifying which networks to advertise to R3. Specifically, only the networks with the first two octets being 172.16 (i.e., 172.16.0.0/16) with a subnet mask of /24 or less will be advertised.

```
R2(config)# ip prefix-list EIGRP-FILTER permit 172.16.0.0/16 le 24
```

c.   Configure a distribute list under the OSPF process to filter routes propagated to R3 using the pre-configured prefix list.

```
R2(config)# router ospf 123
R2(config-router)# distribute-list prefix EIGRP-FILTER out eigrp 1
R2(config-router)# exit
```

Note: If additional router filtering was required, only the EIGRP-FILTER prefix list on R2 would need to be altered.

---

d.   On R3, verify which EIGRP redistributed routes have been learned from R2. Notice how only the 172.16.1.0/24 route is listed because all other routes have subnet masks greater than /24.

```
R3# show ip route ospf | include O E2
O E2     172.16.1.0 [110/100] via 192.168.0.1, 02:07:51, GigabitEthernet0/0/0
```

e.   To observe how a prefix list can be used to filter routes, remove the previously configured prefix list and change the prefix list on R2 to advertise only EIGRP networks with subnet masks of /26 or greater.

```
R2(config)# no ip prefix-list EIGRP-FILTER permit 172.16.0.0/16 le 24
R2(config)# ip prefix-list EIGRP-FILTER permit 172.16.0.0/16 ge 26
```

f.   Verify the change on R3 as shown. Now only the 172.16.1.0/24 route is not listed as all other routes have subnet masks greater than or equal to /26.

```
R3# show ip route ospf | include O E2
O E2     172.16.0.0/30
O E2     172.16.12.0/26
O E2     172.16.13.0/27
O E2     172.16.14.0/28
O E2     172.16.15.0/29
```

g.   Now change the prefix list on R2 to advertise only networks with subnet masks of /28 or less.

```
R2(config)# no ip prefix-list EIGRP-FILTER permit 172.16.0.0/16 ge 26
R2(config)# ip prefix-list EIGRP-FILTER permit 172.16.0.0/16 le 28
```

h.   Verify the output on R3 as shown. Notice how the 172.16.0.0/30 and 172.16.15.0/29 routes are no longer advertised as their subnet masks are greater than /28.

```
R3# show ip route ospf | include O E2
O E2     172.16.1.0/24
O E2     172.16.12.0/26
O E2     172.16.13.0/27
O E2     172.16.14.0/28
```

Part 5: Filter Redistributed Routes using a Route Map

Route maps can also be used to filter routes. A route map works like an access list because it has multiple **deny** and **permit** statements that are read in a sequential order. However, route maps can match and set specific attributes and therefore provide additional options and more flexibility when redistributing routes.

Route maps can be used for:

- **Redistribution** - Route maps provide more options and flexibility to the **redistribute** command.

- **Policy-based routing (PBR)** - PBR allows an administrator to define routing policy other than basic destination-based routing using the routing table. The route map is applied to an interface using the **ip policy route-map** interface configuration command.

- **BGP** - Route maps are the primary tools for implementing BGP policy and allows an administrator to do path control and provide sophisticated manipulation of BGP path attributes. The route map is applied using the BGP **neighbor** router configuration command.

Step 1.   Filter redistributed routes using a route map.

In this step, you will configure R2 to filter the R3 loopback 34 and 35 networks so that they are not redistributed into EIGRP.

a.   On R1, display the current routing table.

```
R1# show ip route eigrp | begin Gateway
Gateway of last resort is not set

      10.0.0.0/24 is subnetted, 1 subnets
D EX    10.10.10.0
           [170/281856] via 172.16.0.1, 02:51:47, GigabitEthernet0/0/0
      192.168.0.0/30 is subnetted, 1 subnets
D EX    192.168.0.0
           [170/281856] via 172.16.0.1, 02:51:47, GigabitEthernet0/0/0
D EX  192.168.3.0/24
           [170/281856] via 172.16.0.1, 02:51:47, GigabitEthernet0/0/0
D EX  192.168.32.0/24
           [170/281856] via 172.16.0.1, 02:51:47, GigabitEthernet0/0/0
D EX  192.168.33.0/24
           [170/281856] via 172.16.0.1, 02:51:47, GigabitEthernet0/0/0
D EX  192.168.34.0/24
           [170/281856] via 172.16.0.1, 02:51:47, GigabitEthernet0/0/0
D EX  192.168.35.0/24
           [170/281856] via 172.16.0.1, 02:51:47, GigabitEthernet0/0/0
```

b. Route maps can be used to filter redistributed traffic in multiple ways. In this step, you will filter and deny the R3 Lo 34 and Lo 35 networks (i.e., 192.168.34.0/28 and 192.168.35.0/29) from being redistributed into the EIGRP routing domain. All other networks connected to R1 will be redistributed.

On R2, create a standard named ACL called **R3-ACL** that identifies the R3 Lo 34 and Lo 35 networks (i.e., 192.168.34.0/28 and 192.168.35.0/29) as shown.

```
R2(config)# ip access-list standard R3-ACL
R2(config-std-nacl)# remark ACL used with the R3-FILTER route map
R2(config-std-nacl)# permit 192.168.34.0 0.0.0.15
R2(config-std-nacl)# permit 192.168.35.0 0.0.0.7
R2(config-std-nacl)# exit
```

c. Configure a route map with a statement that denies traffic based on a match with the named ACL. Then add a permit statement without a match statement to provide an explicit "*permit all*".

```
R2(config)# route-map R3-FILTER deny 10
R2(config-route-map)# description RM filters R3 OSPF routes
R2(config-route-map)# match ip address R3-ACL
R2(config-route-map)# exit
R2(config)# route-map R3-FILTER permit 20
R2(config-route-map)# description RM permits all other R3 OSPF routes
R2(config-route-map)# exit
```

d. Apply this route map to EIGRP by reentering the **redistribute** command using the **route-map** keyword.

```
R2(config)# router eigrp 1
R2(config-router)# redistribute ospf 123 route-map R3-FILTER metric 1000000 100 255 1 1500
```

e. Verify that the two R3 networks are filtered out in the R1 routing table. Notice that the 192.168.34.0/28 and 192.168.35.0/29 networks are no longer in the R1 routing table.

```
R1# show ip route eigrp | begin Gateway
Gateway of last resort is not set


      10.0.0.0/24 is subnetted, 1 subnets
D EX     10.10.10.0 [170/28416] via 172.16.0.1, 00:00:14, GigabitEthernet0/0/0
      192.168.0.0/30 is subnetted, 1 subnets
D EX     192.168.0.0
            [170/28416] via 172.16.0.1, 00:00:14, GigabitEthernet0/0/0
D EX  192.168.3.0/24
            [170/28416] via 172.16.0.1, 00:00:14, GigabitEthernet0/0/0
D EX  192.168.32.0/24
            [170/28416] via 172.16.0.1, 00:00:14, GigabitEthernet0/0/0
D EX  192.168.33.0/24
            [170/28416] via 172.16.0.1, 00:00:14, GigabitEthernet0/0/0
```

Step 2. Filter redistributed routes and set attributes using a route map.

The preceding step was a simple example of using a route map to filter redistributed routes. However, route maps can be used to do much more.

In this step, you will filter the 172.16.13.0/27 network from R1 and change its OSPF metric to 25 instead of 100 like all other redistributed routes. You will also change its metric type from an external type 2 route (i.e., O E2), to an external type 1 route (i.e., O E1).

a. On R3, verify the routing table entry for the routes learned externally identified with the 0 E2 source entry. The 172.16.13.0/27 route will be configured with additional attributes.

```
R3# show ip route ospf | begin Gateway
Gateway of last resort is not set


      10.0.0.0/24 is subnetted, 1 subnets
O IA    10.10.10.0 [110/2] via 192.168.0.1, 03:42:32, GigabitEthernet0/0/0
      172.16.0.0/16 is variably subnetted, 4 subnets, 4 masks
O E2    172.16.1.0/24
           [110/100] via 192.168.0.1, 01:17:56, GigabitEthernet0/0/0
O E2    172.16.12.0/26
           [110/100] via 192.168.0.1, 01:20:12, GigabitEthernet0/0/0
O E2    172.16.13.0/27
           [110/100] via 192.168.0.1, 01:20:12, GigabitEthernet0/0/0
O E2    172.16.14.0/28
           [110/100] via 192.168.0.1, 01:20:12, GigabitEthernet0/0/0
```

b. Although an ACL could be used, this example will use a prefix list. Configure a prefix list identifying the route to be filtered.

```
R2(config)# ip prefix-list R1-PL permit 172.16.13.0/27
```

c. Configure a route map matching the identified route in the prefix list and assign the OSPF metric cost of 25 and change the metric type to External Type 1. Then add a **permit** statement without a **match** statement acting as an explicit *"permit all"*.

```
R2(config)# route-map R1-FILTER permit 10
R2(config-route-map)# description RM filters 172.16.13.0/27
R2(config-route-map)# match ip address prefix-list R1-PL
R2(config-route-map)# set metric 25
R2(config-route-map)# set metric-type type-1
R2(config-route-map)# exit
R2(config)# route-map R1-FILTER permit 20
R2(config-route-map)# description RM permits all other R1 OSPF routes
R2(config-route-map)# exit
```

d. Apply this route map to OSPF by reentering the **redistribute** command using the **route-map** keyword.

```
R2(config)# router ospf 123
R2(config-router)# redistribute eigrp 1 metric 100 subnets route-map R1-FILTER
R2(config-router)# exit
```

e. Verify that the two R3 networks are filtered out in the R1 routing table. Notice that only the 172.16.13.0/27 route is an OSPF External Type 1 route (i.e., O E1) with a cost metric of 26 (i.e., the assigned metric cost of 25 plus the cost of 1 for the R2 to R3 link).

```
R3# show ip route ospf | begin Gateway
Gateway of last resort is not set


      10.0.0.0/24 is subnetted, 1 subnets
O IA    10.10.10.0 [110/2] via 192.168.0.1, 03:48:26, GigabitEthernet0/0/0
      172.16.0.0/16 is variably subnetted, 4 subnets, 4 masks
```

```
O E2    172.16.1.0/24
           [110/100] via 192.168.0.1, 01:23:50, GigabitEthernet0/0/0
O E2    172.16.12.0/26
           [110/100] via 192.168.0.1, 01:26:06, GigabitEthernet0/0/0
O E1    172.16.13.0/27
           [110/26] via 192.168.0.1, 00:00:04, GigabitEthernet0/0/0
O E2    172.16.14.0/28
           [110/100] via 192.168.0.1, 01:26:06, GigabitEthernet0/0/0
```

Router Interface Summary Table

Router Model	Ethernet Interface #1	Ethernet Interface #2	Serial Interface #1	Serial Interface #2
1800	Fast Ethernet 0/0 (F0/0)	Fast Ethernet 0/1 (F0/1)	Serial 0/0/0 (S0/0/0)	Serial 0/0/1 (S0/0/1)
1900	Gigabit Ethernet 0/0 (G0/0)	Gigabit Ethernet 0/1 (G0/1)	Serial 0/0/0 (S0/0/0)	Serial 0/0/1 (S0/0/1)
2801	Fast Ethernet 0/0 (F0/0)	Fast Ethernet 0/1 (F0/1)	Serial 0/1/0 (S0/1/0)	Serial 0/1/1 (S0/1/1)
2811	Fast Ethernet 0/0 (F0/0)	Fast Ethernet 0/1 (F0/1)	Serial 0/0/0 (S0/0/0)	Serial 0/0/1 (S0/0/1)
2900	Gigabit Ethernet 0/0 (G0/0)	Gigabit Ethernet 0/1 (G0/1)	Serial 0/0/0 (S0/0/0)	Serial 0/0/1 (S0/0/1)
4221	Gigabit Ethernet 0/0/0 (G0/0/0)	Gigabit Ethernet 0/0/1 (G0/0/1)	Serial 0/1/0 (S0/1/0)	Serial 0/1/1 (S0/1/1)
4300	Gigabit Ethernet 0/0/0 (G0/0/0)	Gigabit Ethernet 0/0/1 (G0/0/1)	Serial 0/1/0 (S0/1/0)	Serial 0/1/1 (S0/1/1)

Note: To find out how the router is configured, look at the interfaces to identify the type of router and how many interfaces the router has. There is no way to effectively list all the combinations of configurations for each router class. This table includes identifiers for the possible combinations of Ethernet and Serial interfaces in the device. The table does not include any other type of interface, even though a specific router may contain one. An example of this might be an ISDN BRI interface. The string in parenthesis is the legal abbreviation that can be used in Cisco IOS commands to represent the interface.

Device Config Final – Notes

_____

_____

_____

_____

_____

_____

_____

_____

_____

_____

15.1.3 Lab - Path Control Using PBR

Topology

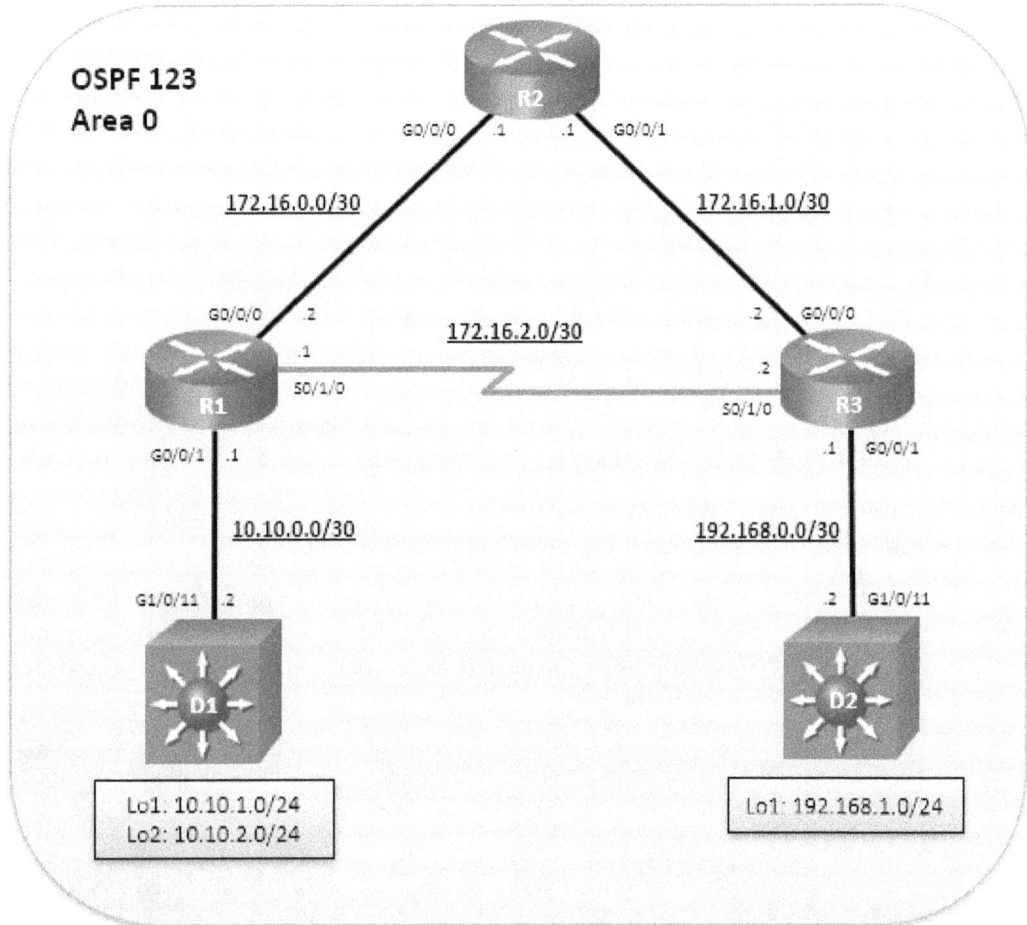

Addressing Table

Device	Interface	IPv4 Address	Subnet Mask
D1	G0/0/11	10.10.0.2	255.255.255.252
	Loopback 1	10.10.1.1	255.255.255.0
	Loopback 2	10.10.2.1	255.255.255.0
R1	G0/0/0	172.16.0.2	255.255.255.252
	G0/0/1	10.10.0.1	255.255.255.252
	S0/1/0	172.16.2.1	255.255.255.252
R2	G0/0/0	172.16.0.1	255.255.255.252
	G0/0/1	172.16.1.1	255.255.255.252
R3	G0/0/0	172.16.1.2	255.255.255.252
	G0/0/1	192.168.0.1	255.255.255.252

Device	Interface	IPv4 Address	Subnet Mask
	S0/1/0	172.16.2.2	255.255.255.252
D2	G0/0/11	192.168.0.2	255.255.255.252
	Loopback 1	192.168.1.1	255.255.255.0

Objectives

In this lab, you will learn how to use policy-based routing to influence path selection.

Part 1: Build the Network and Configure Basic Device Settings

Part 2: Configure and Verify Routing

Part 3: Configure PBR to Provide Path Control

Part 4: Configure Local PBR to Provide Path Control

Background/Scenario

In this scenario, you want to experiment with policy-based routing (PBR) to see how it is implemented and study how it could be used to influence path selection.

Your task is to connect and configure a 3 router and 2 Layer 3 switch OSPF routing domain and verify normal path selection. You will alter the traffic flow for PC2 going to PC3 using PBR. You will also use Cisco IOS IP SLA with PBR to achieve dynamic path control. Finally, you will configure PBR for traffic originating locally on router R1.

Note: This lab is an exercise in configuring distribute lists and redistribution using ALCs, prefix lists, and route maps.

Note: The routers used with CCNP hands-on labs are Cisco 4221 with Cisco IOS XE Release 16.9.4 (universalk9 image). The switches used in the labs are Cisco Catalyst 3650 with Cisco IOS XE Release 16.9.4 (universalk9 image). Other routers, switches, and Cisco IOS versions can be used. Depending on the model and Cisco IOS version, the commands available and output produced might vary from what is shown in the labs. Refer to the Router Interface Summary Table at the end of the lab for the correct interface identifiers.

Note: Make sure that the devices have been erased and have no startup configurations. If you are unsure, contact your instructor.

Required Resources

- 3 Routers (Cisco 4221 with Cisco IOS XE Release 16.9.4 universal image or comparable)
- 2 Switches (Catalyst 3650s with Cisco IOS XE Release 16.9.4 universalk9 image).
- 1 PC (Choice of operating system with terminal emulation program installed)

Instructions

Part 1: Build the Network and Configure Basic Device Settings

In Part 1, you will set up the network topology and configure basic settings and interface addressing on routers.

Step 1.   Cable the network as shown in the topology.

Attach the devices as shown in the topology diagram, and cable as necessary.

Step 2.   Configure basic settings for each device.

a.   Console into each device, enter global configuration mode, and apply the basic settings. The startup configurations for each device are provided below.

Router R1

```
hostname R1
no ip domain lookup
line con 0
 logging sync
 exec-time 0 0
 exit
banner motd # This is R1, Path Control Using PBR #
interface G0/0/0
 description Connection to R2
 ip add 172.16.0.2 255.255.255.252
 no shut
 exit
interface S0/1/0
 description Serial Connection to R3
 ip add 172.16.2.1 255.255.255.252
 no shut
 exit
interface G0/0/1
 description Connection to D1
 ip add 10.10.0.1 255.255.255.252
 no shut
 exit
```

Router R2

```
hostname R2
no ip domain lookup
line con 0
 logging sync
 exec-time 0 0
exit
banner motd # This is R2, Path Control Using PBR #
interface G0/0/0
 description Connection to R1
 ip add 172.16.0.1 255.255.255.252
 no shut
exit
```

```
interface GigabitEthernet0/0/1
 description Connection to R3
 ip address 172.16.1.1 255.255.255.252
 no shut
exit
```

Router R3

```
hostname R3
no ip domain lookup
line con 0
 logging sync
 exec-time 0 0
 exit
banner motd # This is R3, Path Control Using PBR #
interface G0/0/0
 description Connection to R2
 ip add 172.16.1.2 255.255.255.252
 no shut
 exit
interface S0/1/0
 description Serial Connection to R1
 ip add 172.16.2.2 255.255.255.252
 no shut
 exit
interface G0/0/1
 description Connection to D2
 ip add 192.168.0.1 255.255.255.252
 no shut
 exit
```

Switch D1

```
hostname D1
no ip domain lookup
line con 0
exec-timeout 0 0
logging synchronous
exit
banner motd # This is D1, Path Control Using PBR #
interface G1/0/11
 no switchport
 description Connects to R1
 ip address 10.10.0.2 255.255.255.252
 no shut
 exit
interface Loopback 1
 description Interface simulates network
 ip ospf network point-to-point
 ip address 10.10.1.1 255.255.255.0
 exit
interface Loopback 2
 description Interface simulates network
 ip ospf network point-to-point
```

```
 ip address 10.10.2.1 255.255.255.0
 exit
```

Switch D2

```
hostname D2
no ip domain lookup
line con 0
logging sync
exec-time 0 0
exit
banner motd # This is D2, Path Control Using PBR #
interface G1/0/11
 no switchport
 description Connects to R3
 ip address 192.168.0.2 255.255.255.252
 no shut
 exit
interface Loopback 1
 description Interface simulates network
 ip ospf network point-to-point
 ip address 192.168.1.1 255.255.255.0
 exit
```

   b.   Save the running configuration to startup-config.

Part 2: Configure and Verify Routing

In Part 2, you will implement OSPF routing for the routing domain and verify end to end routing.

Step 1.   Configure routing.

In this step, you will configure OSPF.

   a.   On D1, advertise the connected networks using OSPF process ID 123. Also assign D1 the router ID of 1.1.1.2 and set the reference bandwidth to recognize Gigabit Ethernet interfaces.

```
D1(config)# ip routing
D1(config)# router ospf 123
D1(config-router)# router-id 1.1.1.2
D1(config-router)# auto-cost reference-bandwidth 1000
% OSPF: Reference bandwidth is changed.
        Please ensure reference bandwidth is consistent across all routers.
D1(config-router)# network 10.10.0.0 0.0.0.3 area 0
D1(config-router)# network 10.10.1.0 0.0.0.255 area 0
D1(config-router)# network 10.10.2.0 0.0.0.255 area 0
D1(config-router)# end
```

   b.   On R1, advertise the connected networks using OSPF process ID 123. Also assign R1 the router ID of 1.1.1.1 and set the reference bandwidth to recognize Gigabit Ethernet interfaces.

```
R1(config)# router ospf 123
R1(config-router)# router-id 1.1.1.1
R1(config-router)# auto-cost reference-bandwidth 1000
```

```
% OSPF: Reference bandwidth is changed.
        Please ensure reference bandwidth is consistent across all routers.
R1(config-router)# network 10.10.0.0 0.0.0.3 area 0
R1(config-router)# network 172.16.0.0 0.0.0.3 area 0
R1(config-router)# network 172.16.2.0 0.0.0.3 area 0
R1(config-router)# end
R1#
*Feb 19 17:00:40.661: %OSPF-5-ADJCHG: Process 123, Nbr 1.1.1.2 on
GigabitEthernet0/0/1 from LOADING to FULL, Loading Done
```

c. On R2, advertise the connected networks using OSPF process ID 123. Also assign R2 the router ID of 2.2.2.1 and set the reference bandwidth to recognize Gigabit Ethernet interfaces.

```
R2(config)# router ospf 123
R2(config-router)# router-id 2.2.2.1
R2(config-router)# auto-cost reference-bandwidth 1000
% OSPF: Reference bandwidth is changed.
        Please ensure reference bandwidth is consistent across all routers.
*Feb 19 17:02:34.016: %OSPF-6-DFT_OPT: Protocol timers for fast convergence
areEnabled.172.
R2(config-router)# network 172.16.0.0 0.0.0.3 area 0
R2(config-router)# network 172.16.1.0 0.0.0.3 area 0
R2(config-router)# end
R2#
*Feb 19 17:02:42.460: %OSPF-5-ADJCHG: Process 123, Nbr 1.1.1.1 on
GigabitEthernet0/0/0 from LOADING to FULL, Loading Done
```

d. On R3, advertise the connected networks using OSPF process ID 123. Also assign R3 the router ID of 3.3.3.1 and set the reference bandwidth to recognize Gigabit Ethernet interfaces.

```
R3(config)# router ospf 123
R3(config-router)# router-id 3.3.3.1
R3(config-router)# auto-cost reference-bandwidth 1000
% OSPF: Reference bandwidth is changed.
        Please ensure reference bandwidth is consistent across all routers.
R3(config-router)# network 192.168.0.0 0.0.0.3 area 0
R3(config-router)# network 172.16.1.0 0.0.0.3 area 0
R3(config-router)# network 172.16.2.0 0.0.0.3 area 0
R3(config-router)# end
R3#
*Feb 19 17:03:56.362: %OSPF-5-ADJCHG: Process 123, Nbr 1.1.1.1 on Serial0/1/0
from LOADING to FULL, Loading Done
R3#
*Feb 19 17:09:38.978: %OSPF-5-ADJCHG: Process 123, Nbr 2.2.2.1 on
GigabitEthernet0/0/0 from LOADING to FULL, Loading Done
```

e. On D2, advertise the connected networks using OSPF process ID 123. Also assign D2 the router ID of 3.3.3.2 and set the reference bandwidth to recognize Gigabit Ethernet interfaces.

```
D2(config)# ip routing
D2(config)# router ospf 123
D2(config-router)# router-id 3.3.3.2
D2(config-router)# auto-cost reference-bandwidth 1000
```

```
% OSPF: Reference bandwidth is changed.
        Please ensure reference bandwidth is consistent across all routers.
D2(config-router)# network 192.168.0.0 0.0.0.3 area 0
D2(config-router)# network 192.168.1.0 0.0.0.255 area 0
D2(config-router)# end
D2#
*Feb 19 17:29:46.627: %OSPF-5-ADJCHG: Process 123, Nbr 3.3.3.1 on
GigabitEthernet1/0/11 from LOADING to FULL, Loading Done
```

Step 2.   Verify OSPF routing.

   a.   Before configuring PBR, verify the current routing table on all devices. All routing
   tables look accurate.

```
D1# show ip route ospf | begin Gateway
Gateway of last resort is not set


      172.16.0.0/30 is subnetted, 3 subnets
O        172.16.0.0 [110/2] via 10.10.0.1, 00:11:36, GigabitEthernet1/0/11
O        172.16.1.0 [110/3] via 10.10.0.1, 00:10:10, GigabitEthernet1/0/11
O        172.16.2.0 [110/497] via 10.10.0.1, 00:11:36, GigabitEthernet1/0/11
      192.168.0.0/30 is subnetted, 1 subnets
O        192.168.0.0 [110/4] via 10.10.0.1, 00:09:57, GigabitEthernet1/0/11
O     192.168.1.0/24 [110/14] via 10.10.0.1, 00:09:57, GigabitEthernet1/0/11


D2# show ip route ospf | begin Gateway
Gateway of last resort is not set


      10.0.0.0/8 is variably subnetted, 3 subnets, 2 masks
O        10.10.0.0/30 [110/4] via 192.168.0.1, 00:03:55, GigabitEthernet1/0/11
O        10.10.1.0/24 [110/5] via 192.168.0.1, 00:03:55, GigabitEthernet1/0/11
O        10.10.2.0/24 [110/5] via 192.168.0.1, 00:03:55, GigabitEthernet1/0/11
      172.16.0.0/30 is subnetted, 3 subnets
O        172.16.0.0 [110/3] via 192.168.0.1, 00:03:55, GigabitEthernet1/0/11
O        172.16.1.0 [110/2] via 192.168.0.1, 00:03:55, GigabitEthernet1/0/11
O        172.16.2.0 [110/499] via 192.168.0.1, 00:03:55, GigabitEthernet1/0/1


R1# show ip route ospf | begin Gateway
Gateway of last resort is not set


      10.0.0.0/8 is variably subnetted, 4 subnets, 3 masks
O        10.10.1.0/24 [110/11] via 10.10.0.2, 00:12:21, GigabitEthernet0/0/1
O        10.10.2.0/24 [110/11] via 10.10.0.2, 00:12:21, GigabitEthernet0/0/1
      172.16.0.0/16 is variably subnetted, 5 subnets, 2 masks
O        172.16.1.0/30 [110/2] via 172.16.0.1, 00:10:55, GigabitEthernet0/0/0
      192.168.0.0/30 is subnetted, 1 subnets
O        192.168.0.0 [110/3] via 172.16.0.1, 00:10:42, GigabitEthernet0/0/0
O     192.168.1.0/24 [110/13] via 172.16.0.1, 00:10:42, GigabitEthernet0/0/0


R2# show ip route ospf | begin Gateway
Gateway of last resort is not set
```

```
          10.0.0.0/8 is variably subnetted, 5 subnets, 3 masks
O             10.10.0.0/30 [110/2] via 172.16.0.2, 00:12:26, GigabitEthernet0/0/0
O             10.10.1.0/24 [110/12] via 172.16.0.2, 00:12:26, GigabitEthernet0/0/0
O             10.10.2.0/24 [110/12] via 172.16.0.2, 00:12:26, GigabitEthernet0/0/0
          172.16.0.0/16 is variably subnetted, 5 subnets, 2 masks
O             172.16.2.0/30
                  [110/497] via 172.16.0.2, 00:12:26, GigabitEthernet0/0/0
          192.168.0.0/30 is subnetted, 1 subnets
O             192.168.0.0 [110/2] via 172.16.1.2, 00:11:22, GigabitEthernet0/0/1
O          192.168.1.0/24 [110/12] via 172.16.1.2, 00:11:22, GigabitEthernet0/0/1


R3# show ip route ospf | begin Gateway
Gateway of last resort is not set


          10.0.0.0/8 is variably subnetted, 3 subnets, 2 masks
O             10.10.0.0/30 [110/3] via 172.16.1.1, 00:12:56, GigabitEthernet0/0/0
O             10.10.1.0/24 [110/13] via 172.16.1.1, 00:12:56, GigabitEthernet0/0/0
O             10.10.2.0/24 [110/13] via 172.16.1.1, 00:12:56, GigabitEthernet0/0/0
          172.16.0.0/16 is variably subnetted, 5 subnets, 2 masks
O             172.16.0.0/30 [110/2] via 172.16.1.1, 00:12:56, GigabitEthernet0/0/0
O          192.168.1.0/24 [110/11] via 192.168.0.2, 00:12:44, GigabitEthernet0/0/1
```

Step 3.   Verify end-to-end connectivity and path taken.

   a.   From any device, verify connectivity to all configured destinations using the following TCL script. All pings should be successful. Troubleshoot if necessary.

```
tclsh

foreach address {
10.10.0.1
10.10.0.2
10.10.1.1
10.10.2.1
172.16.0.1
172.16.0.2
172.16.1.1
172.16.1.2
172.16.2.1
172.16.2.2
192.168.0.1
192.168.0.2
192.168.1.1
} { ping $address }
```

   b.   On D1, ping the D2 Loopback interface 192.168.1.1 address from the Lo1 interface as shown. The pings should be successful.

```
D1# ping 192.168.1.1 source 10.10.1.1
Type escape sequence to abort.
Sending 5, 100-byte ICMP Echos to 192.168.1.1, timeout is 2 seconds:
Packet sent with a source address of 10.10.1.1
```

```
!!!!!
Success rate is 100 percent (5/5), round-trip min/avg/max = 2/2/3 ms
```

c. Next, identify the path taken to D2 Lo1 interface using the **traceroute** command as shown. Notice that the path taken for the packets sourced from the D1 Lo1 LAN is going through R1 --> R2 --> R3 --> D2.

```
D1# traceroute 192.168.1.1 source 10.10.1.1
Type escape sequence to abort.
Tracing the route to 192.168.1.1
VRF info: (vrf in name/id, vrf out name/id)
  1 10.10.0.1 2 msec 2 msec 2 msec
  2 172.16.0.1 2 msec 2 msec 2 msec
  3 172.16.1.2 2 msec 2 msec 2 msec
  4 192.168.0.2 3 msec *  3 msec
```

Question:

Why is the path not taking the shorter R1 --> R3 --> D2 path?

_____

_____

_____

d. Now **ping** and **traceroute** the D2 Lo1 interface from the D1 Loopback 2 interface as shown. It is also taking the same path.

```
D1# ping 192.168.1.1 source lo 2
Type escape sequence to abort.
Sending 5, 100-byte ICMP Echos to 192.168.1.1, timeout is 2 seconds:
Packet sent with a source address of 10.10.2.1
!!!!!
Success rate is 100 percent (5/5), round-trip min/avg/max = 2/2/4 ms

D1# traceroute 192.168.1.1 source lo 2
Type escape sequence to abort.
Tracing the route to 192.168.1.1
VRF info: (vrf in name/id, vrf out name/id)
  1 10.10.0.1 2 msec 2 msec 1 msec
  2 172.16.0.1 1 msec 2 msec 1 msec
  3 172.16.1.2 2 msec 2 msec 2 msec
  4 192.168.0.2 3 msec *  3 msec
```

e. Display the OSPF routes in the routing table of R1. R1 forwards all packets destined to the 192.168.1.0/24 network out of its G0/0/0 interface to R2.

```
R1# show ip route ospf | begin Gateway
Gateway of last resort is not set

      10.0.0.0/8 is variably subnetted, 4 subnets, 3 masks
O        10.10.1.0/24 [110/2] via 10.10.0.2, 00:19:56, GigabitEthernet0/0/1
O        10.10.2.0/24 [110/2] via 10.10.0.2, 00:19:56, GigabitEthernet0/0/1
      172.16.0.0/16 is variably subnetted, 5 subnets, 2 masks
O        172.16.1.0/30 [110/2] via 172.16.0.1, 00:18:50, GigabitEthernet0/0/0
      192.168.0.0/30 is subnetted, 1 subnets
```

```
O            192.168.0.0 [110/3] via 172.16.0.1, 00:18:37, GigabitEthernet0/0/0
O            192.168.1.0/24 [110/4] via 172.16.0.1, 00:18:33, GigabitEthernet0/0/0
```

f. Display how R1 learned about the 192.168.1.0 network. R1 learned of the network from R2 (i.e., 172.16.0.1) who originally learned it from D2 (i.e., 3.3.3.2).

```
R1# show ip route 192.168.1.0
Routing entry for 192.168.1.0/24
  Known via "ospf 123", distance 110, metric 4, type intra area
  Last update from 172.16.0.1 on GigabitEthernet0/0/0, 00:20:27 ago
  Routing Descriptor Blocks:
  * 172.16.0.1, from 3.3.3.2, 00:20:27 ago, via GigabitEthernet0/0/0
      Route metric is 4, traffic share count is 1
```

Part 3: Configure PBR to Provide Path Control

Recall that route maps can be used for:

- **Redistribution** - Route maps provide more options and flexibility to the **redistribute** command.

- **Policy-based routing (PBR)** - PBR allows an administrator to define routing policy other than basic destination-based routing using the routing table. The route map is applied to an interface using the **ip policy route-map** interface configuration command.

- **BGP** - Route maps are the primary tools for implementing BGP policy and allow an administrator to do path control and provide sophisticated manipulation of BGP path attributes. The route map is applied using the BGP **neighbor** router configuration command.

In this part, you will use PBR to configure source-based IP routing. Specifically, you will override the default IP routing decision based on the OSPF-acquired routing information for selected IP source-to-destination flows and apply a different next-hop router.

Recall that routers normally forward packets to destination addresses based on information in their routing table. By using PBR, you can implement policies that selectively cause packets to take different paths based on source address, protocol type, or application type. Therefore, PBR overrides the router's normal routing behavior.

Configuring PBR involves configuring a route map with **match** and **set** commands and then applying the route map to the interface.

The steps required to implement path control include the following:

- Choose the path control tool to use. Path control tools manipulate or bypass the IP routing table. For PBR, **route-map** commands are used.

- Implement the traffic-matching configuration, specifying which traffic will be manipulated. The **match** commands are used within route maps.

- Define the action for the matched traffic using **set** commands within route maps.

- Apply the route map to incoming traffic.

Step 1.   Configure PBR on R1.

As a test, you will configure the following policy on router R1:

- All traffic sourced from D1 Lo1 LAN must take the R1 --> R2 --> R3 --> D2 path.

- All traffic sourced from D1 Lo2 LAN must take the R1 --> R3 --> D2 path.

a. On R1, create a standard named ACL called **Lo2-ACL** to identify the D1 Loopback 2 (i.e., 10.10.2.0/24) LAN.

```
R1(config)# ip access-list standard Lo2-ACL
R1(config-std-nacl)# remark ACL matches D1 Lo2 traffic
R1(config-std-nacl)# permit 10.10.2.0 0.0.0.255
R1(config-std-nacl)# exit
```

b. Create a route map called **R1-to-R3** that matches Lo2-ACL and sets the next-hop interface to the R3 serial 0/1/0 interface.

```
R1(config)# route-map R1-to-R3 permit
R1(config-route-map)# description RM to forward Lo2 traffic to R3
R1(config-route-map)# match ip address Lo2-ACL
R1(config-route-map)# set ip next-hop 172.16.2.2
R1(config-route-map)# exit
```

c. Apply the R1-to-R3 route map to the G0/0/1 interface using the **ip policy route-map** command.

```
R1(config)# interface g0/0/1
R1(config-if)# ip policy route-map R1-to-R3
R1(config-if)# end
```

d. On R1, display the policy and matches using the **show route-map** command.

```
R1# show route-map
route-map R1-to-R3, permit, sequence 10
  Match clauses:
    ip address (access-lists): Lo2-ACL
  Set clauses:
    ip next-hop 172.16.2.2
  Policy routing matches: 0 packets, 0 bytes
```

Note: There are currently no matches because no packets matching the ACL have passed through R1 G0/0/1.

e. On R1, verify that the R1-to-R3 route map has been applied to the G0/0/1 interface.

```
R1# show ip policy
Interface        Route map
Gi0/0/1          R1-to-R3
```

Step 2.   Test the policy.

Now you are ready to test the policy configured on R1.

a. From D1, test the policy with the **traceroute** command, using D1 Lo1 interface as the source network.

```
D1# traceroute 192.168.1.1 source lo 1
Type escape sequence to abort.
Tracing the route to 192.168.1.1
VRF info: (vrf in name/id, vrf out name/id)
  1 10.10.0.1 2 msec 1 msec 1 msec
  2 172.16.0.1 2 msec 2 msec 2 msec
  3 172.16.1.2 2 msec 4 msec 2 msec
  4 192.168.0.2 3 msec *  3 msec
```

Notice the path taken for the packet sourced from D1 Lo 1 LAN A is still going through R1 --> R2 --> R3 --> D2.

Question:

Why is the traceroute traffic not using the R3 --> R1 path as specified in the R1-to-R3 policy?

_____

_____

b. Now test the policy with the **traceroute** command, using D1 Lo2 interface as the source network. Now the path taken for the packet sourced from D1 Lo 2 LAN is R1 --> R3 --> D2, as expected.

```
D1# traceroute 192.168.1.1 source lo 2
Type escape sequence to abort.
Tracing the route to 192.168.1.1
VRF info: (vrf in name/id, vrf out name/id)
  1 10.10.0.1 2 msec 2 msec 1 msec
  2 172.16.2.2 3 msec 2 msec 3 msec
  3 192.168.0.2 3 msec *  4 msec
```

c. On R1, display the policy and matches using the **show route-map** command.

Note: There are now matches to the policy because packets matching the ACL have passed through R1 G0/0/1 interface. The number of packet and bytes may differ in your implementation.

```
R1# show route-map
route-map R1-to-R3, permit, sequence 10
  Match clauses:
    ip address (access-lists): Lo2-ACL
  Set clauses:
    ip next-hop 172.16.2.2
  Policy routing matches: 5 packets, 210 bytes
```

Part 4: Configure Local PBR to Provide Path Control

How would you policy route packets generated by a router? The answer is to configure Local PBR.

Local PBR is a feature to policy route locally generated traffic. Local PBR policies are applied to the router with the **ip local policy route-map** global config command.

In this part, you will configure R1 to policy route all router generated traffic over the R1 to R3 link.

Step 1.   Configure Local PBR on R1.

a. Verify the path that R1 currently takes without local PBR configured. R1 sends traffic to R2 then R3 and finally D2 as expected.

```
R1# traceroute 192.168.1.1
Type escape sequence to abort.
Tracing the route to 192.168.1.1
VRF info: (vrf in name/id, vrf out name/id)
  1 172.16.0.1 1 msec 1 msec 1 msec
  2 172.16.1.2 1 msec 2 msec 1 msec
  3 192.168.0.2 2 msec *  2 msec
```

b. On R1, create a named extended ACL called **R1-TRAFFIC** which matches all IP generated packets from R1 and destined to the D2 192.162.1.0/24 network.

```
R1(config)# ip access-list extended R1-TRAFFIC
R1(config-ext-nacl)# permit ip any 192.168.1.0 0.0.0.255
R1(config-ext-nacl)# exit
```

c. On R1, create a route map called **LOCAL-PBR** that permits traffic matching the R1-TRAFFIC ACL and redirects it to the R3 172.16.2.2 interface.

```
R1(config)# route-map LOCAL-PBR permit
R1(config-route-map)# match ip address R1-TRAFFIC
R1(config-route-map)# set ip next-hop 172.16.2.2
R1(config-route-map)# exit
```

d. Create a local PBR policy that matches the LOCAL-PBR route map.

```
R1(config)# ip local policy route-map LOCAL-PBR
R1(config)# exit
```

Step 2.   Test Local PBR on R1.

a. Verify the path taken by R1 to reach the 192.168.1.0/24 LAN. The traffic generated by R1 and going to 192.168.1.0/24 is now policy routed directly to R3 (i.e., 172.16.2.2).

```
R1# traceroute 192.168.1.1
Type escape sequence to abort.
Tracing the route to 192.168.1.1
VRF info: (vrf in name/id, vrf out name/id)
  1 172.16.2.2 2 msec 1 msec 2 msec
  2 192.168.0.2 2 msec *  2 msec
```

b. Verify the path taken by R1 to reach other networks. The traffic takes the normal OSPF generated path and is not policy routed.

```
R1# traceroute 192.168.0.2
Type escape sequence to abort.
Tracing the route to 192.168.0.2
VRF info: (vrf in name/id, vrf out name/id)
  1 172.16.0.1 1 msec 1 msec 1 msec
  2 172.16.1.2 1 msec 2 msec 1 msec
  3 192.168.0.2 4 msec *  2 msec
```

c. Verify the route-map counters. The local PBR policy has matched packets.

Note: The number of packets and bytes may differ in your implementation.

```
R1# show route-map
route-map R1-to-R3, permit, sequence 10
  Match clauses:
    ip address (access-lists): Lo2-ACL
  Set clauses:
    ip next-hop 172.16.2.2
  Policy routing matches: 6 packets, 252 bytes
route-map LOCAL-PBR, permit, sequence 10
  Match clauses:
    ip address (access-lists): ICMP-TRAFFIC
  Set clauses:
    ip next-hop 172.16.2.2
  Policy routing matches: 32 packets, 2384 bytes
```

Router Interface Summary Table

Router Model	Ethernet Interface #1	Ethernet Interface #2	Serial Interface #1	Serial Interface #2
1800	Fast Ethernet 0/0 (F0/0)	Fast Ethernet 0/1 (F0/1)	Serial 0/0/0 (S0/0/0)	Serial 0/0/1 (S0/0/1)
1900	Gigabit Ethernet 0/0 (G0/0)	Gigabit Ethernet 0/1 (G0/1)	Serial 0/0/0 (S0/0/0)	Serial 0/0/1 (S0/0/1)
2801	Fast Ethernet 0/0 (F0/0)	Fast Ethernet 0/1 (F0/1)	Serial 0/1/0 (S0/1/0)	Serial 0/1/1 (S0/1/1)
2811	Fast Ethernet 0/0 (F0/0)	Fast Ethernet 0/1 (F0/1)	Serial 0/0/0 (S0/0/0)	Serial 0/0/1 (S0/0/1)
2900	Gigabit Ethernet 0/0 (G0/0)	Gigabit Ethernet 0/1 (G0/1)	Serial 0/0/0 (S0/0/0)	Serial 0/0/1 (S0/0/1)
4221	Gigabit Ethernet 0/0/0 (G0/0/0)	Gigabit Ethernet 0/0/1 (G0/0/1)	Serial 0/1/0 (S0/1/0)	Serial 0/1/1 (S0/1/1)
4300	Gigabit Ethernet 0/0/0 (G0/0/0)	Gigabit Ethernet 0/0/1 (G0/0/1)	Serial 0/1/0 (S0/1/0)	Serial 0/1/1 (S0/1/1)

Note: To find out how the router is configured, look at the interfaces to identify the type of router and how many interfaces the router has. There is no way to effectively list all the combinations of configurations for each router class. This table includes identifiers for the possible combinations of Ethernet and Serial interfaces in the device. The table does not include any other type of interface, even though a specific router may contain one. An example of this might be an ISDN BRI interface. The string in parenthesis is the legal abbreviation that can be used in Cisco IOS commands to represent the interface.

Device Config Final – Notes

15.1.4 Lab - Troubleshoot Route Maps and PBR

Topology

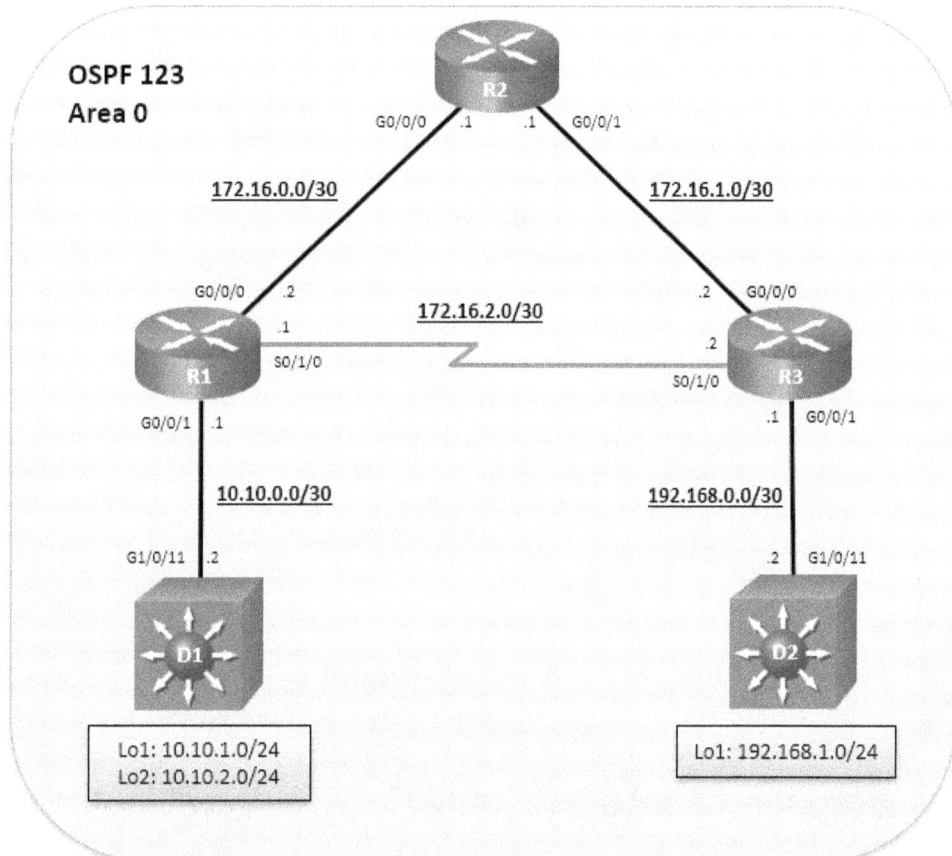

Addressing Table

Device	Interface	IPv4 Address	Subnet Mask
D1	G0/0/11	10.10.0.2	255.255.255.252
	Loopback 1	10.10.1.1	255.255.255.0
	Loopback 2	10.10.2.1	255.255.255.0
R1	G0/0/0	172.16.0.2	255.255.255.252
	G0/0/1	10.10.0.1	255.255.255.252
	S0/1/0	172.16.2.1	255.255.255.252
R2	G0/0/0	172.16.0.1	255.255.255.252
	G0/0/1	172.16.1.1	255.255.255.252
R3	G0/0/0	172.16.1.2	255.255.255.252
	G0/0/1	192.168.0.1	255.255.255.252
	S0/1/0	172.16.2.2	255.255.255.252
D2	G0/0/11	192.168.0.2	255.255.255.252
	Loopback 1	192.168.1.1	255.255.255.0

Objectives

Troubleshoot network issues related to the configuration and operation of PBR using route maps.

Background/Scenario

PBR was recently implemented on R1 and R3. However, there have been problems.

Although the topology has a limited number of routers, you should use the appropriate troubleshooting commands to help find and solve the problems in the three trouble tickets as if this were a much more complex topology with many more routers and networks.

You will be loading configurations with intentional errors onto the network. Your tasks are to FIND the error(s), document your findings and the command(s) or method(s) used to fix them, FIX the issue(s) presented here, and then test the network to ensure both of the following conditions are met:

1. the complaint received in the ticket is resolved

2. full reachability is restored

Note: The routers used with CCNP hands-on labs are Cisco 4221 with Cisco IOS XE Release 16.9.4 (universalk9 image). The switches used in the labs are Cisco Catalyst 3650 with Cisco IOS XE Release 16.9.4 (universalk9 image). Other routers, switches, and Cisco IOS versions can be used. Depending on the model and Cisco IOS version, the commands available and the output produced might vary from what is shown in the labs. Refer to the Router Interface Summary Table at the end of the lab for the correct interface identifiers.

Note: Make sure that the devices have been erased and have no startup configurations. If you are unsure, contact your instructor.

Required Resources

- 3 Routers (Cisco 4221 with Cisco IOS XE Release 16.9.4 universal image or comparable)
- 2 Switches (Cisco 3560 with Cisco IOS XE Release 16.9.4 universal image or comparable)
- Console cables to configure the Cisco IOS devices via the console ports
- Ethernet cables as shown in the topology

Instructions

Part 1: Trouble Ticket 15.1.4.1

Scenario:

The routing table in the OSPF area 0 topology forwards traffic between R1 and R3 via R2 because of the faster Gigabit Ethernet links between R1 and R2 and between R2 and R3. However, corporate policy states that all traffic from the D1 loopback 2 network (i.e., 10.10.2.0/24) should be policy-based routed (PBR) directly to R3 using the R1 to R3 serial link. It was assumed that the policy was working correctly but a recent traceroute from the D1 loopback 2 interface to the D2 loopback 1 (i.e., 192.168.1.0/24) network has revealed otherwise.

```
D1# traceroute 192.168.1.1 source lo2
Type escape sequence to abort.
Tracing the route to 192.168.1.1
```

```
VRF info: (vrf in name/id, vrf out name/id)
  1 10.10.0.1 2 msec 2 msec 2 msec
  2 172.16.0.1 2 msec 2 msec 2 msec
  3 172.16.1.2 2 msec 2 msec 2 msec
  4 192.168.0.2 3 msec *  3 msec
```

Use the commands listed below to load the configuration files for this trouble ticket:

Device	Command
R1	`copy flash:/enarsi/15.1.4.1-r1-config.txt run`
R2	`copy flash:/enarsi/15.1.4.1-r2-config.txt run`
R3	`copy flash:/enarsi/15.1.4.1-r3-config.txt run`
D1	`copy flash:/enarsi/15.1.4.1-d1-config.txt run`
D2	`copy flash:/enarsi/15.1.4.1-d2-config.txt run`

- Traffic from 10.10.2.0/24 going to 192.168.1.0/24 should be routed directly to R3 from R1.

- All other traffic from D1 should be propagated according to the routing table.

- When you have fixed the ticket, change the MOTD on EACH DEVICE using the following command:

  banner motd # This is $(hostname) FIXED from ticket <ticket number> #

- Then save the configuration by issuing the **wri** command (on each device).

- Inform your instructor that you are ready for the next ticket.

- After the instructor approves your solution for this ticket, issue the **reset.now** privileged EXEC command This script will clear your configurations and reload the devices.

Part 2: Trouble Ticket 15.1.4.2

Scenario:

The routing table in the OSPF area 0 topology forwards traffic between R1 and R3 via R2 because of the faster Gigabit Ethernet links between R1 and R2 and between R2 and R3. However, corporate policy states that all traffic from the D1 loopback 2 network (i.e., 10.10.2.0/24) should be policy-based routed (PBR) directly to R3 using the R1 to R3 serial link.

However, a traceroute from the D1 loopback 2 interface to the D2 loopback 1 (i.e., 192.168.1.0/24) network has revealed that traffic is not policy-based routed.

```
D1# traceroute 192.168.1.1 source lo2
Type escape sequence to abort.
Tracing the route to 192.168.1.1
VRF info: (vrf in name/id, vrf out name/id)
  1 10.10.0.1 2 msec 2 msec 2 msec
  2 172.16.0.1 2 msec 2 msec 2 msec
  3 172.16.1.2 2 msec 2 msec 2 msec
  4 192.168.0.2 3 msec *  3 msec
```

Note: This is the same issue as the previous ticket. However, the cause(s) and solution(s) are different.

Use the commands listed below to load the configuration files for this trouble ticket:

Device	Command
R1	copy flash:/enarsi/15.1.4.2-r1-config.txt run
R2	copy flash:/enarsi/15.1.4.2-r2-config.txt run
R3	copy flash:/enarsi/15.1.4.2-r3-config.txt run
D1	copy flash:/enarsi/15.1.4.2-d1-config.txt run
D2	copy flash:/enarsi/15.1.4.2-d2-config.txt run

- Traffic from 10.10.2.0/24 going to 192.168.1.0/24 should be routed directly to R3 from R1.

- All other traffic from D1 should be propagated according to the routing table.

- When you have fixed the ticket, change the MOTD on EACH DEVICE using the following command:

  banner motd # This is $(hostname) FIXED from ticket <ticket number> #

- Then save the configuration by issuing the **wri** command (on each device).

- Inform your instructor that you are ready for the next ticket.

- After the instructor approves your solution for this ticket, issue the **reset.now** privileged EXEC command This script will clear your configurations and reload the devices.

Part 3: Trouble Ticket 15.1.4.3

Scenario:

In this scenario, a local PBR policy was implemented on R3 to route traffic generated for the 10.10.0.0/16 directly to R1. However, a traceroute to 10.10.1.1 displays that it is being forwarded to R2 instead of R1.

```
R3# traceroute 10.10.1.1
Type escape sequence to abort.
Tracing the route to 10.10.1.1
VRF info: (vrf in name/id, vrf out name/id)
  1 172.16.1.1 2 msec 1 msec 1 msec
  2 172.16.0.2 2 msec 1 msec 1 msec
  3 10.10.0.2 4 msec *  2 msec
```

Use the commands listed below to load the configuration files for this trouble ticket:

Device	Command
R1	copy flash:/enarsi/15.1.4.3-r1-config.txt run
R2	copy flash:/enarsi/15.1.4.3-r2-config.txt run
R3	copy flash:/enarsi/15.1.4.3-r3-config.txt run
D1	copy flash:/enarsi/15.1.4.3-d1-config.txt run
D2	copy flash:/enarsi/15.1.4.3-d2-config.txt run

- Traffic generated by R3 going to the 10.10.0.0/16 networks should be sent directly to R1 as identified in the local PBR.

- When you have fixed the ticket, change the MOTD on EACH DEVICE using the following command:

  banner motd # This is $(hostname) FIXED from ticket <ticket number> #

- Then save the configuration by issuing the **wri** command (on each device).

- Inform your instructor that you are ready for the next ticket.

- After the instructor approves your solution for this ticket, issue the **reset.now** privileged EXEC command. This script will clear your configurations and reload the devices.

Router Interface Summary Table

Router Model	Ethernet Interface #1	Ethernet Interface #2	Serial Interface #1	Serial Interface #2
1800	Fast Ethernet 0/0 (F0/0)	Fast Ethernet 0/1 (F0/1)	Serial 0/0/0 (S0/0/0)	Serial 0/0/1 (S0/0/1)
1900	Gigabit Ethernet 0/0 (G0/0)	Gigabit Ethernet 0/1 (G0/1)	Serial 0/0/0 (S0/0/0)	Serial 0/0/1 (S0/0/1)
2801	Fast Ethernet 0/0 (F0/0)	Fast Ethernet 0/1 (F0/1)	Serial 0/1/0 (S0/1/0)	Serial 0/1/1 (S0/1/1)
2811	Fast Ethernet 0/0 (F0/0)	Fast Ethernet 0/1 (F0/1)	Serial 0/0/0 (S0/0/0)	Serial 0/0/1 (S0/0/1)
2900	Gigabit Ethernet 0/0 (G0/0)	Gigabit Ethernet 0/1 (G0/1)	Serial 0/0/0 (S0/0/0)	Serial 0/0/1 (S0/0/1)
4221	Gigabit Ethernet 0/0/0 (G0/0/0)	Gigabit Ethernet 0/0/1 (G0/0/1)	Serial 0/1/0 (S0/1/0)	Serial 0/1/1 (S0/1/1)
4300	Gigabit Ethernet 0/0/0 (G0/0/0)	Gigabit Ethernet 0/0/1 (G0/0/1)	Serial 0/1/0 (S0/1/0)	Serial 0/1/1 (S0/1/1)

Note: To find out how the router is configured, look at the interfaces to identify the type of router and how many interfaces the router has. There is no way to effectively list all the combinations of configurations for each router class. This table includes identifiers for the possible combinations of Ethernet and Serial interfaces in the device. The table does not include any other type of interface, even though a specific router may contain one. An example of this might be an ISDN BRI interface. The string in parenthesis is the legal abbreviation that can be used in Cisco IOS commands to represent the interface.

Device Config Final – Notes

CHAPTER 16

Route Redistribution

16.1.2 Lab - Configure Route Redistribution Between EIGRP and OSPF

Topology

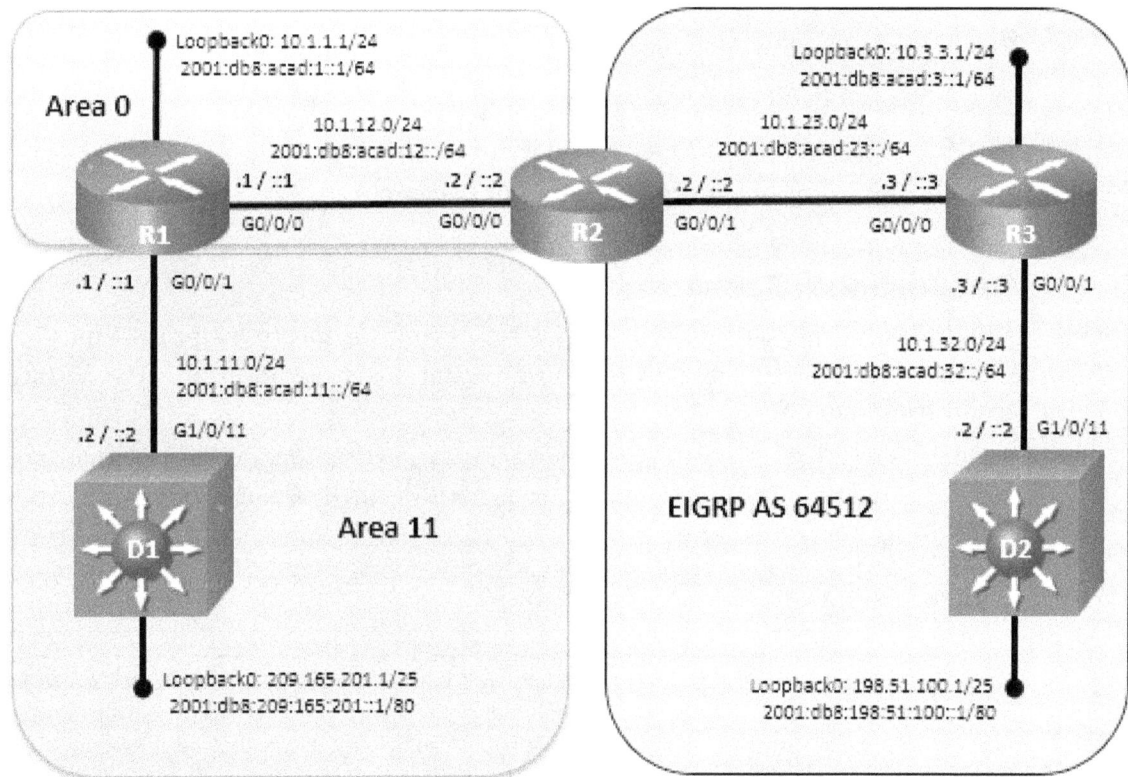

Addressing Table

Device	Interface	IPv4 Address/Mask	IPv6 Address/Prefix	IPv6 Link Local
R1	G0/0/0	10.1.12.1/24	2001:db8:acad:12::1/64	fe80::12:1
	G0/0/1	10.1.11.1/24	2001:db8:acad:11::1/64	fe80::11:1
	Loopback 0	10.1.1.1/24	2001:db8:acad:1::1/64	fe80::1:1
R2	G0/0/0	10.1.12.2/24	2001:db8:acad:12::2/64	fe80::12:2
	G0/0/1	10.1.23.2/24	2001:db8:acad:23::2/64	fe80::23:2
R3	G0/0/0	10.1.23.3/24	2001:db8:acad:23::3/64	fe80::23:3
	G0/0/1	10.1.32.1/24	2001:db8:acad:32::3/64	fe80::32:3
	Loopback 0	10.3.3.3/24	2001:db8:acad:3::3/64	fe80::3:3

Device	Interface	IPv4 Address/Mask	IPv6 Address/Prefix	IPv6 Link Local
D1	G1/0/11	10.1.11.2/24	2001:db8:acad:11::2/64	fe80::11:2
	Loopback 0	209.165.201.1/25	2001:db8:209:165:201::1/80	fe80::209:1
D2	G1/0/11	10.1.32.2/24	2001:db8:acad:32::2/64	fe80::32:2
	Loopback 0	198.51.100.1/25	2001:db8:198:51:100::1/80	fe80::198:1

Objectives

Part 1: Build the Network and Configure Basic Device Settings

Part 2: Verify OSPFv3 AF Neighborships and Routing for IPv4 and IPv6

Part 3: Verify EIGRP Neighborships and Routing for IPv4 and IPv6

Part 4: Configure Redistribution from OSPFv3 to EIGRP

Part 5: Configure Redistribution from EIGRP to OSPFv3

Background/Scenario

In this lab, you will configure redistribution from OSPF into EIGRP for IPv4 and IPv6, and redistribution of EIGRP into OSPF for IPv4 and IPv6. You will also change the metric type for EIGRP routes redistributed into OSPF.

D1, R1 and R2 are configured with OSPFv3 for IPv4 and IPv6 address families, while R2, R3 and D2 are configured with EIGRP using named mode for IPv4 and IPv6 address families.

Note: This lab is an exercise in configuring and verifying two-way route redistribution on R2. Route redistribution in this lab does not reflect networking best practices.

Note: The routers used with CCNP hands-on labs are Cisco 4221 with Cisco IOS XE Release 16.9.4 (universalk9 image). The switches used in the labs are Cisco Catalyst 3650 with Cisco IOS XE Release 16.9.4 (universalk9 image). Other routers, switches, and Cisco IOS versions can be used. Depending on the model and Cisco IOS version, the commands available and the output produced might vary from what is shown in the labs.

Note: Make sure that all the devices have been erased and have no startup configurations. If you are unsure, contact your instructor.

Required Resources

- 3 Routers (Cisco 4221 with Cisco IOS XE Release 16.9.4 universal image or comparable)
- 2 Switches (Cisco 3650 with Cisco IOS XE release 16.9.4 universal image or comparable)
- 1 PC (Choice of operating system with terminal emulation program installed)
- Console cables to configure the Cisco IOS devices via the console ports
- Ethernet cables as shown in the topology

Instructions

Part 1: Build the Network and Configure Basic Device Settings

In Part 1, you will set up the network topology and configure basic settings.

Step 1.   Cable the network as shown in the topology.

Attach the devices as shown in the topology diagram, and cable as necessary.

Step 2.   Configure basic settings for each device.

a.   Console into each device, enter global configuration mode, and apply the basic settings for the lab. Initial configurations for each device are listed below.

Router R1

```
hostname R1
no ip domain lookup
ipv6 unicast-routing
banner motd # R1, Configure Route Redistribution Between EIGRP and OSPF #
line con 0
 exec-timeout 0 0
 logging synchronous
 exit
router ospfv3 1
 router-id 1.1.1.1
exit
interface g0/0/0
 ip address 10.1.12.1 255.255.255.0
 ipv6 address FE80::12:1 link-local
 ipv6 address 2001:DB8:ACAD:12::1/64
 ospfv3 1 ipv6 area 0
 ospfv3 1 ipv4 area 0
 no shutdown
 exit
interface g0/0/1
 ip address 10.1.11.1 255.255.255.0
 ipv6 address fe80::11:1 link-local
 ipv6 address 2001:db8:acad:11::1/64
 ospfv3 1 ipv6 area 11
 ospfv3 1 ipv4 area 11
 no shutdown
 exit
interface loopback 0
 ip address 10.1.1.1 255.255.255.0
 ipv6 address FE80::1:1 link-local
 ipv6 address 2001:DB8:ACAD:1::1/64
 ospfv3 network point-to-point
 ospfv3 1 ipv4 area 0
 ospfv3 1 ipv6 area 0
 no shutdown
 exit
router ospfv3 1
```

```
address-family ipv4 unicast
  passive-interface Loopback0
 exit-address-family
 address-family ipv6 unicast
  passive-interface Loopback0
 exit-address-family
end
```

Router R2

```
hostname R2
no ip domain lookup
ipv6 unicast-routing
banner motd # R2, Configure Route Redistribution Between EIGRP and OSPF #
line con 0
 exec-timeout 0 0
 logging synchronous
 exit
router ospfv3 1
 router-id 2.2.2.2
 address-family ipv4 unicast
 exit-address-family
address-family ipv6 unicast
 exit-address-family
interface g0/0/0
 ip address 10.1.12.2 255.255.255.0
 ipv6 address FE80::12:2 link-local
 ipv6 address 2001:DB8:ACAD:12::2/64
 ospfv3 1 ipv6 area 0
 ospfv3 1 ipv4 area 0
 no shutdown
 exit
interface g0/0/1
 ip address 10.1.23.2 255.255.255.0
 ipv6 address fe80::23:2 link-local
 ipv6 address 2001:db8:acad:23::2/64
 no shutdown
 exit
router eigrp CISCO
address-family ipv4 unicast autonomous-system 64512
  af-interface default
    shutdown
  exit-af-interface
  af-interface GigabitEthernet0/0/1
   no shutdown
  exit-af-interface
  topology base
  exit-af-topology
  network 10.1.23.0 0.0.0.255
  eigrp router-id 2.2.2.2
 exit-address-family
address-family ipv6 unicast autonomous-system 64512
```

```
    af-interface default
     shutdown
    exit-af-interface
    af-interface GigabitEthernet0/0/1
     no shutdown
    exit-af-interface
    topology base
    exit-af-topology
  exit-address-family
 end
```

Router R3

```
hostname R3
no ip domain lookup
ipv6 unicast-routing
banner motd # R3, Configure Route Redistribution Between EIGRP and OSPF #
line con 0
 exec-timeout 0 0
 logging synchronous
 exit
interface g0/0/0
 ip address 10.1.23.3 255.255.255.0
 ipv6 address fe80::23:3 link-local
 ipv6 address 2001:db8:acad:23::3/64
 no shutdown
 exit
interface g0/0/1
 ip address 10.1.32.3 255.255.255.0
 ipv6 address fe80::32:3 link-local
 ipv6 address 2001:db8:acad:32::3/64
 no shutdown
 exit
interface loopback 0
 ip address 10.3.3.3 255.255.255.0
 ipv6 address fe80::3:3 link-local
 ipv6 address 2001:db8:acad:3::3/64
 no shutdown
 exit
router eigrp CISCO
address-family ipv4 unicast autonomous-system 64512
  af-interface default
   shutdown
  exit-af-interface
  af-interface GigabitEthernet0/0/0
   no shutdown
  exit-af-interface
  af-interface GigabitEthernet0/0/1
   no shutdown
  exit-af-interface
  af-interface Loopback0
   no shutdown
```

```
   exit-af-interface
   topology base
   exit-af-topology
   network 10.1.23.0 0.0.0.255
   network 10.1.32.0 0.0.0.255
   network 10.3.3.0 0.0.0.255
   eigrp router-id 3.3.3.3
 exit-address-family
address-family ipv6 unicast autonomous-system 64512
  af-interface default
   shutdown
  exit-af-interface
  af-interface GigabitEthernet0/0/0
   no shutdown
  exit-af-interface
  af-interface GigabitEthernet0/0/1
   no shutdown
  exit-af-interface
  af-interface Loopback0
   no shutdown
  exit-af-interface
  topology base
  exit-af-topology
  eigrp router-id 3.3.3.3
 exit-address-family
end
```

Switch D1

```
hostname D1
no ip domain lookup
ip routing
ipv6 unicast-routing
banner motd # D1, Configure Route Redistribution Between EIGRP and OSPF #
line con 0
 exec-timeout 0 0
 logging synchronous
 exit
router ospfv3 1
 router-id 11.11.11.11
 exit
interface range g1/0/1-24
 shutdown
 exit
interface g1/0/11
 no switchport
 ip address 10.1.11.2 255.255.255.0
 ipv6 address fe80::11:2 link-local
 ipv6 address 2001:db8:acad:11::2/64
 ospfv3 1 ipv6 area 11
 ospfv3 1 ipv4 area 11
 no shutdown
```

```
 exit
interface loopback 0
 ip address 209.165.201.1 255.255.255.128
 ipv6 address fe80::209:1 link-local
 ipv6 address 2001:db8:209:165:201::1/80
 no shutdown
 exit
router ospfv3 1
address-family ipv4 unicast
  passive-interface Loopback0
  default-information originate
 exit-address-family
address-family ipv6 unicast
  passive-interface Loopback0
  default-information originate
 exit-address-family
ip route 0.0.0.0 0.0.0.0 Loopback0
ipv6 route ::/0 Loopback0
exit
```

Switch D2

```
hostname D2
no ip domain lookup
ip routing
ipv6 unicast-routing
banner motd # D2, Configure Route Redistribution Between EIGRP and OSPF #
line con 0
 exec-timeout 0 0
 logging synchronous
 exit
interface range g1/0/1-24
 shutdown
 exit
interface g1/0/11
 no switchport
 ip address 10.1.32.2 255.255.255.0
 ipv6 address fe80::32:2 link-local
 ipv6 address 2001:db8:acad:32::2/64
 no shutdown
 exit
interface loopback 0
 ip address 198.51.100.1 255.255.255.128
 ipv6 address fe80::198:2 link-local
 ipv6 address 2001:db8:198:51:100::1/80
 no shutdown
 router eigrp CISCO
 address-family ipv4 unicast autonomous-system 64512
  af-interface default
   shutdown
  exit-af-interface
  af-interface Loopback0
```

```
                       no shutdown
                       passive-interface
                      exit-af-interface
                      af-interface GigabitEthernet1/0/11
                       no shutdown
                      exit-af-interface
                      topology base
                      exit-af-topology
                      network 10.1.32.0 0.0.0.255
                      network 198.51.100.0 0.0.0.127
                      eigrp router-id 22.22.22.22
                     exit-address-family
                     address-family ipv6 unicast autonomous-system 64512
                      af-interface default
                       shutdown
                      exit-af-interface
                      af-interface Loopback0
                       no shutdown
                       passive-interface
                      exit-af-interface
                      af-interface GigabitEthernet1/0/11
                       no shutdown
                      exit-af-interface
                      topology base
                      exit-af-topology
                      eigrp router-id 22.22.22.22
                     exit-address-family
                     exit
```

b. Set the clock on all devices to UTC time.

c. Save the running configuration to startup-config on all devices.

Part 2: Verify OSPFv3 AF Neighborships and Routing for IPv4 and IPv6

In this part, you will verify that OSPF has established neighbor relationships and routing for IPv4 and IPv6.

Step 1. Verify OSPFv3 AF neighborships on R1.

a. Verify R1 has OSPFv3 neighbors: two neighbors from IPv4 address family and two from IPv6 address family.

```
R1# show ospfv3 neighbor


           OSPFv3 1 address-family ipv4 (router-id 1.1.1.1)


Neighbor ID  Pri  State     Dead Time    Interface ID    Interface
2.2.2.2        1  FULL/BDR  00:00:36     6               GigabitEthernet0/0/0
11.11.11.11    1  FULL/BDR  00:00:31     38              GigabitEthernet0/0/1
```

```
OSPFv3 1 address-family ipv6 (router-id 1.1.1.1)

Neighbor ID  Pri  State     Dead Time  Interface ID  Interface
2.2.2.2       1   FULL/BDR  00:00:39   6             GigabitEthernet0/0/0
11.11.11.11   1   FULL/BDR  00:00:39   38            GigabitEthernet0/0/1
```

  b. The output shows four OSPFv3 neighbors: two neighbors from IPv4 address family and two from IPv6 address family.

Step 2. Verify the IPv4 OSPFv3 routing table on R2.

  a. Verify the OSPFv3 IPv4 routing table on R2. Notice the default route, the intra–area, and inter–area OSPF routes are installed and received from 10.1.12.1, which is R1.

```
R2# show ip route ospfv3 | begin Gateway
Gateway of last resort is 10.1.12.1 to network 0.0.0.0

O*E2  0.0.0.0/0 [110/1] via 10.1.12.1, 02:41:28, GigabitEthernet0/0/0
      10.0.0.0/8 is variably subnetted, 8 subnets, 2 masks
O        10.1.1.0/24 [110/1] via 10.1.12.1, 02:49:12, GigabitEthernet0/0/0
O IA     10.1.11.0/24 [110/2] via 10.1.12.1, 02:44:58, GigabitEthernet0/0/0
```

  b. From R2, ping the Loopback 0 address on D1. The ping should be successful.

```
R2# ping 209.165.201.1
Type escape sequence to abort.
Sending 5, 100-byte ICMP Echos to 209.165.201.1, timeout is 2 seconds:
!!!!!
Success rate is 100 percent (5/5), round-trip min/avg/max = 1/1/2 ms
```

Step 3. Verify IPv6 OSPFv3 routing table on R2.

  a. Verify the OSPFv3 IPv4 routing table on R2. Notice the default route, the intra–area, and inter–area OSPF routes are installed and received from fe80::12:1, which is R1.

```
R2# show ipv6 route ospf
< some output omitted  >
OE2 ::/0 [110/1], tag 1
     via FE80::12:1, GigabitEthernet0/0/0
O   2001:DB8:ACAD:1::/64 [110/2]
     via FE80::12:1, GigabitEthernet0/0/0
OI  2001:DB8:ACAD:11::/64 [110/2]
     via FE80::12:1, GigabitEthernet0/0/0
```

  b. From R2, ping the IPv6 Loopback 0 address on D1. The ping should be successful.

```
R2# ping 2001:db8:209:165:201::1
Type escape sequence to abort.
Sending 5, 100-byte ICMP Echos to 2001:DB8:209:165:201::1, timeout is 2 seconds:
!!!!!
Success rate is 100 percent (5/5), round-trip min/avg/max = 1/1/3 ms
```

  c. The output for the ping in the previous step and this step confirms that R2 has learned OSPFv3 routes for IPv4 and IPv6, including a default route for IPv4 and IPv6. The output also confirms R2 can ping the Loopback 0 address from both IPv4 and IPv6.

Part 3: Verify EIGRP Neighborships and Routing for IPv4 and IPv6

In this part, you will verify that EIGRP has established neighbor relationships and routing for IPv4 and IPv6.

Step 1.   Verify EIGRP for IPv4 neighborships on R3.

Issue the command to verify EIGRP has two IPv4 neighbors, as shown.

```
R3# show ip eigrp neighbors
EIGRP-IPv4 VR(CISCO) Address-Family Neighbors for AS(64512)
H   Address              Interface        Hold Uptime    SRTT   RTO  QSeq
                                          (sec)          (ms)        CntNum
1   10.1.32.2            Gi0/0/1            10 20:13:56      3   100  013
0   10.1.23.2            Gi0/0/0            13 20:31:08      1   100  019
```

Notice the two IPv4 neighbors, 10.1.23.2 and 10.1.32.2.

Step 2.   Verify the EIGRP for IPv6 neighborships on R3.

Issue the command to verify EIGRP has two IPv6 neighbors, as shown.

```
R3# show ipv6 eigrp neighbors
EIGRP-IPv6 VR(CISCO) Address-Family Neighbors for AS(64512)
H   Address              Interface        Hold Uptime    SRTT   RTO  QSeq
                                          (sec)          (ms)        CntNum
1   Link-local address:  Gi0/0/1            13 20:13:20      3   100  09
    FE80::32:2
0   Link-local address:  Gi0/0/0            11 20:32:08      1   100  019
    FE80::23:2
```

Notice the two IPv6 neighbors, fe80::23:2 and fe80::32:2.

Step 3.   Verify EIGRP for IPv4 routing table on R2.

Issue the command to display the EIGRP IPv4 routing table on R2, as shown.

```
R2# show ip route eigrp | begin 10.0
      10.0.0.0/8 is variably subnetted, 8 subnets, 2 masks
D        10.1.32.0/24 [90/15360] via 10.1.23.3, 20:35:38, GigabitEthernet0/0/1
D        10.3.3.0/24 [90/10880] via 10.1.23.3, 20:44:06, GigabitEthernet0/0/1
      198.51.100.0/25 is subnetted, 1 subnets
D        198.51.100.0 [90/16000] via 10.1.23.3, 20:29:04, GigabitEthernet0/0/1
```

Notice three internal EIGRP routes from 10.1.23.3, which is R3.

Step 4.   Verify EIGRP for IPv6 routing table on R2.

Issue the command to display the IPv6 EIGRP routing table on R2, as shown.

```
R2# show ipv6 route eigrp | begin 2001
D   2001:DB8:198:51:100::/80 [90/16000]
      via FE80::23:3, GigabitEthernet0/0/1
D   2001:DB8:ACAD:3::/64 [90/10880]
      via FE80::23:3, GigabitEthernet0/0/1
D   2001:DB8:ACAD:32::/64 [90/15360]
      via FE80::23:3, GigabitEthernet0/0/1
```

The output above confirms R2 has learned EIGRP routes for IPv4 and IPv6.

Part 4: Configure Redistribution from OSPFv3 to EIGRP

Recall that every protocol provides a seed metric at the time of redistribution. By default, when source protocols, such as, OSPF, RIP, and IS-IS, are redistributed into EIGRP, they are given an administrative distance of 170 and a seed metric of infinity. This prevents the installation of the redistributed routes into the EIGRP topology table. The seed metric can be set using the **redistribute** or **default-metric** command. Additionally, when using a route map, the seed metric can be configured using the **set metric** option.

For IPv4, you will set the seed metric using the **redistribute** command and the **default-metric** command.

Step 1.   Redistribute OSPFv3 into EIGRP for IPv4.

In this step we're going to the destination EIGRP AS 64512 to perform redistribution. Since EIGRP is using named mode the **redistribute** command is entered in the address family topology configuration mode, as shown.

```
R2(config)# router eigrp CISCO
R2(config-router)# address-family ipv4 autonomous-system 64512
R2(config-router-af)# topology base
R2(config-router-af-topology)# redistribute ospfv3 1 metric 1000000 10 255 1 1500
R2(config-router-af-topology)# end
```

Step 2.   On D2, verify redistribution of OSPFv3.

Issue the **show ip route eigrp** on D2 to see the external EIGRP routes from OSPFv3.

```
D2# show ip route eigrp | begin Gateway
Gateway of last resort is 10.1.32.3 to network 0.0.0.0

D*EX  0.0.0.0/0 [170/66560] via 10.1.32.3, 00:03:59, GigabitEthernet1/0/11
          10.0.0.0/8 is variably subnetted, 7 subnets, 2 masks
D EX      10.1.1.0/24
              [170/66560] via 10.1.32.3, 00:03:59, GigabitEthernet1/0/11
D EX      10.1.11.0/24
              [170/66560] via 10.1.32.3, 00:03:59, GigabitEthernet1/0/11
D EX      10.1.12.0/24
              [170/66560] via 10.1.32.3, 00:03:59, GigabitEthernet1/0/11
D         10.1.23.0/24
              [90/15360] via 10.1.32.3, 21:20:07, GigabitEthernet1/0/11
D         10.3.3.0/24 [90/10880] via 10.1.32.3, 21:20:07, GigabitEthernet1/0/11
```

Notice the gateway of last resort has been set and D2 has learned four external EIGRP routes which originated from OSPFv3. The OSPFv3 routes are imported into EIGRP as external, D EX routes with an administrative distance of 170, which are higher than the internal EIGRP routes of 90.

Step 3.   Redistribute OSPFv3 into EIGRP for IPv6.

Again, go to the destination protocol to perform redistribution. In this example you will set the seed metric using the **default-metric** command. Both commands are configured in the IPv6 address-family topology base, as shown.

```
R2(config)# router eigrp CISCO
R2(config-router)# address-family ipv6 autonomous-system 64512
R2(config-router-af)# topology base
```

```
R2(config-router-af-topology)# default-metric 1000000 10 255 1 1500
R2(config-router-af-topology)# redistribute ospf 1
```

Note: Do not leave AF topology configuration mode.

In the example above, the seed metric was set using the **default-metric** command.

Notice the **include-connected** option was not configured using the **redistribute ospf 1** command. The **include-connected** command must be set for OSPFv3 IPv6 connected interface on R2, in our example, 2001:db8:acad:12::/64 to be redistributed into EIGRP. With IPv4, connected interfaces are automatically advertised into the routing protocol for connected interfaces the source protocol is advertising. For IPv6, the administrator decides whether the connected subnets are included into redistribution.

Also notice under the EIGRP IPv6 address family, it is not possible to specify OSPFv3 as the source protocol for redistribution. Instead the **ospf** keyword automatically assumes OSPFv3 since the command is entered under the IPv6 address family.

Step 4.   On D2 verify OSPFv3 redistribution for IPv6.

Issue the command to view the IPv6 routing table for EIGRP.

```
D2# show ipv6 route eigrp | begin EX ::
EX  ::/0 [170/66560], tag 1
       via FE80::32:3, GigabitEthernet1/0/11
EX  2001:DB8:ACAD:1::/64 [170/66560]
       via FE80::32:3, GigabitEthernet1/0/11
D   2001:DB8:ACAD:3::/64 [90/10880]
       via FE80::32:3, GigabitEthernet1/0/11
EX  2001:DB8:ACAD:11::/64 [170/66560]
       via FE80::32:3, GigabitEthernet1/0/11
D   2001:DB8:ACAD:23::/64 [90/15360]
       via FE80::32:3, GigabitEthernet1/0/11
```

Notice the three highlighted external routes. The 2001:db8:acad:12::/64 prefix was not redistributed because of the missing **include-connected** keyword.

Step 5.   Redistribute OSPFv3 connected routes into EIGRP for IPv6.

a.   From the EIGRP IPv6 address family topology configuration mode configure redistribution with the same command as the previous step, but this time add **include-connected** as shown.

```
R2(config-router-af-topology)# redistribute ospf 1 include-connected
R2(config-router-af-topology)# end
```

b.   On D2, verify the IPv6 prefixes are being redistributed as before, as well as the connected prefix, which is included and highlighted in the routing table.

```
D2# show ipv6 route eigrp | begin EX ::
EX  ::/0 [170/66560], tag 1
       via FE80::32:3, GigabitEthernet1/0/11
EX  2001:DB8:ACAD:1::/64 [170/66560]
       via FE80::32:3, GigabitEthernet1/0/11
D   2001:DB8:ACAD:3::/64 [90/10880]
       via FE80::32:3, GigabitEthernet1/0/11
EX  2001:DB8:ACAD:11::/64 [170/66560]
       via FE80::32:3, GigabitEthernet1/0/11
```

```
EX  2001:DB8:ACAD:12::/64 [170/66560]
        via FE80::32:3, GigabitEthernet1/0/11
D   2001:DB8:ACAD:23::/64 [90/15360]
        via FE80::32:3, GigabitEthernet1/0/11
```

Part 5: Configure Redistribution from EIGRP for IPv4 into OSPFv3

In this part, you will perform EIGRP for IPv4 redistribution into OSPFv3.

Note: When redistributing into OSPFv2, you must include the **subnets** keyword. The keyword **subnets** is required for classless networks to be advertised. If omitted only classful networks using a classful mask will be redistributed.

Step 1.   On R2, redistribute EIGRP into OSPFv3.

The **redistribute** command is always performed on the destination protocol. Start by accessing the OSPFv3 address family for IPv4. Then redistribute the source protocol, EIGRP 64512 into the destination protocol, as shown.

```
R2(config)# router ospfv3 1
R2(config-router)# address-family ipv4 unicast
R2(config-router-af)# redistribute eigrp 64512
```

Note: Do not leave AF configuration mode.

Step 2.   Verify redistribution on D1.

Issue the **show ip route ospfv3** on D1 to see the external OSPF routes from EIGRP.

```
D1# show ip route ospfv3 | begin Gateway
Gateway of last resort is 0.0.0.0 to network 0.0.0.0

      10.0.0.0/8 is variably subnetted, 7 subnets, 2 masks
O IA    10.1.1.0/24 [110/2] via 10.1.11.1, 02:52:36, GigabitEthernet1/0/11
O IA    10.1.12.0/24 [110/2] via 10.1.11.1, 1d01h, GigabitEthernet1/0/11
O E2    10.1.23.0/24 [110/20] via 10.1.11.1, 00:03:55, GigabitEthernet1/0/11
O E2    10.1.32.0/24 [110/20] via 10.1.11.1, 00:03:55, GigabitEthernet1/0/11
O E2    10.3.3.0/24 [110/20] via 10.1.11.1, 00:03:55, GigabitEthernet1/0/11
      198.51.100.0/25 is subnetted, 1 subnets
O E2    198.51.100.0 [110/20] via 10.1.11.1, 00:03:55, GigabitEthernet1/0/11
```

Notice the highlighted external E2 OSPF routes. By default, external LSAs appear in the routing table marked as E2 with an external cost of 20.

Step 3.   Redistribute EIGRP into OSPFv3 using a type 1.

From the address family configuration mode, modify the **redistribute** command configured in Step 1 to specify an external type 1.

```
R2(config-router-af)# redistribute eigrp 64512 metric-type ?
  1  Set OSPF External Type 1 metrics
  2  Set OSPF External Type 2 metrics

R2(config-router-af)# redistribute eigrp 64512 metric-type 1
R2(config-router-af)# exit
```

Step 4.   Verify redistribution again on D1.

    a.   Issue the **show ip route ospfv3** on D1 to see the external OSPF routes.

```
D1# show ip route ospfv3 | begin Gateway
Gateway of last resort is 0.0.0.0 to network 0.0.0.0

       10.0.0.0/8 is variably subnetted, 7 subnets, 2 masks
O IA    10.1.1.0/24 [110/2] via 10.1.11.1, 03:10:29, GigabitEthernet1/0/11
O IA    10.1.12.0/24 [110/2] via 10.1.11.1, 1d02h, GigabitEthernet1/0/11
O E1    10.1.23.0/24 [110/22] via 10.1.11.1, 00:10:11, GigabitEthernet1/0/11
O E1    10.1.32.0/24 [110/22] via 10.1.11.1, 00:10:11, GigabitEthernet1/0/11
O E1    10.3.3.0/24 [110/22] via 10.1.11.1, 00:10:11, GigabitEthernet1/0/11
       198.51.100.0/25 is subnetted, 1 subnets
O E1    198.51.100.0 [110/22] via 10.1.11.1, 00:10:11, GigabitEthernet1/0/11
```

Notice the highlighted external E1 OSPF routes. These E1 routes have a cost of 22 which includes the default cost of 20 plus the internal cost of 2.

    b.   From D2 ping the Loopback address on D1 using Loopback address of D2. The ping should be successful. This verifies successful two-way redistribution on R2 and end-to-end connectivity for IPv4.

```
D2# ping 209.165.201.1 source loopback 0
Type escape sequence to abort.
Sending 5, 100-byte ICMP Echos to 209.165.201.1, timeout is 2 seconds:
Packet sent with a source address of 198.51.100.1
!!!!!
Success rate is 100 percent (5/5), round-trip min/avg/max = 2/2/3 ms
```

Step 5.   Configure redistribution of EIGRP for IPv6 routes into OSPFv3 using a route map.

Next, you will redistribute EIGRP for IPv6 routes into OSPFv3 using a route map to set the external LSA to a metric type 1, or E1.

    a.   First, you create a route map **named E2O** with a permit statement using a sequence number of 10. Because you are not going to use the **match** command, the default action is to match all. Then you set the metric type to an E1, or m, as shown.

```
R2(config)# route-map E2O permit 10
R2(config-route-map)# set metric-type type-1
R2(config-route-map)# exit
```

    b.   Next, you access the OSPFv3 IPv6 address family. Then you issue the **redistribute** command and specify the route map name. Ensure to add the **include-connected** after the route map name, as shown.

```
R2(config)# router ospfv3 1
R2(config-router)# address-family ipv6
R2(config-router-af)# redistribute eigrp 64512 route-map E2O include-connected
R2(config-router-af)# exit
```

The route map **E2O** will match all redistributed routes, including connected interfaces advertised in EIGRP 64512.

Step 6.   On D1 verify that routes from EIGRP for IPv6 are imported into OSPFv3 with the external metric type 1.

a.   Issue the **show ipv6 route ospf** command on D1 to see the external EIGRP routes. Notice the highlighted external E1 OSPF routes.

```
D1# show ipv6 route ospf
<  output omitted  >
OE1 2001:DB8:198:51:100::/80 [110/22]
     via FE80::11:1, GigabitEthernet1/0/11
OI  2001:DB8:ACAD:1::/64 [110/2]
     via FE80::11:1, GigabitEthernet1/0/11
OE1 2001:DB8:ACAD:3::/64 [110/22]
     via FE80::11:1, GigabitEthernet1/0/11
OI  2001:DB8:ACAD:12::/64 [110/2]
     via FE80::11:1, GigabitEthernet1/0/11
OE1 2001:DB8:ACAD:23::/64 [110/22]
     via FE80::11:1, GigabitEthernet1/0/11
OE1 2001:DB8:ACAD:32::/64 [110/22]
     via FE80::11:1, GigabitEthernet1/0/11
```

b.   From D2, ping the Loopback address on D1 using Loopback address of D2. The ping should be successful. This verifies full successful two-way redistribution on R2 and end-to-end connectivity for IPv6.

```
D2# ping 2001:db8:209:165:201::1 source loopback 0
Type escape sequence to abort.
Sending 5, 100-byte ICMP Echos to 2001:DB8:209:165:201::1, timeout is 2 sec-
onds:
Packet sent with a source address of 2001:DB8:198:51:100::1
!!!!!
Success rate is 100 percent (5/5), round-trip min/avg/max = 2/4/9 ms
```

Reflection Questions

1.   What is the difference between an external OSPF E2 and E1?

2.   What are three ways to set a seed metric during redistribution?

3.   What is the default action in a route map if you do not include the **match** command?

Router Interface Summary Table

Router Model	Ethernet Interface #1	Ethernet Interface #2	Serial Interface #1	Serial Interface #2
1800	Fast Ethernet 0/0 (F0/0)	Fast Ethernet 0/1 (F0/1)	Serial 0/0/0 (S0/0/0)	Serial 0/0/1 (S0/0/1)
1900	Gigabit Ethernet 0/0 (G0/0)	Gigabit Ethernet 0/1 (G0/1)	Serial 0/0/0 (S0/0/0)	Serial 0/0/1 (S0/0/1)
2801	Fast Ethernet 0/0 (F0/0)	Fast Ethernet 0/1 (F0/1)	Serial 0/1/0 (S0/1/0)	Serial 0/1/1 (S0/1/1)
2811	Fast Ethernet 0/0 (F0/0)	Fast Ethernet 0/1 (F0/1)	Serial 0/0/0 (S0/0/0)	Serial 0/0/1 (S0/0/1)
2900	Gigabit Ethernet 0/0 (G0/0)	Gigabit Ethernet 0/1 (G0/1)	Serial 0/0/0 (S0/0/0)	Serial 0/0/1 (S0/0/1)
4221	Gigabit Ethernet 0/0/0 (G0/0/0)	Gigabit Ethernet 0/0/1 (G0/0/1)	Serial 0/1/0 (S0/1/0)	Serial 0/1/1 (S0/1/1)
4300	Gigabit Ethernet 0/0/0 (G0/0/0)	Gigabit Ethernet 0/0/1 (G0/0/1)	Serial 0/1/0 (S0/1/0)	Serial 0/1/1 (S0/1/1)

Note: To find out how the router is configured, look at the interfaces to identify the type of router and how many interfaces the router has. There is no way to effectively list all the combinations of configurations for each router class. This table includes identifiers for the possible combinations of Ethernet and Serial interfaces in the device. The table does not include any other type of interface, even though a specific router may contain one. An example of this might be an ISDN BRI interface. The string in parenthesis is the legal abbreviation that can be used in Cisco IOS commands to represent the interface.

Device Config Final – Notes

16.1.3 Lab - Configure Route Redistribution Within the Same Interior Gateway Protocol

Topology

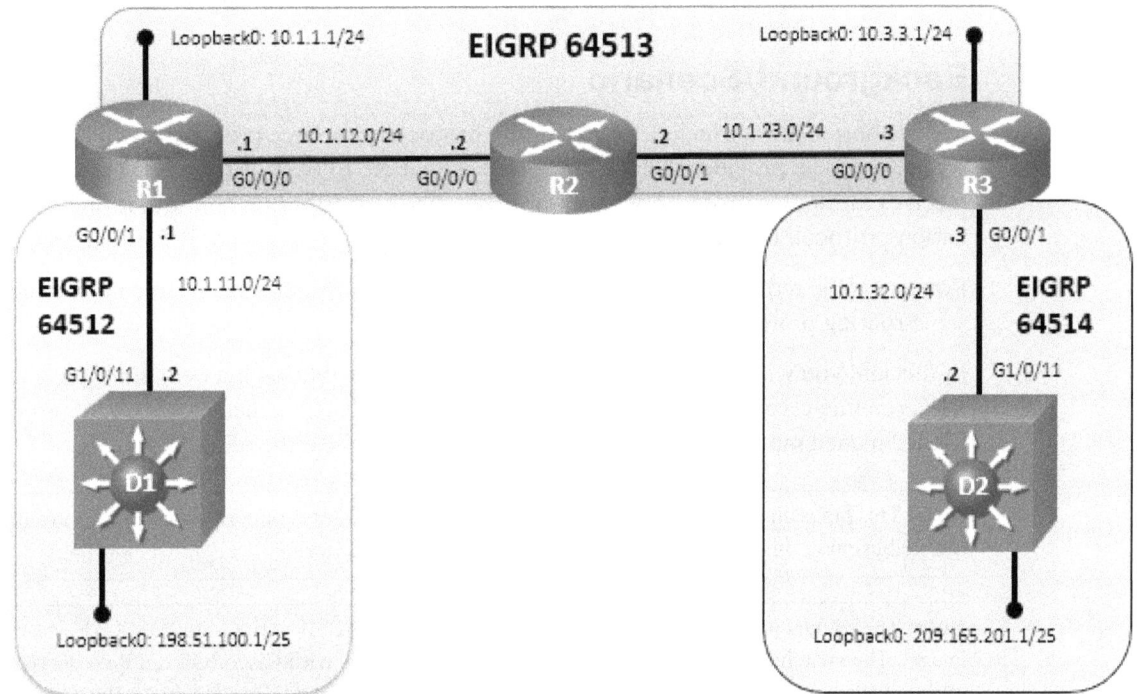

Addressing Table

Device	Interface	IP Address	Subnet Mask
R1	G0/0/0	10.1.12.1	255.255.255.0
	G0/0/1	10.1.11.1	255.255.255.0
	Loopback 0	10.1.1.1	255.255.255.0
R2	G0/0/0	10.1.12.2	255.255.255.0
	G0/0/1	10.1.23.2	255.255.255.0
R3	G0/0/0	10.1.23.3	255.255.255.0
	G0/0/1	10.1.32.1	255.255.255.0
	Loopback 0	10.3.3.3	255.255.255.0
D1	G1/0/11	10.1.11.2	255.255.255.0
	Loopback 0	198.51.100.1	255.255.255.128
D2	G1/0/11	10.1.32.2	255.255.255.0
	Loopback 0	209.165.201.1	255.255.255.128

Objectives

Part 1: Build the Network and Configure Basic Device Settings

Part 2: Configure and Verify Two-Way Redistribution on R1

Part 3: Configure and Verify Two-Way Redistribution on R3

Part 4: Filter and Verify Redistribution using a Distribute List and Prefix List

Background/Scenario

Redistribution always includes two routing protocols: a source protocol and a destination protocol. The source protocol provides the network prefixes that are to be redistributed. The destination protocol receives the source protocol network prefixes. The redistribution configuration exists under the destination protocol. Examples of source protocols are static, connected, RIP, EIGRP, OSPF, IS-IS, and BGP.

Routes can be redistributed between different routing protocols or between different processes of the same routing protocol.

In this lab, you will configure mutual or two-way redistribution between multiple EIGRP processes. R1 is running classic mode EIGRP for AS 64512 and EIGRP named mode for AS 64513. R3 is running EIGRP named mode for AS 64513 and classic mode EIGRP for AS 64514.

Note: This lab is an exercise in configuring and verifying two-way route redistribution on routers R1 and R3. Route redistribution in this lab does not reflect networking best practices.

Note: The routers used with CCNP hands-on labs are Cisco 4221s with Cisco IOS XE Release 16.9.4 (universalk9 image). The switches used in the labs are Cisco Catalyst 3650s with Cisco IOS XE Release 16.9.4 (universalk9 image). Other routers, switches, and Cisco IOS versions can be used. Depending on the model and Cisco IOS version, the commands available and the output produced might vary from what is shown in the labs.

Note: Make sure that all the devices have been erased and have no startup configurations. If you are unsure, contact your instructor.

Required Resources

- 3 Routers (Cisco 4221 with Cisco IOS XE Release 16.9.4 universal image or comparable)
- 2 Switches (Cisco 3650 with Cisco IOS XE release 16.9.4 universal image or comparable)
- 1 PC (Choice of operating system with terminal emulation program installed)
- Console cables to configure the Cisco IOS devices via the console ports
- Ethernet cables as shown in the topology

Instructions

Part 1: Build the Network and Configure Basic Device Settings

In Part 1, you will set up the network topology and configure basic settings.

Step 1.   Cable the network as shown in the topology.

Attach the devices as shown in the topology diagram, and cable as necessary.

Step 2.   Configure basic settings for each device.

a.   Console into each device, enter global configuration mode, and apply the basic settings for the lab. Initial configurations for each device are listed below.

Router R1

```
hostname R1
no ip domain lookup
banner motd # R1, Configure Route Redistribution Within the Same Interior
Gateway Protocol #
line con 0
 exec-timeout 0 0
 logging synchronous
 exit
interface g0/0/0
 ip address 10.1.12.1 255.255.255.0
 no shutdown
 exit
interface g0/0/1
 ip address 10.1.11.1 255.255.255.0
 no shutdown
 exit
interface loopback 0
 ip address 10.1.1.1 255.255.255.0
 no shutdown
 exit
router eigrp 64512
  eigrp router-id 1.1.1.1
  network 10.1.11.0 0.0.0.255
  exit
router eigrp CISCO
 address-family ipv4 unicast autonomous-system 64513
  eigrp router-id 1.1.1.1
  network 10.1.1.0 0.0.0.255
  network 10.1.12.0 0.0.0.255
  exit
end
```

Router R2

```
hostname R2
no ip domain lookup
banner motd # R2, Configure Route Redistribution Within the Same Interior
Gateway Protocol #
line con 0
 exec-timeout 0 0
 logging synchronous
 exit
interface g0/0/0
```

```
    ip address 10.1.12.2 255.255.255.0
    no shutdown
    exit
  interface g0/0/1
    ip address 10.1.23.2 255.255.255.0
    no shutdown
    exit
  router eigrp CISCO
   address-family ipv4 unicast autonomous-system 64513
   eigrp router-id 2.2.2.2
     network 10.1.12.0 0.0.0.255
     network 10.1.23.0 0.0.0.255
   end
```

Router R3

```
hostname R3
no ip domain lookup
banner motd # R3, Configure Route Redistribution Within the Same Interior
Gateway Protocol #
line con 0
  exec-timeout 0 0
  logging synchronous
  exit
interface g0/0/0
  ip address 10.1.23.1 255.255.255.0
  no shutdown
  exit
interface g0/0/1
  ip address 10.1.32.1 255.255.255.0
  no shutdown
  exit
interface loopback 0
  ip address 10.3.3.1 255.255.255.0
  no shutdown
  exit
router eigrp 64514
   eigrp router-id 3.3.3.3
   network 10.1.32.0 0.0.0.255
   exit
router eigrp CISCO
  address-family ipv4 unicast autonomous-system 64513
   eigrp router-id 3.3.3.3
   network 10.1.23.0 0.0.0.255
   network 10.3.3.0 0.0.0.255
   exit
end
```

Switch D1

```
hostname D1
no ip domain lookup
ip routing
banner motd # D1, Configure Route Redistribution Within the Same Interior
```

```
Gateway Protocol #
line con 0
 exec-timeout 0 0
 logging synchronous
 exit
interface range g1/0/1-24
 shutdown
 exit
interface g1/0/11
 no switchport
 ip address 10.1.11.2 255.255.255.0
 no shutdown
 exit
interface loopback 0
 ip address 198.51.100.1 255.255.255.128
 no shutdown
 exit
router eigrp 64512
 eigrp router-id 11.11.11.11
 network 10.1.11.0 0.0.0.255
 network 198.51.100.0 0.0.0.127
end
```

Switch D2

```
hostname D2
no ip domain lookup
ip routing
banner motd # D2, Configure Route Redistribution Within the Same Interior
Gateway Protocol #
line con 0
 exec-timeout 0 0
 logging synchronous
 exit
interface range g1/0/1-24
 shutdown
 exit
interface g1/0/11
 no switchport
 ip address 10.1.32.2 255.255.255.0
 no shutdown
 exit
interface loopback 0
 ip address 209.165.201.1 255.255.255.128
 no shutdown
 exit
router eigrp 64514
 eigrp router-id 22.22.22.22
 network 10.1.32.0 0.0.0.255
 redistribute static
 eigrp stub static
exit
```

```
ip route 0.0.0.0 0.0.0.0 Loopback0
end
```

    b. Set the clock on all devices to UTC time.

    c. Save the running configuration to startup-config on all devices.

Step 3. Verify EIGRP neighborships on R1.

    a. Verify that R1 has two EIGRP neighbor relationships. One EIGRP neighbor is from named mode AS 64513. The other neighbor is from EIGRP classic mode AS 64512.

```
R1# show ip eigrp neighbors
EIGRP-IPv4 Neighbors for AS(64512)
H   Address              Interface       Hold Uptime   SRTT   RTO  QSeq
                                         (sec)         (ms)        CntNum
0   10.1.11.2            Gi0/0/1           10 00:02:10    3   100  02
EIGRP-IPv4 VR(CISCO) Address-Family Neighbors for AS(64513)
H   Address              Interface       Hold Uptime   SRTT   RTO  QSeq
                                         (sec)         (ms)        CntNum
0   10.1.12.2            Gi0/0/0           11 00:03:38    2   100  07
```

    b. Next, verify that R1 has learned internal routes form each EIGRP neighbor. Issue the **show ip route eigrp** command for AS 64512, as shown. Notice the gateway of last resort is not set and the internal EIGRP route is from D1 Loopback 0.

```
R1# show ip route eigrp 64512 | begin Gateway
Gateway of last resort is not set
      198.51.100.0/25 is subnetted, 1 subnets
D        198.51.100.0
           [90/130816] via 10.1.11.2, 04:24:45, GigabitEthernet0/0/1
```

    c. Issue the **show ip route eigrp** command for EIGRP named mode, as shown. Notice the gateway of last resort is not set and the internal EIGRP routes are from R1 and R3.

```
R2# show ip route eigrp 64513 | begin Gateway
Gateway of last resort is not set
      10.0.0.0/8 is variably subnetted, 6 subnets, 2 masks
D        10.1.1.0/24 [90/10880] via 10.1.12.1, 00:10:57, GigabitEthernet0/0/0
D        10.3.3.0/24 [90/10880] via 10.1.23.1, 00:00:21, GigabitEthernet0/0/1
```

    d. Verify that R3 has two EIGRP neighbor relationships. First, issue the **show ip eigrp neighbors detail** command. Notice that neighbor 10.1.32.2 is an EIGRP stub neighbor advertising static routes. Notice that R3 is using EIGRP named mode for neighbor 10.1.23.2.

```
R3# show ip eigrp neighbors detail
EIGRP-IPv4 Neighbors for AS(64514)
H   Address            Interface      Hold Uptime  SRTT   RTO  Q  Seq
                                      (sec)        (ms)       Cnt Num
0   10.1.32.2          Gi0/0/1          11 07:40:16    1   100  0  3
    Version 25.0/2.0, Retrans: 2, Retries: 0, Prefixes: 1
    < some output omitted >

    Stub Peer Advertising (STATIC ) Routes
    Suppressing queries
```

```
EIGRP-IPv4 VR(CISCO) Address-Family Neighbors for AS(64513)
H   Address              Interface         Hold Uptime    SRTT   RTO  Q Seq
                                           (sec)          (ms)        CntNum
0   10.1.23.2            Gi0/0/0             13 07:36:50    1    100  0 39
    Version 23.0/2.0, Retrans: 1, Retries: 0, Prefixes: 2
< some output omitted >
```

e. Issue the **show ip route eigrp** command for EIGRP named mode, as shown. Notice that the two internal EIGRP routes are from AS 64513.

```
R3# show ip route eigrp 64513 | begin Gateway
Gateway of last resort is 10.1.32.2 to network 0.0.0.0

      10.0.0.0/8 is variably subnetted, 8 subnets, 2 masks
D        10.1.1.0/24 [90/16000] via 10.1.23.2, 00:34:20, GigabitEthernet0/0/0
D        10.1.12.0/24 [90/15360] via 10.1.23.2, 00:34:20, GigabitEthernet0/0/0
```

f. Issue the **show ip route eigrp** command for AS 64514, as shown. Notice that R3 has learned a default from the EIGRP stub neighbor.

```
R3# show ip route eigrp 64514 | begin Gateway
Gateway of last resort is 10.1.32.2 to network 0.0.0.0

D*EX  0.0.0.0/0 [170/130816] via 10.1.32.2, 07:47:51, GigabitEthernet0/0/1
```

g. From R3, ping the Loopback 0 interface on D2. The ping should be successful.

```
R3# ping 209.165.201.1
Type escape sequence to abort.
Sending 5, 100-byte ICMP Echos to 209.165.201.1, timeout is 2 seconds:
!!!!!
Success rate is 100 percent (5/5), round-trip min/avg/max = 1/1/2 ms
```

h. From R3, ping the Loopback 0 interface on D2 and source the ping from Loopback 0, as shown. The ping should not be successful.

```
R3# ping 209.165.201.1 source 10.3.3.1
Type escape sequence to abort.
Sending 5, 100-byte ICMP Echos to 209.165.201.1, timeout is 2 seconds:
Packet sent with a source address of 10.3.3.1
.....
Success rate is 0 percent (0/5)
```

Part 2: Configure Two-Way Redistribution on R1

In this part of the lab, you will perform EIGRP-to-EIGRP redistribution on R1. Remember that every protocol provides a seed metric at the time of redistribution. By default, source protocols, such as OSPF, RIP, IS-IS redistributed into EIGRP are given an administrative distance of 170 and a seed metric of infinity. This prevents the installation of the redistributed routes into the EIGRP topology table. However, if an EIGRP AS redistributes into another EIGRP AS, all the path metrics are preserved and included during redistribution. Therefore, setting an EIGRP seed metric is not required with performing EIGRP–to–EIGRP redistribution.

When performing redistribution, the **router** command defines the destination protocol and the **redistribute** command identifies the source protocol. For example:

```
Router(config)# router eigrp 5 !<--destination protocol
Router(config-router)# redistribute eigrp 10 !<--source protocol
```

In our example, the destination protocol is EIGRP AS 5 and the source protocol is EIGRP 10. This results in all EIGRP routes from AS 10 being redistributed into EIGRP AS 5.

Step 1.   Redistribute EIGRP 64513 into EIGRP 64512.

In this step, you are going to the destination EIGRP AS 64512 to perform redistribution. The source EIGRP AS is 64513.

```
R1(config)# router eigrp 64512
R1(config-router)# redistribute eigrp 64513
R1(config-router)# exit
```

Step 2.   On D1 verify one-way redistribution.

Issue the **show ip route eigrp** command on D1 to see the external EIGRP routes from AS 64513. Notice that the external EIGRP routes all originated from AS 64513. Also, notice that a gateway of last resort has not been set.

```
D1# show ip route eigrp | begin Gateway
Gateway of last resort is not set

      10.0.0.0/8 is variably subnetted, 6 subnets, 2 masks
D EX    10.1.1.0/24
           [170/130816] via 10.1.11.1, 00:07:40, GigabitEthernet1/0/11
D EX    10.1.12.0/24
           [170/3072] via 10.1.11.1, 00:07:40, GigabitEthernet1/0/11
D EX    10.1.23.0/24
           [170/3328] via 10.1.11.1, 00:07:40, GigabitEthernet1/0/11
D EX    10.3.3.0/24 [170/3353] via 10.1.11.1, 00:07:40, GigabitEthernet1/0/11
```

Step 3.   Redistribute EIGRP 64512 into EIGRP 64513.

Next, go to the destination AS 64513 to perform redistribution. The source EIGRP AS is 64512. To redistribute using an EIGRP named instance, you need to access the topology base, as shown.

```
R1(config)# router eigrp CISCO
R1(config-router)# address-family ipv4 unicast autonomous-system 64513
R1(config-router-af)# topology base
R1(config-router-af-topology)# redistribute eigrp 64512
R1(config-router-af-topology)# end
```

Step 4.   On R3 verify two-way redistribution on R1.

Issue the **show ip route eigrp 64513 | section D EX command** on D3 to see the external EIGRP routes from AS 64512.

```
R3# show ip route eigrp 64513 | section D EX
D EX    10.1.11.0/24
           [170/20480] via 10.1.23.2, 00:12:47, GigabitEthernet0/0/0
D EX    198.51.100.0
           [170/2580480] via 10.1.23.2, 00:12:47, GigabitEthernet0/0/0
```

Part 3: Configure Two-Way Redistribution on R3

In this part of the lab, you will perform EIGRP-to-EIGRP redistribution on R3. Remember a seed metric is not required.

Step 1.   Redistribute EIGRP 64513 into EIGRP 64514.

The **redistribute** command is always performed on the destination protocol. Start by accessing the EIGRP process 64514. Then redistribute the source protocol, EIGRP 64513, into the destination protocol, as shown.

```
R3(config)# router eigrp 64514
R3(config-router)# redistribute eigrp 64513
R3(config-router)# end
```

Step 2.   Verify redistribution on D2.

Issue the **show ip route eigrp** command on D2 to see the external EIGRP routes from AS 64513. Notice that the highlighted external EIGRP routes originated from AS 64512 and the other four external EIGRP prefixes originated from AS 64513.

```
D2# show ip route eigrp | begin Gateway
Gateway of last resort is 0.0.0.0 to network 0.0.0.0

      10.0.0.0/8 is variably subnetted, 7 subnets, 2 masks
D EX    10.1.1.0/24 [170/3353] via 10.1.32.1, 01:34:57, GigabitEthernet1/0/11
D EX    10.1.11.0/24
           [170/3584] via 10.1.32.1, 01:34:57, GigabitEthernet1/0/11
D EX    10.1.12.0/24
           [170/3328] via 10.1.32.1, 01:34:57, GigabitEthernet1/0/11
D EX    10.1.23.0/24
           [170/3072] via 10.1.32.1, 01:34:57, GigabitEthernet1/0/11
D EX    10.3.3.0/24
           [170/130816] via 10.1.32.1, 01:34:57, GigabitEthernet1/0/11
      198.51.100.0/25 is subnetted, 1 subnets
D EX    198.51.100.0
           [170/131584] via 10.1.32.1, 01:34:57, GigabitEthernet1/0/11
```

Step 3.   Redistribute EIGRP 64514 into EIGRP 64513 on R3.

Next, go to the destination AS 64513 to perform redistribution. To redistribute using an EIGRP named mode you need to access the topology base. Then, the source AS 64514 is specified using the **redistribute** command, as shown.

```
R3(config)# router eigrp CISCO
R3(config-router)# address-family ipv4 unicast autonomous-system 64513
R3(config-router-af)# topology base
R3(config-router-af-topology)# redistribute eigrp 64514
R3(config-router-af-topology)# end
```

Step 4.   On D1 verify two-way redistribution on R3 and end-to-end connectivity.

a.   Issue the **show ip route eigrp** command on D1 to see the external EIGRP routes. Notice that both of the highlighted external EIGRP routes originated from AS 64514.

```
D1# show ip route eigrp | begin Gateway
Gateway of last resort is 10.1.11.1 to network 0.0.0.0

D*EX  0.0.0.0/0 [170/131584] via 10.1.11.1, 00:01:28, GigabitEthernet1/0/11
      10.0.0.0/8 is variably subnetted, 7 subnets, 2 masks
D EX    10.1.1.0/24
           [170/130816] via 10.1.11.1, 04:48:32, GigabitEthernet1/0/11
```

```
D EX     10.1.12.0/24
               [170/3072] via 10.1.11.1, 04:48:32, GigabitEthernet1/0/11
D EX     10.1.23.0/24
               [170/3328] via 10.1.11.1, 04:48:32, GigabitEthernet1/0/11
D EX     10.1.32.0/24
               [170/3584] via 10.1.11.1, 00:01:28, GigabitEthernet1/0/11
D EX     10.3.3.0/24 [170/3353] via 10.1.11.1, 04:48:32, GigabitEthernet1/0/11
```

b. Next, from D1 ping the 209.165.201.1 address on D2 using the Loopback 0 address on D1. The ping should be successful. This verifies full end-to-end connectivity and successful two-way redistribution on R1 and R3.

```
D1# ping 209.165.201.1 source 198.51.100.1
Type escape sequence to abort.
Sending 5, 100-byte ICMP Echos to 209.165.201.1, timeout is 2 seconds:
Packet sent with a source address of 198.51.100.1
!!!!!
Success rate is 100 percent (5/5), round-trip min/avg/max = 2/2/4 ms
```

Part 4: Filter and Verify Redistribution using a Distribute List and Prefix List

In this part of the lab, we will filter specific EIGRP prefixes being redistributed into AS 64512 on R1 and advertised to D1. Note that the **redistribute** command cannot directly reference a prefix list, but a route map can refer to a prefix list using the **match** command. In our example, we will bind the prefix list using a distribute list.

Step 1.   Create a prefix list named FILTER and specify the action for each statement.

Only allow the default route, as well as the Loopback 0 address on R1 and R3, to be sent to D1 using the prefix list name FILTER, as shown on R1. Notice the **permit** statement allows prefixes to be advertised. The last statement, sequence 20 filters all other prefixes. If not explicitly set, the deny statement is implied. This is similar to using an ACL.

```
R1(config)# ip prefix-list FILTER seq 5 permit 0.0.0.0/0
R1(config)# ip prefix-list FILTER seq 10 permit 10.1.1.0/24
R1(config)# ip prefix-list FILTER seq 15 permit 10.3.3.0/24
R1(config)# ip prefix-list FILTER seq 20 deny 0.0.0.0/0 le 32
```

Step 2.   Apply the IP prefix-list using the distribute-list command in EIGRP 64512.

Next, apply the prefix-list FILTER to the distribute-list which filters routing advertisements to D1. The **out** keyword in the **distribute-list** command specifies that subnets matching prefix list FILTER will be filtered as the routing updates exit the GigabitEthernet0/0/1 interface toward D1. Using the keyword **in** would filter routes entering the routing table.

```
R1(config)# router eigrp 64512
R1(config-router)# distribute-list prefix FILTER out GigabitEthernet0/0/1
R1(config-router)# end
```

Step 3.   Verify route filtering.

a. Issue the **show ip prefix-list detail** command on R1 to verify the hit count for each sequence in the prefix list. Notice in our example the default route has 2 hits while the deny statement has 10 hits.

```
R1# show ip prefix-list detail
Prefix-list with the last deletion/insertion: FILTER
```

```
                  ip prefix-list FILTER:
                      count: 4, range entries: 1, sequences: 5 - 20, refcount: 4
                      seq 5 permit 0.0.0.0/0 (hit count: 2, refcount: 1)
                      seq 10 permit 10.1.1.0/24 (hit count: 6, refcount: 1)
                      seq 15 permit 10.3.3.0/24 (hit count: 4, refcount: 2)
                      seq 20 deny 0.0.0.0/0 le 32 (hit count: 10, refcount: 0)
```

b. Issue the **show ip route eigrp** command on D1 to see the external EIGRP routes. Notice the smaller routing table on D1.

```
D1# show ip route eigrp | begin Gateway
Gateway of last resort is 10.1.11.1 to network 0.0.0.0

D*EX   0.0.0.0/0 [170/131584] via 10.1.11.1, 01:22:55, GigabitEthernet1/0/11
          10.0.0.0/8 is variably subnetted, 4 subnets, 2 masks
D EX      10.1.1.0/24
              [170/130816] via 10.1.11.1, 00:31:39, GigabitEthernet1/0/11
D EX      10.3.3.0/24 [170/3353] via 10.1.11.1, 00:31:27, GigabitEthernet1/0/11
```

c. From D1 ping the 209.165.201.1 address on D2 using the Loopback 0 address on D1. The ping should be successful. This verifies full end-to-end connectivity and successful redistribution on R1 and R3, as well as route filtering on R1.

```
D1# ping 209.165.201.1 source 198.51.100.1
Type escape sequence to abort.
Sending 5, 100-byte ICMP Echos to 209.165.201.1, timeout is 2 seconds:
Packet sent with a source address of 198.51.100.1
!!!!!
Success rate is 100 percent (5/5), round-trip min/avg/max = 2/2/3 ms
```

Reflection Questions

1. Why is a seed metric not required when redistributing EIGRP into another EIGRP process?

_____

_____

2. What other source protocol(s), other than EIGRP does not require a seed metric defined for redistribution into EIGRP?

_____

_____

3. Which EIGRP prefixes were filtered on R1 and not sent to D1?

_____

_____

Router Interface Summary Table

Router Model	Ethernet Interface #1	Ethernet Interface #2	Serial Interface #1	Serial Interface #2
1800	Fast Ethernet 0/0 (F0/0)	Fast Ethernet 0/1 (F0/1)	Serial 0/0/0 (S0/0/0)	Serial 0/0/1 (S0/0/1)
1900	Gigabit Ethernet 0/0 (G0/0)	Gigabit Ethernet 0/1 (G0/1)	Serial 0/0/0 (S0/0/0)	Serial 0/0/1 (S0/0/1)
2801	Fast Ethernet 0/0 (F0/0)	Fast Ethernet 0/1 (F0/1)	Serial 0/1/0 (S0/1/0)	Serial 0/1/1 (S0/1/1)
2811	Fast Ethernet 0/0 (F0/0)	Fast Ethernet 0/1 (F0/1)	Serial 0/0/0 (S0/0/0)	Serial 0/0/1 (S0/0/1)
2900	Gigabit Ethernet 0/0 (G0/0)	Gigabit Ethernet 0/1 (G0/1)	Serial 0/0/0 (S0/0/0)	Serial 0/0/1 (S0/0/1)
4221	Gigabit Ethernet 0/0/0 (G0/0/0)	Gigabit Ethernet 0/0/1 (G0/0/1)	Serial 0/1/0 (S0/1/0)	Serial 0/1/1 (S0/1/1)
4300	Gigabit Ethernet 0/0/0 (G0/0/0)	Gigabit Ethernet 0/0/1 (G0/0/1)	Serial 0/1/0 (S0/1/0)	Serial 0/1/1 (S0/1/1)

Note: To find out how the router is configured, look at the interfaces to identify the type of router and how many interfaces the router has. There is no way to effectively list all the combinations of configurations for each router class. This table includes identifiers for the possible combinations of Ethernet and Serial interfaces in the device. The table does not include any other type of interface, even though a specific router may contain one. An example of this might be an ISDN BRI interface. The string in parenthesis is the legal abbreviation that can be used in Cisco IOS commands to represent the interface.

Device Config Final – Notes

16.1.4 Lab - Implement Route Redistribution Between Multiple Protocols

Topology

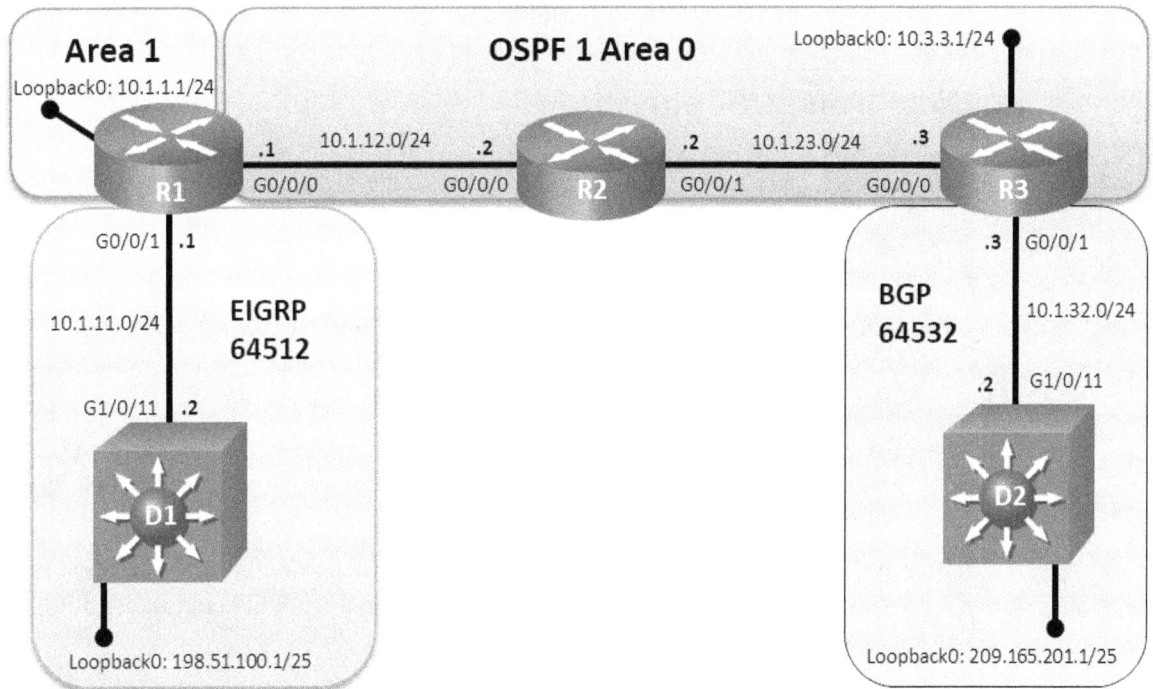

Addressing Table

Device	Interface	IP Address	Subnet Mask
R1	G0/0/0	10.1.12.1	255.255.255.0
	G0/0/1	10.1.11.1	255.255.255.0
	Loopback 0	10.1.1.1	255.255.255.0
R2	G0/0/0	10.1.12.2	255.255.255.0
	G0/0/1	10.1.23.2	255.255.255.0
R3	G0/0/0	10.1.23.3	255.255.255.0
	G0/0/1	10.1.32.3	255.255.255.0
	Loopback 0	10.3.3.3	255.255.255.0
D1	G1/0/11	10.1.11.2	255.255.255.0
	Loopback 0	198.51.100.1	255.255.255.128
D2	G1/0/11	10.1.32.2	255.255.255.0
	Loopback 0	209.165.201.1	255.255.255.128

Objectives

Part 1: Build the Network and Configure Basic Device Settings

Part 2: Configure and Verify Two-Way Redistribution on R1

Part 3: Configure and Verify Two-Way Redistribution on R3

Part 4: Filter and Verify Redistribution using a Route Map

Background/Scenario

Every routing protocol has a unique redistribution behavior. The default redistribution behavior for EIGRP, OSPF, and BGP is as follows:

- External routes redistributed into EIGRP have a seed metric of infinity and EIGRP routes set with infinity are not installed into the EIGRP topology table.

- External routes redistributed into OSPF by default, are Type 2 (E2) external. Routes sourced from BGP will have a seed metric of 1, while other routing protocols will have a seed metric of 20. Only classful networks are redistributed, not subnets.

- External routes redistributed into BGP have the origin set to **incomplete (?)**, the multi-exit discriminator (MED) is set to the IGP metric and the weight is set to 32,768. By default, BGP does not redistribute internal BGP routes.

In this lab, you will configure mutual or two-way redistribution between multiple EIGRP and OSPF on R1. Then you will configure two-way redistribution between OSPF and BGP on R3. Finally, a route map will be used to selectively redistribute routes.

Note: This lab is an exercise in configuring and verifying two-way route redistribution on routers R1 and R3. Route redistribution in this lab does not reflect networking best practices.

Note: The routers used with CCNP hands-on labs are Cisco 4221 with Cisco IOS XE Release 16.9.4 (universalk9 image). The switches used in the labs are Cisco Catalyst 3650 with Cisco IOS XE Release 16.9.4 (universalk9 image). Other routers, switches, and Cisco IOS versions can be used. Depending on the model and Cisco IOS version, the commands available and the output produced might vary from what is shown in the labs.

Note: Make sure that all the devices have been erased and have no startup configurations. If you are unsure, contact your instructor.

Required Resources

- 3 Routers (Cisco 4221 with Cisco IOS XE Release 16.9.4 universal image or comparable)

- 2 Switches (Cisco 3650 with Cisco IOS XE release 16.9.4 universal image or comparable)

- 1 PC (Windows with terminal emulation program)

- Console cables to configure the Cisco IOS devices via the console ports

- Ethernet cables as shown in the topology

Instructions

Part 1: Build the Network and Configure Basic Device Settings

In Part 1, you will set up the network topology and configure basic settings.

Step 1.   Cable the network as shown in the topology.

Attach the devices as shown in the topology diagram, and cable as necessary.

Step 2.   Configure basic settings for each device.

a.   Console into each device, enter global configuration mode, and apply the basic settings for the lab. Initial configurations for each device are listed below.

Router R1

```
hostname R1
no ip domain lookup
banner motd # R1, Configure BGP Route Redistribution #
line con 0
 exec-timeout 0 0
 logging synchronous
 exit
interface g0/0/0
 ip address 10.1.12.1 255.255.255.0
 no shutdown
 exit
interface g0/0/1
 ip address 10.1.11.1 255.255.255.0
 no shutdown
 exit
interface loopback 0
 ip address 10.1.1.1 255.255.255.0
 ip ospf network point-to-point
 ip ospf cost 15
 no shutdown
 exit
router eigrp 64512
  eigrp router-id 1.1.1.1
  network 10.1.11.0 0.0.0.255
  exit
router ospf 1
  router-id 1.1.1.1
  network 10.1.1.0 0.0.0.255 area 1
  network 10.1.12.0 0.0.0.255 area 0
  exit
end
```

Router R2

```
hostname R2
no ip domain lookup
banner motd # R2, Configure BGP Route Redistribution #
line con 0
```

```
 exec-timeout 0 0
 logging synchronous
 exit
interface g0/0/0
 ip address 10.1.12.2 255.255.255.0
 no shutdown
 exit
interface g0/0/1
 ip address 10.1.23.2 255.255.255.0
 no shutdown
 exit
router ospf 1
  router-id 2.2.2.2
  network 10.1.12.0 0.0.0.255 area 0
  network 10.1.23.0 0.0.0.255 area 0
  end
```

Router R3

```
hostname R3
no ip domain lookup
banner motd # R3, Configure BGP Route Redistribution #
line con 0
 exec-timeout 0 0
 logging synchronous
 exit
interface g0/0/0
 ip address 10.1.23.3 255.255.255.0
 no shutdown
 exit
interface g0/0/1
 ip address 10.1.32.3 255.255.255.0
 no shutdown
 exit
interface loopback 0
 ip address 10.3.3.1 255.255.255.0
 ip ospf network point-to-point
 no shutdown
 exit
router ospf 1
  router-id 3.3.3.3
  network 10.3.3.0 0.0.0.255 area 0
  network 10.1.23.0 0.0.0.255 area 0
router bgp 64532
 bgp router-id 3.3.3.3
 no bgp default ipv4-unicast
 neighbor 10.1.32.2 remote-as 64532
 address-family ipv4
  neighbor 10.1.32.2 activate
  neighbor 10.1.32.2 next-hop-self
exit-address-family
end
```

Switch D1

```
hostname D1
no ip domain lookup
ip routing
banner motd # D1, Configure BGP Route Redistribution #
line con 0
 exec-timeout 0 0
 logging synchronous
 exit
interface range g1/0/1-24
 shutdown
 exit
interface g1/0/11
 no switchport
 ip address 10.1.11.2 255.255.255.0
 no shutdown
 exit
interface loopback 0
 ip address 198.51.100.1 255.255.255.128
 no shutdown
 exit
router eigrp 64512
 eigrp router-id 11.11.11.11
 network 10.1.11.0 0.0.0.255
 network 198.51.100.0 0.0.0.127
end
```

Switch D2

```
hostname D2
no ip domain lookup
ip routing
banner motd # D2, Configure BGP Route Redistribution #
line con 0
 exec-timeout 0 0
 logging synchronous
 exit
interface range g1/0/1-24
 shutdown
 exit
interface g1/0/11
 no switchport
 ip address 10.1.32.2 255.255.255.0
 no shutdown
 exit
interface loopback 0
 ip address 209.165.201.1 255.255.255.128
 no shutdown
 exit
router bgp 64532
 bgp router-id 22.22.22.22
 no bgp default ipv4-unicast
```

```
neighbor 10.1.32.3 remote-as 64532
address-family ipv4
 network 209.165.201.0 mask 255.255.255.128
 neighbor 10.1.32.3 activate
exit-address-family
end
```

b. Set the clock on all devices to UTC time.

c. Save the running configuration to startup-config on all devices.

Step 3. Verify EIGRP on R1.

a. Verify that R1 has one EIGRP neighbor with D1.

```
R1# show ip eigrp neighbors
EIGRP-IPv4 Neighbors for AS(64512)
H   Address              Interface          Hold Uptime    SRTT   RTO  QSeq
                                            (sec)          (ms)        CntNum
0   10.1.11.2            Gi0/0/1            10 00:04:08    3      100  032
```

b. Next, issue the **show ip route eigrp** command, as shown, and notice the internal EIGRP route is from D1, Loopback 0.

```
R1# show ip route eigrp | begin Gateway
Gateway of last resort is not set
      198.51.100.0/25 is subnetted, 1 subnets
D        198.51.100.0
              [90/130816] via 10.1.11.2, 00:07:43, GigabitEthernet0/0/1
```

Step 4. Verify OSPF on R1.

a. Verify that R1 has one OSPF neighbor with R2.

```
R1# show ip ospf neighbor

Neighbor ID     Pri   State      Dead Time   Address        Interface
2.2.2.2          1    FULL/BDR   00:00:39    10.1.12.2      GigabitEthernet0/0/0
```

b. Next, on R1 issue the **show ip route ospf** command, as shown. Notice the two OSPF intra–area routes.

```
R1# show ip route ospf | begin Gateway
Gateway of last resort is not set
      10.0.0.0/8 is variably subnetted, 8 subnets, 2 masks
O        10.1.23.0/24 [110/2] via 10.1.12.2, 00:35:32, GigabitEthernet0/0/0
O        10.3.3.0/24 [110/3] via 10.1.12.2, 00:35:32, GigabitEthernet0/0/0
```

Step 5. Verify OSPF on R3.

a. Verify that R3 has one OSPF neighbor with R2 using the **show ip ospf neighbor** command.

```
R3# show ip ospf neighbor

Neighbor ID     Pri   State      Dead Time   Address        Interface
2.2.2.2          1    FULL/DR    00:00:36    10.1.23.2      GigabitEthernet0/0/0
```

b. Next, issue the **show ip route ospf** command, as shown. Notice the first route is an OSPF inter–area route from Area 1 on R1 with an OSPF cost of 17. Notice the other route is an OSPF intra–area prefix with an OSPF cost of 2.

```
R3# show ip route ospf | begin Gateway
Gateway of last resort is not set

      10.0.0.0/8 is variably subnetted, 6 subnets, 2 masks
O IA    10.1.1.0/24 [110/17] via 10.1.23.2, 06:07:43, GigabitEthernet0/0/0
O       10.1.12.0/24 [110/2] via 10.1.23.2, 06:16:33, GigabitEthernet0/0/0
```

Step 6.   Verify BGP on R3.

a.   Issue the **show bgp ipv4 unicast neighbors** command, as shown. Notice the "established" BGP peer at 10.1.32.2, D2.

```
R3# show bgp ipv4 unicast neighbors | include BGP
BGP neighbor is 10.1.32.2,  remote AS 64532, internal link
  BGP version 4, remote router ID 22.22.22.22
  BGP state = Established, up for 00:11:11
  < some output omitted >
```

b.   Next, issue the **show bgp ipv4 unicast** command and notice the 209.165.201.0/25 prefix is learned via internal BGP (iBGP).

```
R3# show bgp ipv4 unicast | begin Network
     Network          Next Hop          Metric LocPrf Weight Path
*>i  209.165.201.0/25 10.1.32.2              0    100      0 i
```

Part 2: Configure Two-Way Redistribution on R1

In this part of the lab, you will perform mutual EIGRP-to-OSPF and OSPF-to-EIGRP redistribution on R1.

Step 1.   Redistribute EIGRP 64512 into OSPF.

By default, EIGRP routes redistributed into OSPF will be seen as external Type 2 (E2) routes. In this step, you will change the external Type 2 (E2) routes to external Type 1 (E1) routes and specify the **subnets** keyword.

Note: Best practice suggests always entering the keyword **subnets**. However, depending on the IOS version, the keyword **subnets** may automatically be appended to the **redistribute** command in OSPFv2.

```
R1(config)# router ospf 1
R1(config-router)# redistribute eigrp 64512 metric-type 1 subnets
R1(config-router)# exit
```

Step 2.   Verify one-way redistribution on R3.

a.   Issue the **show ip route ospf** command on R3 to see the external OSPF routes are Type 1 with a cost of 22. Both E1 routes originated from EIGRP AS 64512.

```
R3# show ip route ospf | begin Gateway
Gateway of last resort is not set

      10.0.0.0/8 is variably subnetted, 9 subnets, 2 masks
O IA    10.1.1.0/24 [110/17] via 10.1.23.2, 08:19:58, GigabitEthernet0/0/0
O E1    10.1.11.0/24 [110/22] via 10.1.23.2, 01:21:42, GigabitEthernet0/0/0
O       10.1.12.0/24 [110/2] via 10.1.23.2, 08:28:48, GigabitEthernet0/0/0
      198.51.100.0/25 is subnetted, 1 subnets
O E1    198.51.100.0 [110/22] via 10.1.23.2, 01:21:42, GigabitEthernet0/0/0
```

Step 3. Redistribute OSPF into EIGRP 64512.

By default, external EIGRP routes are given an administrative distance of 170 and a seed metric of infinity, which prevents the installation of the redistributed routes into the EIGRP topology table. This default path metric can be changed from infinity to a specific value for bandwidth, delay, reliability, load, and maximum transmission unit (MTU).

Redistribute OSPF into EIGRP 64512 and set the EIGRP K values as shown.

```
R1(config)# router eigrp 64512
R1(config-router)# redistribute ospf 1 metric 1000000 10 255 1 1500
R1(config-router)# end
```

Step 4. Verify Two-Way Redistribution on D1.

Issue the **show ip route eigrp | begin Gateway command** on D1 to see four external EIGRP routes from OSPF.

```
D1# show ip route eigrp | begin Gateway
Gateway of last resort is not set
      10.0.0.0/8 is variably subnetted, 6 subnets, 2 masks
D EX     10.1.1.0/24 [170/5376] via 10.1.11.1, 00:00:12, GigabitEthernet1/0/11
D EX     10.1.12.0/24
             [170/5376] via 10.1.11.1, 00:00:12, GigabitEthernet1/0/11
D EX     10.1.23.0/24
             [170/5376] via 10.1.11.1, 00:00:12, GigabitEthernet1/0/11
D EX     10.3.3.0/24 [170/5376] via 10.1.11.1, 00:00:12, GigabitEthernet1/0/11
```

Part 3: Configure Two-Way Redistribution on R3

In this part of the lab you will perform OSPF-to-BGP and BGP-to-OSPF redistribution on R3.

Step 1. Redistribute OSPF into BGP.

By default, when configuring redistribution of OSPF into BGP without any keywords, only OSPF intra-area and inter-area routes are redistributed. Your BGP configuration determines where the **redistribute** command is entered. When using an address family, the **redistribute** command is entered in the address family configuration mode, otherwise it is entered under the BGP process.

In this lab, the **redistribute** command is configured under the BGP IPv4 address family as shown. Notice that no additional keywords or sub-commands are configured.

```
R3(config)# router bgp 64532
R3(config-router)# address-family ipv4
R3(config-router-af)# redistribute ospf 1
R3(config-router-af)# exit
```

Step 2. Verify redistribution on D2.

a. Issue the **show bgp ipv4 unicast** command on D2 to see the default behavior of OSPF being redistributed into BGP. Notice that only intra-area and inter-area routes are redistributed. All routes redistributed into BGP have the origin code set to incomplete (?) and the weight set to 32,768. Additionally, the MED / Metric value was set based on the OSPF cost on R3. The 10.1.1.0/24 was the inter-area route and shows a Metric of 17. The 10.1.12.0/24 route was an intra-area route with a Metric of 2. The 10.1.23.0/24 and 10.3.3.0/24 routes were directly connected via OSPF and show a Metric of 0.

```
D2# show bgp ipv4 unicast | begin Origin
Origin codes: i - IGP, e - EGP, ? - incomplete
RPKI validation codes: V valid, I invalid, N Not found

        Network         Next Hop        Metric LocPrf Weight Path
 *>i 10.1.1.0/24        10.1.23.2           17    100      0 ?
 *>i 10.1.12.0/24       10.1.23.2            2    100      0 ?
 *>i 10.1.23.0/24       0.0.0.0              0    100      0 ?
 *>i 10.3.3.0/24        0.0.0.0              0    100      0 ?
 *>  209.165.201.0/25 10.1.32.2              0         32768 i
```

b. Next, on R3 issue the **redistribute** command again, and add the keyword **match** to redistribute internal and external Type 1 OSPF routes into BGP.

```
R3(config)# router bgp 64532
R3(config-router)# address-family ipv4
R3(config-router-af)# redistribute ospf 1 match internal external 1
R3(config-router-af)# exit
```

c. Issue the **show bgp ipv4 unicast** command on D2, as shown, to see the two external OSPF routes redistributed into BGP. Notice the metric of 22 and origin code of incomplete (?). Remember that both prefixes originated in EIGRP AS 64512.

```
D2# show bgp ipv4 unicast | begin Network
        Network         Next Hop        Metric LocPrf Weight Path
 *>i 10.1.1.0/24        10.1.32.3           17    100      0 ?
 *>i 10.1.11.0/24       10.1.32.3           22    100      0 ?
 *>i 10.1.12.0/24       10.1.32.3            2    100      0 ?
 *>i 10.1.23.0/24       10.1.32.3            0    100      0 ?
 *>i 10.3.3.0/24        10.1.32.3            0    100      0 ?
 *>i 198.51.100.0/25    10.1.32.3           22    100      0 ?
 *>  209.165.201.0/25 0.0.0.0               0         32768 i
```

Step 3.   Redistribute BGP into OSPF.

In this step of the lab, you will redistribute BGP into OSPF.

Note: BGP is designed to support a large routing table, whereas, IGPs are not. Redistribution of BGP into an IGP on a router with a larger BGP routing table (for example the internet table with 800,000 plus routes) should use selective route redistribution. Otherwise the IGP can become unstable in the routing domain, which can lead to packet loss.

When redistributing BGP into OSPF, internal BGP routes are not redistributed, by default.

On R3 redistribute BGP into OSPF as shown, adding the **subnets** keyword and leaving the default OSPF external Type 2.

```
R3(config)# router ospf 1
R3(config-router)# redistribute bgp 64532 subnets
R3(config-router)# exit
```

Step 4.   Verify redistribution of BGP into OSPF on R1.

Issue the **show ip route ospf** command on R1. Notice the internal BGP prefix 209.165.201.0/25 was not redistributed and missing from the OSPF routing table. Normally only external BGP routes are redistributed. However, in this lab there are no eBGP routes.

```
R1# show ip route ospf | begin Gateway
Gateway of last resort is not set
```

```
            10.0.0.0/8 is variably subnetted, 9 subnets, 2 masks
O              10.1.23.0/24 [110/2] via 10.1.12.2, 03:49:00, GigabitEthernet0/0/0
O              10.3.3.0/24 [110/3] via 10.1.12.2, 03:49:00, GigabitEthernet0/0/0
```

Step 5.   Allow iBGP routes to be redistributed into OSPF.

To allow internal BGP routes to be redistributed into OSPF requires the **bgp redistribute-internal** command. This command is issued within the BGP address family process for IPv4, as shown.

```
R3(config)# router bgp 64532
R3(config-router)# address-family ipv4
R3(config-router-af)# bgp redistribute-internal
R3(config-router-af)# end
```

Step 6.   Verify redistribution of iBGP into OSPF on R1.

a.   Issue the **show ip route ospf** command on R1 to see the OSPF external Type 2 (E2) route from BGP. Notice the default seed metric of 1 for BGP routes redistributed into OSPF.

```
R1# show ip route ospf | begin Gateway
Gateway of last resort is not set

        10.0.0.0/8 is variably subnetted, 9 subnets, 2 masks
O          10.1.23.0/24 [110/2] via 10.1.12.2, 04:12:12, GigabitEthernet0/0/0
O          10.3.3.0/24 [110/3] via 10.1.12.2, 04:12:12, GigabitEthernet0/0/0
        209.165.201.0/25 is subnetted, 1 subnets
O E2      209.165.201.0 [110/1] via 10.1.12.2, 00:00:06, GigabitEthernet0/0/0
```

b.   From D2 ping the 198.51.100.1 address on D1 using the Loopback 0 address on D2. The ping should be successful. This verifies full end-to-end connectivity and successful redistribution on R1 and R3.

```
D2# ping 198.51.100.1 source 209.165.201.1
Type escape sequence to abort.
Sending 5, 100-byte ICMP Echos to 198.51.100.1, timeout is 2 seconds:
Packet sent with a source address of 209.165.201.1
!!!!!
Success rate is 100 percent (5/5), round-trip min/avg/max = 2/2/4 ms
```

Part 4: Filter and Verify Redistribution using a Prefix List and Route Map

In this part of the lab, you will use a prefix list and router map on R3 to filter specific OSPF prefixes from being redistributed into BGP.

Step 1.   Create a prefix list named LOOPBACK and specify the action for each statement.

Permit only the Loopback addresses on D1, R1 and R3, as shown. The last sequence 20 statement filters all other prefixes. If not explicitly set, the deny statement is implied similar to using an ACL.

```
R3(config)# ip prefix-list LOOPBACK seq 5 permit 198.51.100.0/25
R3(config)# ip prefix-list LOOPBACK seq 10 permit 10.1.1.0/24
R3(config)# ip prefix-list LOOPBACK seq 15 permit 10.3.3.0/24
R3(config)# ip prefix-list LOOPBACK seq 20 deny 0.0.0.0/0 le 32
```

Step 2. Apply the IP prefix list using a route map.

Create a route map named OSPF-into-BGP. Next, apply the prefix-list LOOPBACK to the route map which allows redistribution of prefixes into BGP. Any prefixes matching the named prefix list LOOPBACK with a permit statement, will be redistributed into BGP.

```
R3(config)# route-map OSPF-into-BGP permit 10
R3(config-route-map)# match ip address prefix-list LOOPBACK
R3(config-route-map)# exit
```

Step 3. Apply the route map to the redistribute command.

Apply the route map named OSPF-into-BGP at the end of the redistribute command, as shown.

```
R3(config)# router bgp 64532
R3(config-router)# address-family ipv4
R3(config-router-af)# redistribute ospf 1 match internal external 1 route-map OSPF-into-BGP
R3(config-router-af)# end
```

Step 4. Verify redistribution filtering.

a. Issue the **show ip prefix-list detail** command on R3 to verify the hit count for each sequence in the prefix list. Notice in our example each Loopback address has 2 hits, while the deny statement has 6 hits.

```
R3# show ip prefix-list detail
Prefix-list with the last deletion/insertion: LOOPBACK
ip prefix-list LOOPBACK:
   count: 4, range entries: 1, sequences: 5 - 20, refcount: 3
   seq 5 permit 198.51.100.0/25 (hit count: 2, refcount: 1)
   seq 10 permit 10.1.1.0/24 (hit count: 2, refcount: 1)
   seq 15 permit 10.3.3.0/24 (hit count: 2, refcount: 2)
   seq 20 deny 0.0.0.0/0 le 32 (hit count: 6, refcount: 1)
```

b. Issue the **show bgp ipv4 unicast** command to verify filtering of OSPF prefixes into BGP. Notice only the Loopback addresses on D1, R1 and R3 are redistributed into BGP.

```
D2# show bgp ipv4 unicast | begin Network
     Network          Next Hop          Metric LocPrf Weight Path
 *>i 10.1.1.0/24      10.1.32.3             17    100      0 ?
 *>i 10.3.3.0/24      10.1.32.3              0    100      0 ?
 *>i 198.51.100.0/25  10.1.32.3             22    100      0 ?
 *>  209.165.201.0/25 0.0.0.0                0         32768 i
```

c. From D2 ping the 198.51.100.1 address on D1. The ping should not be successful.

```
D2# ping 198.51.100.1
Type escape sequence to abort.
Sending 5, 100-byte ICMP Echos to 198.51.100.1, timeout is 2 seconds:
.....
Success rate is 0 percent (0/5)
```

d. From D2 ping the 198.51.100.1 address on D1 using the Loopback 0 address on D2. The ping should be successful. This verifies full end-to-end connectivity and successful redistribution on R1 and R3 as well as redistribution filtering on R3 using a prefix list and route map.

```
D2# ping 198.51.100.1 source 209.165.201.1
Type escape sequence to abort.
```

```
Sending 5, 100-byte ICMP Echos to 198.51.100.1, timeout is 2 seconds:
Packet sent with a source address of 209.165.201.1
!!!!!
Success rate is 100 percent (5/5), round-trip min/avg/max = 2/2/4 ms
```

Reflection Questions

1. Why does the ping to 198.51.100.1 fail when you do not specify the source Loopback 209.165.201.1 on D2?

   _____

   _____

   _____

2. By default, routes redistributed into BGP have the origin code, weight, and MED have which values?

   _____

   _____

   _____

3. By default, which OSFP prefixes are redistributed into BGP using the **redistribute ospf 1** command?

   _____

   _____

   _____

4. Redistributed routes into OSPF have a metric of 20, with the exception of redistributed BGP routes which have a seed metric of _____ ?

Router Interface Summary Table

Router Model	Ethernet Interface #1	Ethernet Interface #2	Serial Interface #1	Serial Interface #2
1800	Fast Ethernet 0/0 (F0/0)	Fast Ethernet 0/1 (F0/1)	Serial 0/0/0 (S0/0/0)	Serial 0/0/1 (S0/0/1)
1900	Gigabit Ethernet 0/0 (G0/0)	Gigabit Ethernet 0/1 (G0/1)	Serial 0/0/0 (S0/0/0)	Serial 0/0/1 (S0/0/1)
2801	Fast Ethernet 0/0 (F0/0)	Fast Ethernet 0/1 (F0/1)	Serial 0/1/0 (S0/1/0)	Serial 0/1/1 (S0/1/1)
2811	Fast Ethernet 0/0 (F0/0)	Fast Ethernet 0/1 (F0/1)	Serial 0/0/0 (S0/0/0)	Serial 0/0/1 (S0/0/1)
2900	Gigabit Ethernet 0/0 (G0/0)	Gigabit Ethernet 0/1 (G0/1)	Serial 0/0/0 (S0/0/0)	Serial 0/0/1 (S0/0/1)
4221	Gigabit Ethernet 0/0/0 (G0/0/0)	Gigabit Ethernet 0/0/1 (G0/0/1)	Serial 0/1/0 (S0/1/0)	Serial 0/1/1 (S0/1/1)
4300	Gigabit Ethernet 0/0/0 (G0/0/0)	Gigabit Ethernet 0/0/1 (G0/0/1)	Serial 0/1/0 (S0/1/0)	Serial 0/1/1 (S0/1/1)

Note: To find out how the router is configured, look at the interfaces to identify the type of router and how many interfaces the router has. There is no way to effectively list all the combinations of configurations for each router class. This table includes identifiers for the possible combinations of Ethernet and Serial interfaces in the device. The table does not include any other type of interface, even though a specific router may contain one. An example of this might be an ISDN BRI interface. The string in parenthesis is the legal abbreviation that can be used in Cisco IOS commands to represent the interface.

Device Config Final – Notes

CHAPTER 17

Troubleshooting Redistribution

17.1.2 Lab - Troubleshoot Redistribution

Objectives

Troubleshoot network issues related to redistribution.

Background/Scenario

In this topology D1, R1, and R2 are implementing OSPFv2. D1 is redistributing Loopback 0 into OSPFv2 area 10. R2 and R3 are BGP neighbors in AS 64512. You will be loading configurations with intentional errors or missing configurations onto the network. Your tasks are to FIND the error(s), document your findings and the command(s) or method(s) used to fix them, FIX the issue(s) presented here, and then test the network to ensure both of the following conditions are met:

1. the complaint received in the ticket is resolved

2. full reachability is restored

Note: The routers used with CCNP hands-on labs are Cisco 4221 with Cisco IOS XE Release 16.9.4 (universalk9 image). The switches used in the labs are Cisco Catalyst 3650 with Cisco IOS XE Release 16.9.4 (universalk9 image). Other routers, switches, and Cisco IOS versions can be used. Depending on the model and Cisco IOS version, the commands available and the output produced might vary from what is shown in the labs. Refer to the Router Interface Summary Table at the end of the lab for the correct interface identifiers.

Note: Make sure that the devices have been erased and have no startup configurations. If you are unsure, contact your instructor.

Required Resources

- 3 Routers (Cisco 4221 with Cisco IOS XE Release 16.9.4 universal image or comparable)

- 2 Switches (Cisco 3560 with Cisco IOS XE Release 16.9.4 universal image or comparable)

- 1 PC (Choice of operating system with terminal emulation program installed)

- Console cables to configure the Cisco IOS devices via the console ports

- Ethernet cables as shown in the topology

Instructions

Part 1: Trouble Ticket 17.1.2.1

Topology

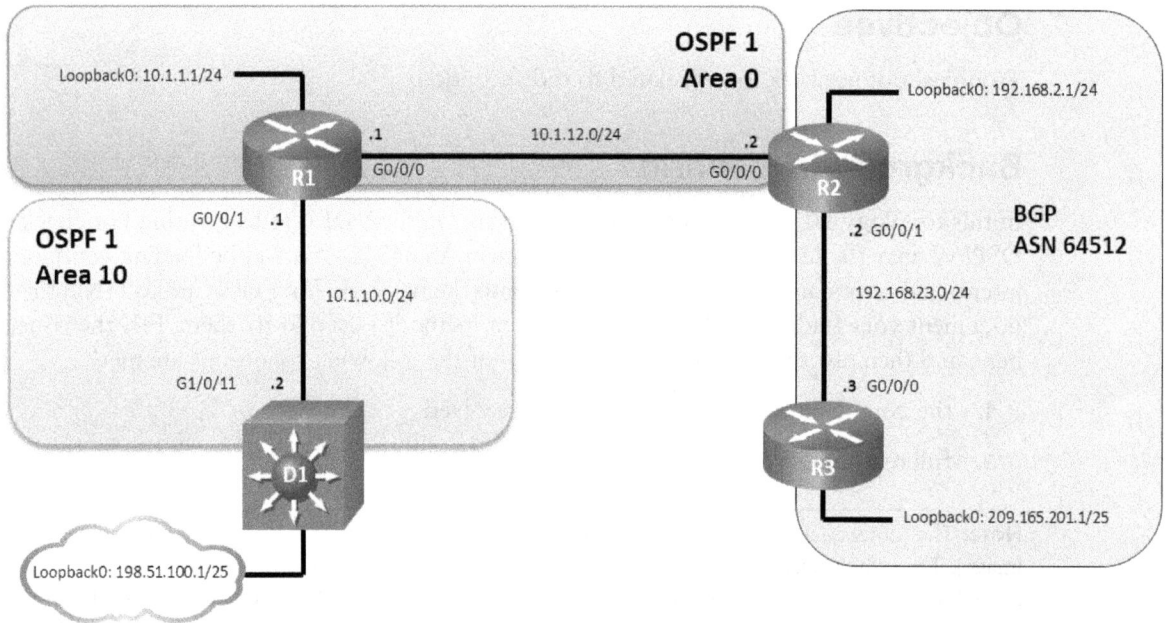

Addressing Table

Device	Interface	IP Address	Subnet Mask
R1	G0/0/0	10.1.12.1	255.255.255.0
	G0/0/1	10.1.10.1	255.255.255.0
	Loopback 0	10.1.1.1	255.255.255.0
R2	G0/0/0	10.1.12.2	255.255.255.0
	G0/0/1	192.168.23.2	255.255.255.0
	Loopback 0	192.168.2.1	255.255.255.0
R3	G0/0/0	192.168.23.3	255.255.255.0
	Loopback 0	209.165.201.1	255.255.255.128
D1	G1/0/11	10.1.10.2	255.255.255.0
	Loopback 0	198.51.100.1	255.255.255.128

Scenario

During a routine maintenance window, router R2 was replaced and upgraded to support gigabit interfaces. Instead of modifying the previous R2 configuration to support gigabit interfaces and loading the
configuration file on R2, the network engineer decided to type the IOS commands on R2 from memory. As a result, only partial redistribution is occurring between OSPFv2 and BGP. The network engineer
requested your help in diagnosing and resolving the issue(s) to restore full connectivity.

Use the commands listed below to load the configuration files for this trouble ticket:

Device	Command
R1	`copy flash:/enarsi/17.1.2.1-r1-config.txt run`
R2	`copy flash:/enarsi/17.1.2.1-r2-config.txt run`
R3	`copy flash:/enarsi/17.1.2.1-r3-config.txt run`
D1	`copy flash:/enarsi/17.1.2.1-d1-config.txt run`

- Passwords on all devices are **cisco12345**. If a username is required, use **admin**.
- After you have fixed the ticket, change the MOTD on EACH DEVICE using the following command

  banner motd # This is $(hostname) FIXED from ticket <ticket number> #

- Save the configuration by issuing the **wri** command (on each device).
- Inform your instructor that you are ready for the next ticket.
- After the instructor approves your solution for this ticket, issue the **reset.now** privileged EXEC command. This script will clear your configurations and reload the devices.

Part 2: Trouble Ticket 17.1.2.2

Topology

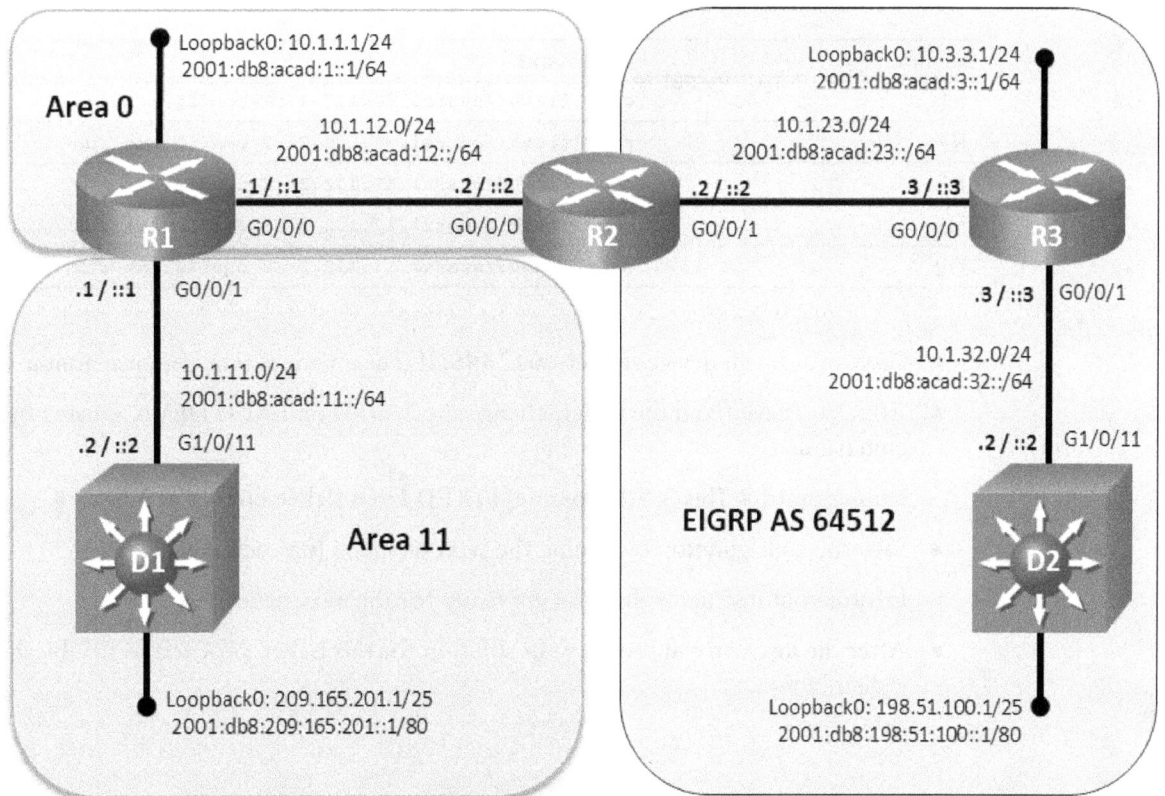

Addressing Table

Device	Interface	IPv4 Address/Mask	IPv6 Address/Prefix	IPv6 Link Local
R1	G0/0/0	10.1.12.1/24	2001:db8:acad:12::1/64	fe80::12:1
	G0/0/1	10.1.11.1/24	2001:db8:acad:11::1/64	fe80::11:1
	Loopback 0	10.1.1.1/24	2001:db8:acad:1::1/64	fe80::1:1
R2	G0/0/0	10.1.12.2/24	2001:db8:acad:12::2/64	fe80::12:2
	G0/0/1	10.1.23.2/24	2001:db8:acad:23::2/64	fe80::23:2
R3	G0/0/0	10.1.23.3/24	2001:db8:acad:23::3/64	fe80::23:3
	G0/0/1	10.1.32.1/24	2001:db8:acad:32::3/64	fe80::32:3
	Loopback 0	10.3.3.3/24	2001:db8:acad:3::3/64	fe80::3:3
D1	G1/0/11	10.1.11.2/24	2001:db8:acad:11::2/64	fe80::11:2
	Loopback 0	209.165.201.1/25	2001:db8:209:165:201::1/80	fe80::209:1
D2	G1/0/11	10.1.32.2/24	2001:db8:acad:32::2/64	fe80::32:2
	Loopback 0	198.51.100.1/25	2001:db8:198:51:100::1/80	fe80::198:1

Scenario

R2 is performing route redistribution between OSPFv3 AF and Named EIGRP for IPv4 and IPv6. During a routine maintenance window, router R2 was upgraded. As a result, redistribution is not occurring for IPv4 between OSPFv3 and named EIGRP.

Use the commands listed below to load the configuration files for this trouble ticket:

Device	Command
R1	`copy flash:/enarsi/17.1.2.2-r1-config.txt run`
R2	`copy flash:/enarsi/17.1.2.2-r2-config.txt run`
R3	`copy flash:/enarsi/17.1.2.2-r3-config.txt run`
D1	`copy flash:/enarsi/17.1.2.2-d1-config.txt run`
D2	`copy flash:/enarsi/17.1.2.2-d2-config.txt run`

- Passwords on all devices are **cisco12345**. If a username is required, use **admin**.
- After you have fixed the ticket, change the MOTD on EACH DEVICE using the following command:

  banner motd # This is $(hostname) FIXED from ticket <ticket number> #

- Save the configuration by issuing the **wri** command (on each device).
- Inform your instructor that you are ready for the next ticket.
- After the instructor approves your solution for this ticket, proceed to troubleshoot IPv6 redistribution.

Part 3: Trouble Ticket 17.1.2.3

Topology

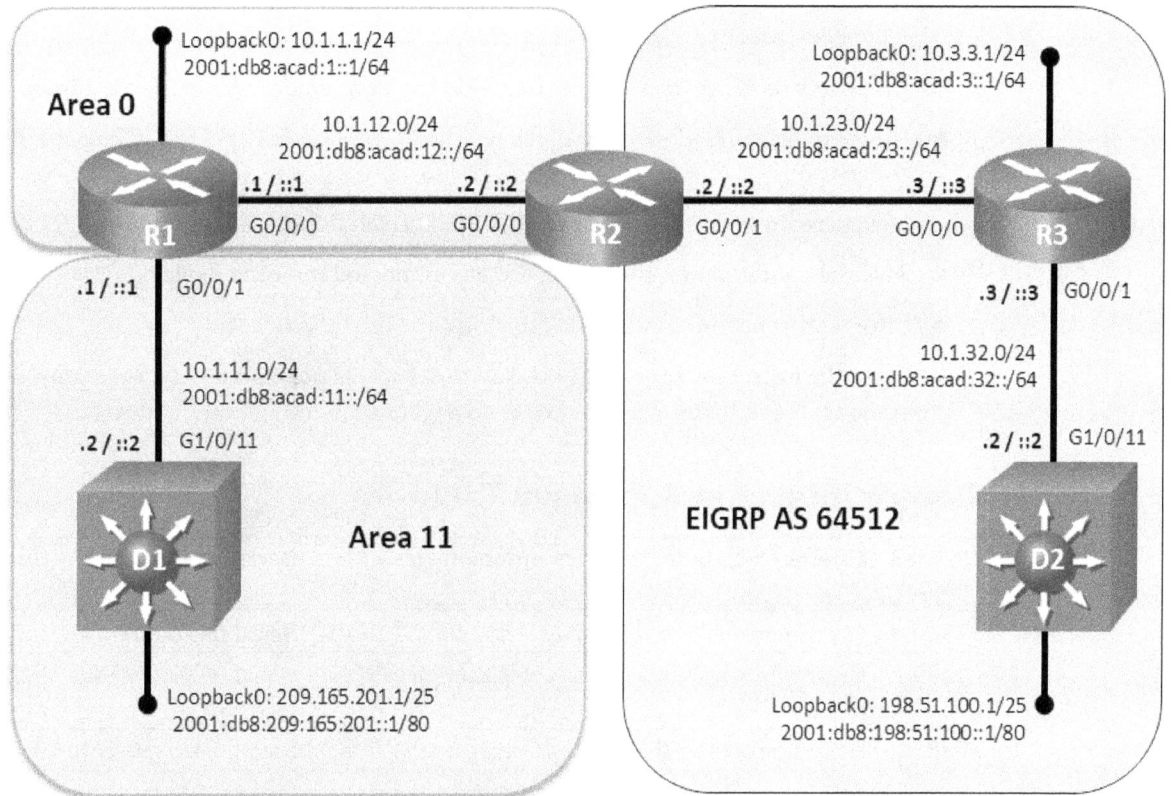

Addressing Table

Device	Interface	IPv4 Address/Mask	IPv6 Address/Prefix	IPv6 Link Local
R1	G0/0/0	10.1.12.1/24	2001:db8:acad:12::1/64	fe80::12:1
	G0/0/1	10.1.11.1/24	2001:db8:acad:11::1/64	fe80::11:1
	Loopback 0	10.1.1.1/24	2001:db8:acad:1::1/64	fe80::1:1
R2	G0/0/0	10.1.12.2/24	2001:db8:acad:12::2/64	fe80::12:2
	G0/0/1	10.1.23.2/24	2001:db8:acad:23::2/64	fe80::23:2
R3	G0/0/0	10.1.23.3/24	2001:db8:acad:23::3/64	fe80::23:3
	G0/0/1	10.1.32.1/24	2001:db8:acad:32::3/64	fe80::32:3
	Loopback 0	10.3.3.3/24	2001:db8:acad:3::3/64	fe80::3:3
D1	G1/0/11	10.1.11.2/24	2001:db8:acad:11::2/64	fe80::11:2
	Loopback 0	209.165.201.1/25	2001:db8:209:165:201::1/80	fe80::209:1
D2	G1/0/11	10.1.32.2/24	2001:db8:acad:32::2/64	fe80::32:2
	Loopback 0	198.51.100.1/25	2001:db8:198:51:100::1/80	fe80::198:1

Scenario

R2 is performing route redistribution between OSPFv3 AF and Named EIGRP for IPv4 and IPv6. During a routine maintenance window router R2 was upgraded. As a result, IPv6 has limited connectivity and only partial redistribution is occurring between OSPFv3 and named EIGRP for IPv6.

Use the previously loaded 17.1.2.2 configuration files for trouble ticket 17.1.2.3:

- Passwords on all devices are **cisco12345**. If a username is required, use **admin**.

- After you have fixed the ticket, change the MOTD on EACH DEVICE using the following command

  banner motd # This is $(hostname) FIXED from ticket <ticket number> #

- Save the configuration by issuing the **wri** command (on each device).

- Inform your instructor that you are ready for the next ticket.

- After the instructor approves your solution for this ticket, issue the **reset.now** privileged EXEC command. This script will clear your configurations and reload the devices.

Router Interface Summary Table

Router Model	Ethernet Interface #1	Ethernet Interface #2	Serial Interface #1	Serial Interface #2
1800	Fast Ethernet 0/0 (F0/0)	Fast Ethernet 0/1 (F0/1)	Serial 0/0/0 (S0/0/0)	Serial 0/0/1 (S0/0/1)
1900	Gigabit Ethernet 0/0 (G0/0)	Gigabit Ethernet 0/1 (G0/1)	Serial 0/0/0 (S0/0/0)	Serial 0/0/1 (S0/0/1)
2801	Fast Ethernet 0/0 (F0/0)	Fast Ethernet 0/1 (F0/1)	Serial 0/1/0 (S0/1/0)	Serial 0/1/1 (S0/1/1)
2811	Fast Ethernet 0/0 (F0/0)	Fast Ethernet 0/1 (F0/1)	Serial 0/0/0 (S0/0/0)	Serial 0/0/1 (S0/0/1)
2900	Gigabit Ethernet 0/0 (G0/0)	Gigabit Ethernet 0/1 (G0/1)	Serial 0/0/0 (S0/0/0)	Serial 0/0/1 (S0/0/1)
4221	Gigabit Ethernet 0/0/0 (G0/0/0)	Gigabit Ethernet 0/0/1 (G0/0/1)	Serial 0/1/0 (S0/1/0)	Serial 0/1/1 (S0/1/1)
4300	Gigabit Ethernet 0/0/0 (G0/0/0)	Gigabit Ethernet 0/0/1 (G0/0/1)	Serial 0/1/0 (S0/1/0)	Serial 0/1/1 (S0/1/1)

Note: To find out how the router is configured, look at the interfaces to identify the type of router and how many interfaces the router has. There is no way to effectively list all the combinations of configurations for each router class. This table includes identifiers for the possible combinations of Ethernet and Serial interfaces in the device. The table does not include any other type of interface, even though a specific router may contain one. An example of this might be an ISDN BRI interface. The string in parenthesis is the legal abbreviation that can be used in Cisco IOS commands to represent the interface.

Device Config Final – Notes

CHAPTER 18

VRF, MPLS, and MPLS Layer 3 VPNs

18.1.2 Lab - Implement VRF-Lite

Topology

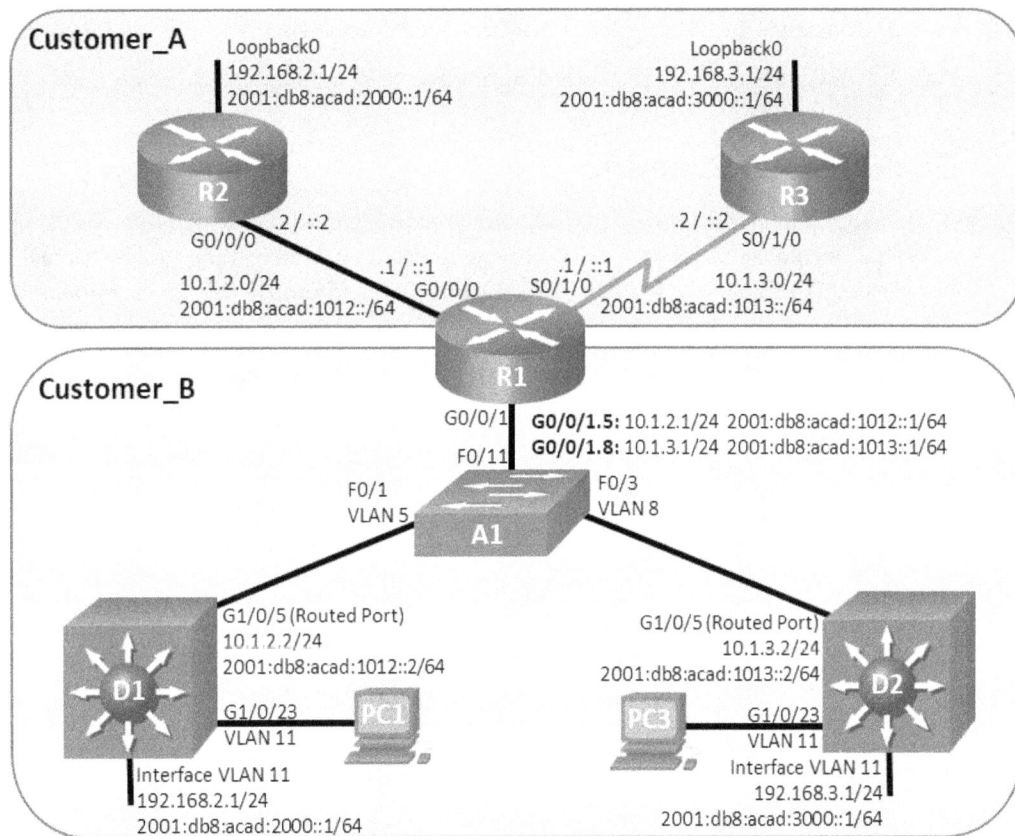

Addressing Table

Device	Interface	IPv4 Address	IPv6 Address	IPv6 Link-Local
R1	G0/0/0	10.1.2.1/24	2001:db8:acad:1012::1/64	fe80::1:1
	G0/0/1.5	10.1.2.1/24	2001:db8:acad:1012::1/64	fe80::1:2
	G0/0/1.8	10.1.3.1/24	2001:db8:acad:1013::1/64	fe80::1:4
	S0/1/0	10.1.3.1/25	2001:db8:acad:1013::1/64	fe80::1:2
R2	G0/0/0	10.2.3.2/24	2001:db8:acad:1023::2/64	fe80::2:1
	Loopback0	192.168.2.1/24	2001:db8:acad:2000::1/64	fe80::2:2
R3	S0/1/0	10.1.3.3/25	2001:db8:acad:1013::3/64	fe80::3:1
	Loopback0	192.168.3.1/27	2001:db8:acad:3000::1/64	fe80::3:2
D1	G1/0/5	10.1.2.2/24	2001:db8:acad:1012::2/64	fe80::d1:1
	VLAN 11	192.168.2.1/24	2001:db8:acad:2000::2/64	fe80::d1:2

Device	Interface	IPv4 Address	IPv6 Address	IPv6 Link-Local
D2	G1/0/5	10.1.3.2/24	2001:db8:acad:1013::2/64	fe80::d2:1
	VLAN 11	192.168.3.1/24	2001:db8:acad:3000::1/64	fe80::d2:2

Objectives

Part 1: Build the Network and Configure Basic Device Settings

Part 2: Configure and Verify VRF and Interface Addressing

Part 3: Configure and Verify Static Routing for Reachability Inside Each VRF

Background/Scenario

By default, all interfaces on a router are included in the global routing table. Service providers must be able to virtualize the router, thus creating multiple, virtual routing tables. Virtual Routing and Forwarding (VRF) can do just that. VRF-Lite is VRF without the MPLS component.

In this lab, you will work on R1, playing the part of a service provider router, as it supports two customers who have the same addressing scheme configured. Your task is to deploy VRF-Lite and static routing so that the customers have full reachability within their network.

Note: This lab is an exercise in developing, deploying, and verifying VRF-Lite, and does not reflect networking best practices.

Note: The routers and switches used with CCNP hands-on labs are Cisco 4221 and Cisco 3650, both with Cisco IOS XE Release 16.9.4 (universalk9 image), and Cisco 2960+ with IOS release 15.2 (lanbase image). Other routers, switches, and Cisco IOS versions can be used. Depending on the model and Cisco IOS version, the commands available and the output produced might vary from what is shown in the labs.

Note: Ensure that the routers and switches have been erased and have no startup configurations. If you are unsure contact your instructor.

Note: The PCs used in this lab do not require addressing. They are needed to bring interface VLAN 11 up.

Required Resources

- 3 Routers (Cisco 4221 with Cisco IOS XE Release 16.9.4 universal image or comparable)
- 2 Switches (Cisco 3650 with Cisco IOS XE release 16.9.4 universal image or comparable)
- 1 Switch (Cisco 2960+ with Cisco IOS release 15.2 lanbase image or comparable)
- 2 PCs (Windows with a terminal emulation program, such as Tera Term)
- Console cables to configure the Cisco IOS devices via the console ports
- Ethernet and serial cables as shown in the topology

Part 1: Build the Network and Configure Basic Device Settings

In Part 1, you will set up the network topology and configure basic settings on all devices.

Step 1.   Cable the network as shown in the topology.

Attach the devices as shown in the topology diagram, and cable as necessary.

Step 2.   Configure basic settings for each device.

    a.   Console into each device, enter global configuration mode, and apply the basic settings. A command list for each device using the following startup configurations.

Router R1

```
enable
configure terminal
hostname R1
no ip domain lookup
ipv6 unicast-routing
banner motd # R1, Implement VRF-Lite #
line con 0
 exec-timeout 0 0
 logging synchronous
 exit
line vty 0 4
 privilege level 15
 password cisco123
 exec-timeout 0 0
 logging synchronous
 login
 cxit
```

Router R2

```
enable
configure terminal
hostname R2
no ip domain lookup
ipv6 unicast-routing
banner motd # R2, Implement VRF-Lite #
line con 0
 exec-timeout 0 0
 logging synchronous
 exit
line vty 0 4
 privilege level 15
 password cisco123
 exec-timeout 0 0
 logging synchronous
 login
 exit
interface g0/0/0
 ip address 10.1.2.2 255.255.255.0
 ipv6 address fe80::2:1 link-local
 ipv6 address 2001:db8:acad:1012::2/64
```

```
  no shutdown
  exit
interface loopback 0
  ip address 192.168.2.1 255.255.255.0
  ipv6 address fe80::2:2 link-local
  ipv6 address 2001:db8:acad:2000::1/64
  no shutdown
  exit
ip route 0.0.0.0 0.0.0.0 g0/0/0 10.1.2.1
ipv6 route ::/0 g0/0/0 2001:db8:acad:1012::1
```

Router R3

```
enable
configure terminal
hostname R3
no ip domain lookup
ipv6 unicast-routing
banner motd # R3, Implement VRF-Lite #
line con 0
  exec-timeout 0 0
  logging synchronous
  exit
line vty 0 4
  privilege level 15
  password cisco123
  exec-timeout 0 0
  logging synchronous
  login
  exit
interface s0/1/0
  ip address 10.1.3.2 255.255.255.0
  ipv6 address fe80::3:1 link-local
  ipv6 address 2001:db8:acad:1013::2/64
  no shutdown
  exit
interface loopback 0
  ip address 192.168.3.1 255.255.255.0
  ipv6 address fe80::3:2 link-local
  ipv6 address 2001:db8:acad:3000::1/64
  no shutdown
  exit
ip route 0.0.0.0 0.0.0.0 s0/1/0 10.1.3.1
ipv6 route ::/0 s0/1/0 2001:db8:acad:1013::1
```

Switch D1

```
enable
configure terminal
hostname D1
no ip domain lookup
ip routing
ipv6 unicast-routing
banner motd # D1, Implement VRF-Lite #
```

```
line con 0
 exec-timeout 0 0
 logging synchronous
 exit
line vty 0 4
 privilege level 15
 password cisco123
 exec-timeout 0 0
 logging synchronous
 login
 exit
interface range g1/0/1-24, g1/1/1-4, g0/0
 shutdown
 exit
interface g1/0/5
 no switchport
 ip address 10.1.2.2 255.255.255.0
 ipv6 address fe80::d1:1 link-local
 ipv6 address 2001:db8:acad:1012::2/64
 no shutdown
 exit
vlan 11
 name LOCAL_VLAN
 exit
interface vlan 11
 ip address 192.168.2.1 255.255.255.0
 ipv6 address fe80::d1:2 link-local
 ipv6 address 2001:db8:acad:2000::1/64
 no shutdown
 exit
interface g1/0/23
 switchport mode access
 switchport access vlan 11
 no shutdown
 exit
ip route 0.0.0.0 0.0.0.0 g1/0/5 10.1.2.1
ipv6 route ::/0 g1/0/5 2001:db8:acad:1012::1
```

Switch D2

```
enable
configure terminal
hostname D2
no ip domain lookup
ip routing
ipv6 unicast-routing
banner motd # D2, Implement VRF-Lite #
line con 0
 exec-timeout 0 0
 logging synchronous
 exit
line vty 0 4
```

```
 privilege level 15
 password cisco123
 exec-timeout 0 0
 logging synchronous
 login
 exit
interface range g1/0/1-24, g1/1/1-4, g0/0
 shutdown
 exit
interface g1/0/5
 no switchport
 ip address 10.1.3.2 255.255.255.0
 ipv6 address fe80::d2:1 link-local
 ipv6 address 2001:db8:acad:1013::2/64
 no shutdown
 exit
vlan 11
 name LOCAL_VLAN
 exit
interface vlan 11
 ip address 192.168.3.1 255.255.255.0
 ipv6 address fe80::d2:2 link-local
 ipv6 address 2001:db8:acad:3000::1/64
 no shutdown
 exit
interface g1/0/23
 switchport mode access
 switchport access vlan 11
 no shutdown
 exit
ip route 0.0.0.0 0.0.0.0 g1/0/5 10.1.3.1
ipv6 route ::/0 g1/0/5 2001:db8:acad:1013::1
```

Switch A1

```
enable
configure terminal
hostname A1
no ip domain lookup
banner motd # A1, Implement VRF-Lite #
line con 0
 exec-timeout 0 0
 logging synchronous
 exit
line vty 0 4
 privilege level 15
 password cisco123
 exec-timeout 0 0
 logging synchronous
 login
 exit
interface range f0/1-24, g0/1-2
```

```
 shutdown
 exit
vlan 5
 name D1
 exit
vlan 8
 name D2
 exit
interface f0/11
 switchport mode trunk
 switchport nonegotiate
 no shutdown
 exit
interface f0/1
 switchport mode access
 switchport access vlan 5
 no shutdown
 exit
interface f0/3
 switchport mode access
 switchport access vlan 8
 no shutdown
```

   b.   Set the clock on each router to UTC time.

   c.   Save the running configuration to startup-config.

Part 2: Configure and Verify VRF and Interface Addressing

In Part 2, you will configure and verify VRF-Lite on R1. The other devices, R2, R3, D1, D2, and A1 require no additional configuration. Once again, the configuration being used here is not meant to represent best practice, but to assess your ability to complete the required configurations.

Step 1.   On R1, create the required VRFs.

   a.   Create the Customer_A and Customer_B VRFs, and initialize them for both IPv4 and IPv6. The VRF names are case sensitive.

```
R1(config)# vrf definition Customer_A
R1(config-vrf)# address-family ipv4
R1(config-vrf-af)# address-family ipv6
R1(config-vrf-af)# exit
R1(config-vrf)# vrf definition Customer_B
R1(config-vrf)# address-family ipv4
R1(config-vrf-af)# address-family ipv6
R1(config-vrf-af)# exit
```

   b.   Configure interfaces G0/0/0 and S0/1/0 for the Customer_A network.

```
R1(config)# interface g0/0/0
R1(config-if)# vrf forwarding Customer_A
R1(config-if)# ip address 10.1.2.1 255.255.255.0
R1(config-if)# ipv6 address fe80::1:1 link-local
R1(config-if)# ipv6 address 2001:db8:acad:1012::1/64
R1(config-if)# no shutdown
```

```
R1(config-if)# exit
R1(config)# interface s0/1/0
R1(config-if)# vrf forwarding Customer_A
R1(config-if)# ip address 10.1.3.1 255.255.255.0
R1(config-if)# ipv6 address fe80::1:4 link-local
R1(config-if)# ipv6 address 2001:db8:acad:1013::1/64
R1(config-if)# no shutdown
R1(config-if)# exit
```

    c. Configure R1 interface G0/0/1 to support the Customer_B network. G0/0/1 will be performing inter-VLAN routing between VLANs 5 and 8.

```
R1(config)# interface g0/0/1
R1(config-if)# no shutdown
R1(config-if)# exit
R1(config)# interface g0/0/1.5
R1(config-subif)# encapsulation dot1q 5
R1(config-subif)# vrf forwarding Customer_B
R1(config-subif)# ip address 10.1.2.1 255.255.255.0
R1(config-subif)# ipv6 address fe80::1:2 link-local
R1(config-subif)# ipv6 address 2001:db8:acad:1012::1/64
R1(config-subif)# exit
R1(config)# interface g0/0/1.8
R1(config-subif)# encapsulation dot1q 8
R1(config-subif)# vrf forwarding Customer_B
R1(config-subif)# ip address 10.1.3.1 255.255.255.0
R1(config-subif)# ipv6 address fe80::1:3 link-local
R1(config-subif)# ipv6 address 2001:db8:acad:1013::1/64
R1(config-subif)# end
```

Step 2.   Verify the VRF-Lite configuration.

    a. Verify the interface assignments using the **show ip vrf interfaces** command.

```
R1# show ip vrf interfaces
Interface            IP-Address       VRF
Protocol
Gi0/0/0              10.1.2.1         Customer_A              up
Se0/1/0              10.1.3.1         Customer_A              up
Gi0/0/1.5            10.1.2.1         Customer_B              up
Gi0/0/1.8            10.1.3.1         Customer_B              up
```

    b. Verify the VRF routing tables with the **show ip route vrf** *vrf_name* and **show ipv6 route vrf** *vrf_name* command.

```
R1# show ip route vrf Customer_A | begin Gateway
Gateway of last resort is not set


      10.0.0.0/8 is variably subnetted, 4 subnets, 2 masks
C        10.1.2.0/24 is directly connected, GigabitEthernet0/0/0
L        10.1.2.1/32 is directly connected, GigabitEthernet0/0/0
C        10.1.3.0/24 is directly connected, Serial0/1/0
L        10.1.3.1/32 is directly connected, Serial0/1/0
```

```
R1# show ipv6 route vrf Customer_B
IPv6 Routing Table - Customer_B - 5 entries
Codes: C - Connected, L - Local, S - Static, U - Per-user Static route
<output omitted>
      a - Application
C   2001:DB8:ACAD:1012::/64 [0/0]
      via GigabitEthernet0/0/1.5, directly connected
L   2001:DB8:ACAD:1012::1/128 [0/0]
      via GigabitEthernet0/0/1.5, receive
C   2001:DB8:ACAD:1013::/64 [0/0]
      via GigabitEthernet0/0/1.8, directly connected
L   2001:DB8:ACAD:1013::1/128 [0/0]
      via GigabitEthernet0/0/1.8, receive
L   FF00::/8 [0/0]
      via Null0, receive
```

c. Verify next-hop reachability within each vrf with the **ping vrf** *vrf_name* **address** command.

```
R1# ping vrf Customer_A 10.1.2.2
Type escape sequence to abort.
Sending 5, 100-byte ICMP Echos to 10.1.2.2, timeout is 2 seconds:
.!!!!
Success rate is 80 percent (4/5), round-trip min/avg/max = 1/1/1 ms
R1# ping vrf Customer_A 2001:db8:acad:1012::2
Type escape sequence to abort.
Sending 5, 100-byte ICMP Echos to 2001:DB8:ACAD:1012::2, timeout is 2 seconds:
!!!!!
Success rate is 100 percent (5/5), round-trip min/avg/max = 1/2/10 ms
R1# ping vrf Customer_A 10.1.3.2
Type escape sequence to abort.
Sending 5, 100-byte ICMP Echos to 10.1.3.2, timeout is 2 seconds:
!!!!!
Success rate is 100 percent (5/5), round-trip min/avg/max = 2/2/3 ms
R1# ping vrf Customer_A 2001:db8:acad:1013::2
Type escape sequence to abort.
Sending 5, 100-byte ICMP Echos to 2001:DB8:ACAD:1013::2, timeout is 2 seconds:
!!!!!
Success rate is 100 percent (5/5), round-trip min/avg/max = 2/2/3 ms
```

Part 3: Configure and Verify Static Routing for Reachability Inside Each VRF

In Part 3, you will configure static routing so that all networks are reachable within their respective VRFs. At the end of this part, R1 should be able to successfully source a ping from interface loopback0 to R3 interface loopback0, and D1 should be able to successfully source a ping from interface VLAN 11 to D2 interface VLAN 11. Once again, the way these networks are being implemented is not meant to represent best practice, but to assess your ability to complete the required configurations.

Step 1.   Verify that distant networks are not reachable within each VRF.

In this step, you will check to make sure that distant networks are not reachable from R1 within each VRF.

a.   On R1, issue the commands **ping vrf Customer_A 192.168.2.1** and **ping vrf Customer_A 192.168.3.1**. Neither should succeed.

```
R1# ping vrf Customer_A 192.168.2.1
Type escape sequence to abort.
Sending 5, 100-byte ICMP Echos to 192.168.2.1, timeout is 2 seconds:
.....
Success rate is 0 percent (0/5)
R1# ping vrf Customer_A 192.168.3.1
Type escape sequence to abort.
Sending 5, 100-byte ICMP Echos to 192.168.3.1, timeout is 2 seconds:
.....
Success rate is 0 percent (0/5)
```

b.   On R1, issue the commands **ping vrf Customer_A 2001:db8:acad:2000::1** and **ping vrf Customer_A 2001:db8:acad:3000::1**. Neither should succeed.

```
R1# ping vrf Customer_A 2001:db8:acad:2000::1
Type escape sequence to abort.
Sending 5, 100-byte ICMP Echos to 2001:DB8:ACAD:2000::1, timeout is 2 seconds:

% No valid route for destination
Success rate is 0 percent (0/1)
R1# ping vrf Customer_A 2001:db8:acad:3000::1
Type escape sequence to abort.
Sending 5, 100-byte ICMP Echos to 2001:DB8:ACAD:3000::1, timeout is 2 seconds:

% No valid route for destination
Success rate is 0 percent (0/1)
```

c.   On R1, issue the commands **ping vrf Customer_B 192.168.2.1** and **ping vrf Customer_B 192.168.3.1**. Neither should succeed.

```
R1# ping vrf Customer_B 192.168.2.1
Type escape sequence to abort.
Sending 5, 100-byte ICMP Echos to 192.168.2.1, timeout is 2 seconds:
.....
Success rate is 0 percent (0/5)
R1# ping vrf Customer_B 192.168.3.1
Type escape sequence to abort.
Sending 5, 100-byte ICMP Echos to 192.168.3.1, timeout is 2 seconds:
.....
Success rate is 0 percent (0/5)
```

d.   On R1, issue the commands **ping vrf Customer_B 2001:db8:acad:2000::1** and **ping vrf Customer_B 2001:db8:acad:3000::1**. Neither should succeed.

```
R1# ping vrf Customer_B 2001:db8:acad:2000::1
Type escape sequence to abort.
Sending 5, 100-byte ICMP Echos to 2001:DB8:ACAD:2000::1, timeout is 2 seconds:
```

```
% No valid route for destination
Success rate is 0 percent (0/1)


R1# ping vrf Customer_B 2001:db8:acad:3000::1
Type escape sequence to abort.
Sending 5, 100-byte ICMP Echos to 2001:DB8:ACAD:3000::1, timeout is 2 seconds:


% No valid route for destination
Success rate is 0 percent (0/1)
```

Step 2. Configure static routing at R1 for each VRF.

In this step, you will configure R1 so that it can reach distant networks in each VRF. The neighbor systems (D1, D2, R2, and R3) have static routes already configured, so as soon as you correctly install these static routes, there will be full reachability within each VRF.

a. On R1, create static routes for the distant networks in the Customer_A VRF using the **ip route vrf** *vrf_name destination_network next-hop* command.

```
R1(config)# ip route vrf Customer_A 192.168.2.0 255.255.255.0 g0/0/0 10.1.2.2
R1(config)# ip route vrf Customer_A 192.168.3.0 255.255.255.0 s0/1/0 10.1.3.2
R1(config)# ipv6 route vrf Customer_A 2001:db8:acad:2000::/64 g0/0/0
2001:db8:acad:1012::2
R1(config)# ipv6 route vrf Customer_A 2001:db8:acad:3000::/64 s0/1/0
2001:db8:acad:1013::2
```

b. Use the example above to correctly configure fully specified static routes for the Customer_B network.

_____

_____

_____

_____

_____

Step 3. Verify full reachability within each VRF.

a. On R2, ping the IPv4 and IPv6 addresses of R3 interface Loopback0 using a source address of R2 interface Loopback0. All pings should be successful.

```
R2# ping 192.168.3.1 source loopback0
Type escape sequence to abort.
Sending 5, 100-byte ICMP Echos to 192.168.3.1, timeout is 2 seconds:
Packet sent with a source address of 192.168.2.1
!!!!!
Success rate is 100 percent (5/5), round-trip min/avg/max = 2/2/3 ms
R2# ping 2001:db8:acad:3000::1 source loopback0
Type escape sequence to abort.
Sending 5, 100-byte ICMP Echos to 2001:DB8:ACAD:3000::1, timeout is 2 seconds:
Packet sent with a source address of 2001:DB8:ACAD:2000::1
!!!!!
Success rate is 100 percent (5/5), round-trip min/avg/max = 2/2/2 ms
```

b. On D1, ping the IPv4 and IPv6 addresses of D2 interface VLAN 11 using a source address of D1 interface VLAN 11. All pings should be successful.

```
D1# ping 192.168.3.1 source vlan11

Type escape sequence to abort.
Sending 5, 100-byte ICMP Echos to 192.168.3.1, timeout is 2 seconds:
Packet sent with a source address of 192.168.2.1
!!!!!
Success rate is 100 percent (5/5), round-trip min/avg/max = 1/4/9 ms


D1# ping 2001:db8:acad:3000::1 source vlan11

Type escape sequence to abort.
Sending 5, 100-byte ICMP Echos to 2001:DB8:ACAD:3000::1, timeout is 2 seconds:
Packet sent with a source address of 2001:DB8:ACAD:2000::1
!!!!!
Success rate is 100 percent (5/5), round-trip min/avg/max = 0/5/17 ms
```

Router Interface Summary Table

Router Model	Ethernet Interface #1	Ethernet Interface #2	Serial Interface #1	Serial Interface #2
1800	Fast Ethernet 0/0 (F0/0)	Fast Ethernet 0/1 (F0/1)	Serial 0/0/0 (S0/0/0)	Serial 0/0/1 (S0/0/1)
1900	Gigabit Ethernet 0/0 (G0/0)	Gigabit Ethernet 0/1 (G0/1)	Serial 0/0/0 (S0/0/0)	Serial 0/0/1 (S0/0/1)
2801	Fast Ethernet 0/0 (F0/0)	Fast Ethernet 0/1 (F0/1)	Serial 0/1/0 (S0/1/0)	Serial 0/1/1 (S0/1/1)
2811	Fast Ethernet 0/0 (F0/0)	Fast Ethernet 0/1 (F0/1)	Serial 0/0/0 (S0/0/0)	Serial 0/0/1 (S0/0/1)
2900	Gigabit Ethernet 0/0 (G0/0)	Gigabit Ethernet 0/1 (G0/1)	Serial 0/0/0 (S0/0/0)	Serial 0/0/1 (S0/0/1)
4221	Gigabit Ethernet 0/0/0 (G0/0/0)	Gigabit Ethernet 0/0/1 (G0/0/1)	Serial 0/1/0 (S0/1/0)	Serial 0/1/1 (S0/1/1)
4300	Gigabit Ethernet 0/0/0 (G0/0/0)	Gigabit Ethernet 0/0/1 (G0/0/1)	Serial 0/1/0 (S0/1/0)	Serial 0/1/1 (S0/1/1)

Note: To find out how the router is configured, look at the interfaces to identify the type of router and how many interfaces the router has. There is no way to effectively list all the combinations of configurations for each router class. This table includes identifiers for the possible combinations of Ethernet and Serial interfaces in the device. The table does not include any other type of interface, even though a specific router may contain one. An example of this might be an ISDN BRI interface. The string in parenthesis is the legal abbreviation that can be used in Cisco IOS commands to represent the interface.

Device Config Final – Notes

19.1.2 Lab - Implement a GRE Tunnel

Topology

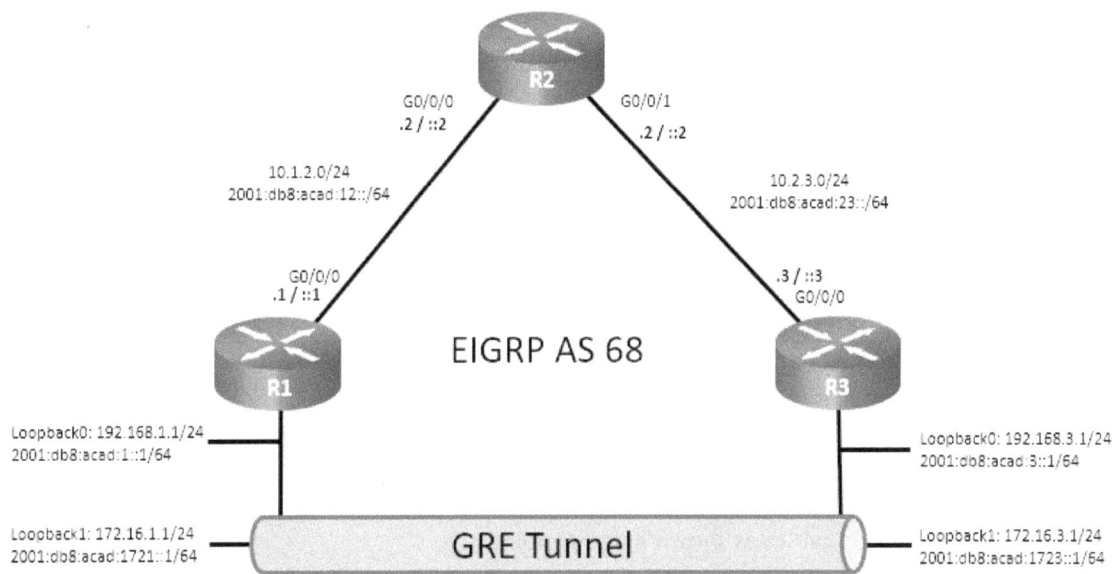

Addressing Table

Device	Interface	IPv4 Address	IPv6 Address	IPv6 Link-Local
R1	G0/0/0	10.1.2.1/24	2001:db8:acad:12::1/64	fe80::1:1
	Loopback 0	192.168.1.1/24	2001:db8:acad:1::1/64	fe80::1:2
	Loopback 1	172.16.1.1/24	2001:db8:acad:1721::1/64	fe80::1:3
R2	G0/0/0	10.1.2.2/24	2001:db8:acad:12::2/64	fe80::2:1
	G0/0/1	10.2.3.2/24	2001:db8:acad:23::2/64	fe80::2:1
R3	G0/0/0	10.2.3.3/24	2001:db8:acad:23::3/64	fe80::3:1
	Loopback 0	192.168.3.1/24	2001:db8:acad:3::1/64	fe80::3:2
	Loopback 1	172.16.3.1/24	2001:db8:acad:1723::1/64	fe80::3:3

Objectives

Part 1: Build the Network and Configure Basic Device Settings

Part 2: Configure and Verify GRE Tunnels with Static Routing

Part 3: Configure and Verify GRE Tunnels with Dynamic Routing

Background/Scenario

Overlay networks allow you to insert flexibility into existing topologies. An existing physical topology is referred to as an underlay network. Generic Routing Encapsulation (GRE) protocol, which was originally developed by Cisco, is a very useful tool that allows you to create overlay networks to support many different purposes. GRE is very flexible and works with IPv4 and IPv6 in an underlay network. In this lab you will deploy basic GRE tunnels over both IPv4 and IPv6 underlay networks.

Note: This lab is an exercise in configuring and verifying various implementations of GRE tunnels and does not reflect networking best practices.

Note: The routers used with CCNP hands-on labs are Cisco 4221 with Cisco IOS XE Release 16.9.4 (universalk9 image). Other routers and Cisco IOS versions can be used. Depending on the model and Cisco IOS version, the commands available and the output produced might vary from what is shown in the labs.

Note: Ensure that the routers' startup configurations have been erased and the devices reloaded if necessary. If you are unsure contact your instructor.

Required Resources

- 3 Routers (Cisco 4221 with Cisco IOS XE Release 16.9.4 universal image or comparable)
- 1 PC (Choice of operating system with a terminal emulation program installed)
- Console cables to configure the Cisco IOS devices via the console ports
- Ethernet cables as shown in the topology

Instructions

Part 1: Build the Network and Configure Basic Device Settings

In Part 1, you will set up the network topology and configure basic settings.

Step 1. Cable the network as shown in the topology.

Attach the devices as shown in the topology diagram, and cable as necessary.

Step 2. Configure basic settings for each switch.

a. Console into each router, enter global configuration mode, and apply the basic settings for the lab. Initial configurations for each device are listed below.

Router R1

```
hostname R1
no ip domain lookup
ipv6 unicast-routing
banner motd # R1, Implement a GRE Tunnel #
line con 0
 exec-timeout 0 0
 logging synchronous
 exit
line vty 0 4
```

```
 privilege level 15
 password cisco123
 exec-timeout 0 0
 logging synchronous
 login
 exit
router eigrp EIGRP-IPv4_GRE_LAB
 address-family ipv4 unicast autonomous-system 68
  eigrp router-id 1.1.1.1
  network 10.1.2.0 255.255.255.0
  network 192.168.1.0 255.255.255.0
  network 172.16.1.0 255.255.255.0
  exit
 exit
router eigrp EIGRP-IPv6_GRE_LAB
 address-family ipv6 unicast autonomous-system 68
  eigrp router-id 1.1.1.1
  exit
 exit
interface g0/0/0
 ip address 10.1.2.1 255.255.255.0
 ipv6 address fe80::1:1 link-local
 ipv6 address 2001:db8:acad:12::1/64
 no shutdown
 exit
interface loopback 0
 ip address 192.168.1.1 255.255.255.0
 ipv6 address fe80::1:2 link-local
 ipv6 address 2001:db8:acad:1::1/64
 no shutdown
 exit
interface loopback 1
 ip address 172.16.1.1 255.255.255.0
 ipv6 address fe80::1:3 link-local
 ipv6 address 2001:db8:acad:1721::1/64
 exit
```

Router R2

```
hostname R2
no ip domain lookup
ipv6 unicast-routing
banner motd # R2, Implement a GRE Tunnel #
line con 0
 exec-timeout 0 0
 logging synchronous
 exit
line vty 0 4
 privilege level 15
 password cisco123
 exec-timeout 0 0
 logging synchronous
```

```
  login
 exit
router eigrp EIGRP-IPv4_GRE_LAB
 address-family ipv4 unicast autonomous-system 68
  eigrp router-id 2.2.2.2
  network 10.1.2.0 255.255.255.0
  network 10.2.3.0 255.255.255.0
  exit
 exit
router eigrp EIGRP-IPv6_GRE_LAB
 address-family ipv6 unicast autonomous-system 68
  eigrp router-id 2.2.2.2
  exit
 exit
interface g0/0/0
 ip address 10.1.2.2 255.255.255.0
 ipv6 address fe80::2:1 link-local
 ipv6 address 2001:db8:acad:12::2/64
 no shutdown
 exit
interface g0/0/1
 ip address 10.2.3.2 255.255.255.0
 ipv6 address fe80::2:2 link-local
 ipv6 address 2001:db8:acad:23::2/64
 no shutdown
 exit
```

Router R3

```
hostname R3
no ip domain lookup
ipv6 unicast-routing
banner motd # R3, Implement a GRE Tunnel #
line con 0
 exec-timeout 0 0
 logging synchronous
 exit
line vty 0 4
 privilege level 15
 password cisco123
 exec-timeout 0 0
 logging synchronous
 login
 exit
router eigrp EIGRP-IPv4_GRE_LAB
 address-family ipv4 unicast autonomous-system 68
  eigrp router-id 3.3.3.3
  network 10.2.3.0 255.255.255.0
  network 192.168.3.1 255.255.255.0
  network 172.16.3.0 255.255.255.0
  exit
```

```
      exit
router eigrp EIGRP-IPv6_GRE_LAB
 address-family ipv6 unicast autonomous-system 68
  eigrp router-id 3.3.3.3
  exit
 exit
interface g0/0/0
 ip address 10.2.3.3 255.255.255.0
 ipv6 address fe80::3:1 link-local
 ipv6 address 2001:db8:acad:23::3/64
 no shutdown
 exit
interface loopback 0
 ip address 192.168.3.1 255.255.255.0
 ipv6 address fe80::3:2 link-local
 ipv6 address 2001:db8:acad:3::1/64
 no shutdown
 exit
interface loopback 1
 ip address 172.16.3.1 255.255.255.0
 ipv6 address fe80::3:3 link-local
 ipv6 address 2001:db8:acad:1723::1/64
 no shutdown
 exit
```

b. Set the clock on each device to UTC time.

c. Save the running configuration to startup-config.

Part 2: Configure and Verify GRE Tunnels with Static Routing

In Part 2, you will configure and verify a GRE tunnel between R1 and R3. You will use static routes for overlay reachability and dynamic routing for underlay reachability. You will configure two tunnels, one for IPv4 traffic and one for IPv6 traffic. GRE tunnels are extremely flexible, and there are many options for implementation beyond what is being done in this lab.

Step 1.   Verify reachability between R1 and R3.

a. From R1, ping R3 interface Loopback 0 using IPv4. All pings should be successful.

```
R1# ping 192.168.3.1
Type escape sequence to abort.
Sending 5, 100-byte ICMP Echos to 192.168.3.1, timeout is 2 seconds:
!!!!!
Success rate is 100 percent (5/5), round-trip min/avg/max = 1/1/1 ms
```

b. From R1, ping R3 interface Loopback 0 using IPv6. All pings should be successful.

```
R1# ping 2001:db8:acad:3::1
Type escape sequence to abort.
Sending 5, 100-byte ICMP Echos to 2001:DB8:ACAD:3::1, timeout is 2 seconds:
!!!!!
Success rate is 100 percent (5/5), round-trip min/avg/max = 1/2/7 ms
```

Step 2.   Create an IPv4-based GRE tunnel between R1 and R3.

    a.   On R1, create interface Tunnel 0, by specifying the IP address 100.100.100.1/30, a tunnel source of Loopback0, and a tunnel destination of 192.168.3.1.

```
R1(config)# interface tunnel 0
R1(config-if)# ip address 100.100.100.1 255.255.255.252
R1(config-if)# tunnel source loopback 0
R1(config-if)# tunnel destination 192.168.3.1
R1(config-if)# exit
```

    b.   On R1, create a static route to 172.16.3.0/24 via interface Tunnel 0.

```
R1(config)# ip route 172.16.3.0 255.255.255.0 tunnel 0
```

    c.   On R3, create interface Tunnel 0, by specifying the IP address 100.100.100.2/30, a tunnel source of Loopback0, and a tunnel destination of 192.168.1.1.

```
R3(config)# interface tunnel 0
R3(config-if)# ip address 100.100.100.2 255.255.255.252
R3(config-if)# tunnel source loopback 0
R3(config-if)# tunnel destination 192.168.1.1
R3(config-if)# exit
```

    d.   On R3, create a static route to 172.16.1.0/24 via interface Tunnel 0.

```
R3(config)# ip route 172.16.1.0 255.255.255.0 tunnel 0
```

    e.   On R1, issue the **show interface tunnel 0** command and examine the output.

```
R1# show interface tunnel 0
Tunnel0 is up, line protocol is up
  Hardware is Tunnel
  Internet address is 100.100.100.1/30
  MTU 9976 bytes, BW 100 Kbit/sec, DLY 50000 usec,
     reliability 255/255, txload 1/255, rxload 1/255
  Encapsulation TUNNEL, loopback not set
  Keepalive not set
  Tunnel linestate evaluation up
  Tunnel source 192.168.1.1 (Loopback0), destination 192.168.3.1
   Tunnel Subblocks:
      src-track:
         Tunnel0 source tracking subblock associated with Loopback0
          Set of tunnels with source Loopback0, 1 member (includes iterators),
on interface <OK>
  Tunnel protocol/transport GRE/IP
    Key disabled, sequencing disabled
    Checksumming of packets disabled
  Tunnel TTL 255, Fast tunneling enabled
  Tunnel transport MTU 1476 bytes
  Tunnel transmit bandwidth 8000 (kbps)
  Tunnel receive bandwidth 8000 (kbps)
  Last input never, output never, output hang never
  Last clearing of "show interface" counters 00:02:45
  Input queue: 0/375/0/0 (size/max/drops/flushes); Total output drops: 0
  Queueing strategy: fifo
  Output queue: 0/0 (size/max)
  5 minute input rate 0 bits/sec, 0 packets/sec
```

```
    5 minute output rate 0 bits/sec, 0 packets/sec
       0 packets input, 0 bytes, 0 no buffer
       Received 0 broadcasts (0 IP multicasts)
       0 runts, 0 giants, 0 throttles
       0 input errors, 0 CRC, 0 frame, 0 overrun, 0 ignored, 0 abort
       0 packets output, 0 bytes, 0 underruns
       0 output errors, 0 collisions, 0 interface resets
       0 unknown protocol drops
       0 output buffer failures, 0 output buffers swapped out
```

f. From R1, **ping 172.16.3.1**. The pings should be successful.

Step 3. Create an IPv6-based GRE tunnel between R1 and R3.

a. On R1, create interface Tunnel 1, by specifying the IPv6 address 2001:db8:ffff::1/64, a tunnel source of Loopback 0, a tunnel destination of 2001:db8:acad:3::1, and the tunnel mode GRE IPv6.

```
R1(config)# interface tunnel 1
R1(config-if)# ipv6 address 2001:db8:ffff::1/64
R1(config-if)# tunnel source loopback 0
R1(config-if)# tunnel destination 2001:db8:acad:3::1
R1(config-if)# tunnel mode gre ipv6
R1(config-if)# exit
```

b. On R1, create a static route to 2001:db8:acad:1723::/64 via interface Tunnel 1.

```
R1(config)# ipv6 route 2001:db8:acad:1723::/64 tunnel 1
```

c. On R3, create interface Tunnel 1, by specifying the IPv6 address 1002:db8:ffff::2/64, a tunnel source of Loopback 0, a tunnel destination of 2001:db8:acad:1::1 and the tunnel mode of GRE IPv6.

```
R3(config)# interface tunnel 1
R3(config-if)# ipv6 address 2001:db8:ffff::2/64
R3(config-if)# tunnel source loopback 0
R3(config-if)# tunnel destination 2001:db8:acad:1::1
R3(config-if)# tunnel mode gre ipv6
R3(config-if)# exit
```

d. On R3, create a static route to 2001:db8:acad:1721::/64 via interface Tunnel 1.

```
R3(config)# ipv6 route 2001:db8:acad:1721::/64 tunnel 1
```

e. On R1, issue the **show interface tunnel 1** command and examine the output.

```
R1# show interface tunnel 1
Tunnel1 is up, line protocol is up
  Hardware is Tunnel
  MTU 1456 bytes, BW 100 Kbit/sec, DLY 50000 usec,
     reliability 255/255, txload 255/255, rxload 255/255
  Encapsulation TUNNEL, loopback not set
  Keepalive not set
  Tunnel linestate evaluation up
  Tunnel source 2001:DB8:ACAD:1::1 (Loopback0), destination 2001:DB8:ACAD:3::1
   Tunnel Subblocks:
       src-track:
          Tunnel1 source tracking subblock associated with Loopback0
           Set of tunnels with source Loopback0, 2 members (includes
iterators),on interface <OK>
```

```
Tunnel protocol/transport GRE/IPv6
  Key disabled, sequencing disabled
  Checksumming of packets disabled
Tunnel TTL 255
Path MTU Discovery, ager 10 mins, min MTU 1280
Tunnel transport MTU 1456 bytes
Tunnel transmit bandwidth 8000 (kbps)
Tunnel receive bandwidth 8000 (kbps)
Last input 00:00:31, output 00:01:01, output hang never
Last clearing of "show interface" counters 00:06:58
Input queue: 0/375/0/0 (size/max/drops/flushes); Total output drops: 0
Queueing strategy: fifo
Output queue: 0/0 (size/max)
5 minute input rate 367000 bits/sec, 395 packets/sec
5 minute output rate 367000 bits/sec, 395 packets/sec
   246335 packets input, 28574884 bytes, 0 no buffer
   Received 0 broadcasts (0 IP multicasts)
   0 runts, 0 giants, 0 throttles
   0 input errors, 0 CRC, 0 frame, 0 overrun, 0 ignored, 0 abort
   246336 packets output, 28575000 bytes, 0 underruns
   0 output errors, 0 collisions, 0 interface resets
   0 unknown protocol drops
   0 output buffer failures, 0 output buffers swapped out
```

   f.   From R1, **ping 2001:db8:acad:1723::1**. The pings should be successful.

Part 3: Configure and Verify GRE Tunnels with Dynamic Routing

In Part 3, you will configure and verify GRE tunnels between R1 and R3. You will use dynamic routing for overlay reachability and static routing for underlay reachability. You will configure two tunnels, one for IPv4 traffic and one for IPv6 traffic.

Step 1.   Remove the tunnel interfaces and dynamic routing on R1 and R3.

   a.   Issue the **no interface tunnel 0** and **no interface tunnel 1** command on R1 and R3. This will also remove the static routes.

   b.   On R1, R2, and R3, remove EIGRP with the **no router eigrp EIGRP-IPv4_GRE_LAB** and **no router eigrp EIGRP-IPv6_GRE_LAB** commands.

Step 2.   Replace the EIGRP configuration on R1, R2, and R3 with static routing.

   a.   On R1 and R3, create IPv4 and IPv6 static default routes that point to R2.

   _____

   _____

   _____

   _____

   _____

b. On R2, create IPv4 and IPv6 static routes that point to the R1 and R3 Loopback 0 networks.

```
R2(config)# ip route 192.168.1.0 255.255.255.0 10.1.2.1
R2(config)# ip route 192.168.3.0 255.255.255.0 10.2.3.3
R2(config)# ipv6 route 2001:db8:acad:1::/64 2001:db8:acad:12::1
R2(config)# ipv6 route 2001:db8:acad:3::/64 2001:db8:acad:23::3
```

c. Verify that R1 can reach Loopback 0 on R3 with pings that use the source address of the R1 Loopback 0 address.

```
R1# ping 192.168.3.1 source loopback 0
Type escape sequence to abort.
Sending 5, 100-byte ICMP Echos to 192.168.3.1, timeout is 2 seconds:
Packet sent with a source address of 192.168.1.1
!!!!!
Success rate is 100 percent (5/5), round-trip min/avg/max = 1/1/2 ms

R1# ping 2001:db8:acad:3::1 source loopback 0
Type escape sequence to abort.
Sending 5, 100-byte ICMP Echos to 2001:DB8:ACAD:3::1, timeout is 2 seconds:
Packet sent with a source address of 2001:DB8:ACAD:1::1
!!!!!
Success rate is 100 percent (5/5), round-trip min/avg/max = 1/1/1 ms
```

Step 3. Create an IPv4-based GRE tunnel between R1 and R3.

a. On R1, create interface Tunnel 0, by specifying the IP address 100.100.100.1/30, bandwidth of 4000 kbps, a tunnel source of Loopback 0, and a tunnel destination of 192.168.3.1.

```
R1(config)# interface tunnel 0
R1(config-if)# ip address 100.100.100.1 255.255.255.252
R1(config-if)# bandwidth 4000
R1(config-if)# ip mtu 1400
R1(config-if)# tunnel source loopback 0
R1(config-if)# tunnel destination 192.168.3.1
R1(config-if)# exit
```

b. On R1, configure classic EIGRP for IPv4 with router-id 1.1.1.1 and AS 68. The network statements should include Tunnel 0 and Loopback 1.

```
R1(config)# router eigrp 68
R1(config-router)# eigrp router-id 1.1.1.1
R1(config-router)# network 100.100.100.0 255.255.255.252
R1(config-router)# network 172.16.1.0 255.255.255.0
R1(config-router)# exit
```

c. On R3, create interface Tunnel 0, by specifying the IP address 100.100.100.2/30, bandwidth of 4000 kbps, a tunnel source of Loopback0, and a tunnel destination of 192.168.1.1.

```
R3(config)# interface tunnel 0
R3(config-if)# ip address 100.100.100.2 255.255.255.252
R3(config-if)# bandwidth 4000
R3(config-if)# ip mtu 1400
```

```
R3(config-if)# tunnel source loopback 0
R3(config-if)# tunnel destination 192.168.1.1
R3(config-if)# exit
```

d. On R3, configure classic EIGRP for IPv4 with router-id 3.3.3.3 and AS 68. The network statements should include Tunnel 0 and Loopback 1.

```
R3(config)# router eigrp 68
R3(config-router)# eigrp router-id 3.3.3.3
R3(config-router)# network 100.100.100.0 255.255.255.252
R3(config-router)# network 172.16.3.0 255.255.255.0
R3(config-router)# end
```

e. On R1, issue the show interface tunnel 0 command and examine the output.

```
R1# show interface tunnel 0
Tunnel0 is up, line protocol is up
  Hardware is Tunnel
  Internet address is 100.100.100.1/30
  MTU 9976 bytes, BW 4000 Kbit/sec, DLY 50000 usec,
     reliability 255/255, txload 1/255, rxload 1/255
  Encapsulation TUNNEL, loopback not set
  Keepalive not set
  Tunnel linestate evaluation up
  Tunnel source 192.168.1.1 (Loopback0), destination 192.168.3.1
   Tunnel Subblocks:
      src-track:
         Tunnel0 source tracking subblock associated with Loopback0
          Set of tunnels with source Loopback0, 1 member (includes iterators),
on interface <OK>
  Tunnel protocol/transport GRE/IP
    Key disabled, sequencing disabled
    Checksumming of packets disabled
  Tunnel TTL 255, Fast tunneling enabled
  Tunnel transport MTU 1476 bytes
  Tunnel transmit bandwidth 8000 (kbps)
  Tunnel receive bandwidth 8000 (kbps)
  Last input 00:00:01, output 00:00:04, output hang never
  Last clearing of "show interface" counters 00:06:11
  Input queue: 0/375/0/0 (size/max/drops/flushes); Total output drops: 0
  Queueing strategy: fifo
  Output queue: 0/0 (size/max)
  5 minute input rate 0 bits/sec, 0 packets/sec
  5 minute output rate 0 bits/sec, 0 packets/sec
     23 packets input, 2064 bytes, 0 no buffer
     Received 0 broadcasts (0 IP multicasts)
     0 runts, 0 giants, 0 throttles
     0 input errors, 0 CRC, 0 frame, 0 overrun, 0 ignored, 0 abort
     58 packets output, 6784 bytes, 0 underruns
     0 output errors, 0 collisions, 0 interface resets
     0 unknown protocol drops
     0 output buffer failures, 0 output buffers swapped out
```

f. On R1, issue the **show ip route eigrp | begin Gateway** command and verify that 172.16.3.0/24 appears in the routing table as an EIGRP route.

```
R1# show ip route eigrp | begin Gateway
Gateway of last resort is 10.1.2.2 to network 0.0.0.0


      172.16.0.0/16 is variably subnetted, 3 subnets, 2 masks
D        172.16.3.0/24 [90/2048000] via 100.100.100.2, 00:02:01, Tunnel0
```

g. From R1, **ping 172.16.3.1**. The pings should be successful.

Step 4. Create an IPv6-based GRE tunnel between R1 and R3.

a. On R1, create interface Tunnel 1, by specifying the IPv6 address 2001:db8:ffff::1/64, bandwidth of 4000 kbps, a tunnel source of Loopback 0, and a tunnel destination of 2001:db8:acad:3::1.

```
R1(config)# interface tunnel 1
R1(config-if)# ipv6 address 2001:db8:ffff::1/64
R1(config-if)# bandwidth 4000
R1(config-if)# tunnel source loopback 0
R1(config-if)# tunnel destination 2001:db8:acad:3::1
R1(config-if)# tunnel mode gre ipv6
R1(config-if)# exit
```

b. On R1, configure classic EIGRP for IPv6 with router-id 1.1.1.1 and AS 68. Add the **ipv6 eigrp 68** command to the Tunnel 1 and Loopback 1 interfaces.

```
R1(config)# ipv6 router eigrp 68
R1(config-rtr)# eigrp router-id 1.1.1.1
R1(config-rtr)# exit
R1(config)# interface tunnel 1
R1(config-if)# ipv6 eigrp 68
R1(config-if)# exit
R1(config)# interface loopback 1
R1(config-if)# ipv6 eigrp 68
R1(config-if)# end
```

c. On R3, create interface Tunnel 1, by specifying the IPv6 address 1002:db8:ffff::2/64, bandwidth of 4000 kbps, a tunnel source of Loopback 0, and a tunnel destination of 2001:db8:acad:1::1.

```
R3(config)# interface tunnel 1
R3(config-if)# ipv6 address 2001:db8:ffff::2/64
R3(config-if)# bandwidth 4000
R3(config-if)# tunnel source loopback 0
R3(config-if)# tunnel destination 2001:db8:acad:1::1
R3(config-if)# tunnel mode gre ipv6
R3(config-if)# exit
```

d. On R3, configure classic EIGRP for IPv6 with router-id 3.3.3.3 and AS 68. Add the **ipv6 eigrp 68** command to the Tunnel 1 and Loopback 1 interfaces.

```
R3(config)# ipv6 router eigrp 68
R3(config-rtr)# eigrp router-id 3.3.3.3
R3(config-rtr)# exit
R3(config)# interface tunnel 1
R3(config-if)# ipv6 eigrp 68
R3(config-if)# exit
```

```
R3(config)# interface loopback 1
R3(config-if)# ipv6 eigrp 68
R3(config-if)# exit
```

e. On R1, issue the **show interface tunnel 1** command and examine the output.

```
R1# show interface tunnel 1
Tunnel1 is up, line protocol is up
  Hardware is Tunnel
  MTU 1456 bytes, BW 4000 Kbit/sec, DLY 50000 usec,
     reliability 255/255, txload 1/255, rxload 1/255
  Encapsulation TUNNEL, loopback not set
  Keepalive not set
  Tunnel linestate evaluation up
  Tunnel source 2001:DB8:ACAD:1::1 (Loopback0), destination 2001:DB8:ACAD:3::1
   Tunnel Subblocks:
      src-track:
         Tunnel1 source tracking subblock associated with Loopback0
          Set of tunnels with source Loopback0, 2 members (includes
iterators),on interface <OK>
  Tunnel protocol/transport GRE/IPv6
    Key disabled, sequencing disabled
    Checksumming of packets disabled
  Tunnel TTL 255
  Path MTU Discovery, ager 10 mins, min MTU 1280
  Tunnel transport MTU 1456 bytes
  Tunnel transmit bandwidth 8000 (kbps)
  Tunnel receive bandwidth 8000 (kbps)
  Last input 00:00:09, output 00:00:04, output hang never
  Last clearing of "show interface" counters 00:04:20
  Input queue: 0/375/0/0 (size/max/drops/flushes); Total output drops: 0
  Queueing strategy: fifo
  Output queue: 0/0 (size/max)
  5 minute input rate 0 bits/sec, 0 packets/sec
  5 minute output rate 0 bits/sec, 0 packets/sec
     31 packets input, 4048 bytes, 0 no buffer
     Received 0 broadcasts (0 IP multicasts)
     0 runts, 0 giants, 0 throttles
     0 input errors, 0 CRC, 0 frame, 0 overrun, 0 ignored, 0 abort
     46 packets output, 5864 bytes, 0 underruns
     0 output errors, 0 collisions, 0 interface resets
     0 unknown protocol drops
     0 output buffer failures, 0 output buffers swapped out
```

f. On R1, issue the **show ipv6 route eigrp | section D** command and verify that 2001:db8:acad:1723::/64 appears in the routing table as an OSPF route.

```
R1# show ipv6 route eigrp | section D
        I2 - ISIS L2, IA - ISIS interarea, IS - ISIS summary, D - EIGRP
        EX - EIGRP external, ND - ND Default, NDp - ND Prefix, DCE - Destination
        NDr - Redirect, RL - RPL, O - OSPF Intra, OI - OSPF Inter
D   2001:DB8:ACAD:1723::/64 [90/2048000]
      via FE80::2FC:BAFF:FE94:29B0, Tunnel1
```

g. From R1, ping 2001:db8:acad:1723::1. The pings should be successful.

Router Interface Summary Table

Router Model	Ethernet Interface #1	Ethernet Interface #2	Serial Interface #1	Serial Interface #2
1800	Fast Ethernet 0/0 (F0/0)	Fast Ethernet 0/1 (F0/1)	Serial 0/0/0 (S0/0/0)	Serial 0/0/1 (S0/0/1)
1900	Gigabit Ethernet 0/0 (G0/0)	Gigabit Ethernet 0/1 (G0/1)	Serial 0/0/0 (S0/0/0)	Serial 0/0/1 (S0/0/1)
2801	Fast Ethernet 0/0 (F0/0)	Fast Ethernet 0/1 (F0/1)	Serial 0/1/0 (S0/1/0)	Serial 0/1/1 (S0/1/1)
2811	Fast Ethernet 0/0 (F0/0)	Fast Ethernet 0/1 (F0/1)	Serial 0/0/0 (S0/0/0)	Serial 0/0/1 (S0/0/1)
2900	Gigabit Ethernet 0/0 (G0/0)	Gigabit Ethernet 0/1 (G0/1)	Serial 0/0/0 (S0/0/0)	Serial 0/0/1 (S0/0/1)
4221	Gigabit Ethernet 0/0/0 (G0/0/0)	Gigabit Ethernet 0/0/1 (G0/0/1)	Serial 0/1/0 (S0/1/0)	Serial 0/1/1 (S0/1/1)
4300	Gigabit Ethernet 0/0/0 (G0/0/0)	Gigabit Ethernet 0/0/1 (G0/0/1)	Serial 0/1/0 (S0/1/0)	Serial 0/1/1 (S0/1/1)

Note: To find out how the router is configured, look at the interfaces to identify the type of router and how many interfaces the router has. There is no way to effectively list all the combinations of configurations for each router class. This table includes identifiers for the possible combinations of Ethernet and Serial interfaces in the device. The table does not include any other type of interface, even though a specific router may contain one. An example of this might be an ISDN BRI interface. The string in parenthesis is the legal abbreviation that can be used in Cisco IOS commands to represent the interface.

Device Config Final – Notes

_____

_____

_____

_____

_____

_____

_____

_____

_____

_____

_____

_____

_____

_____

19.1.3 Lab - Implement a DMVPN Phase 1 Hub-to-Spoke Topology

Topology

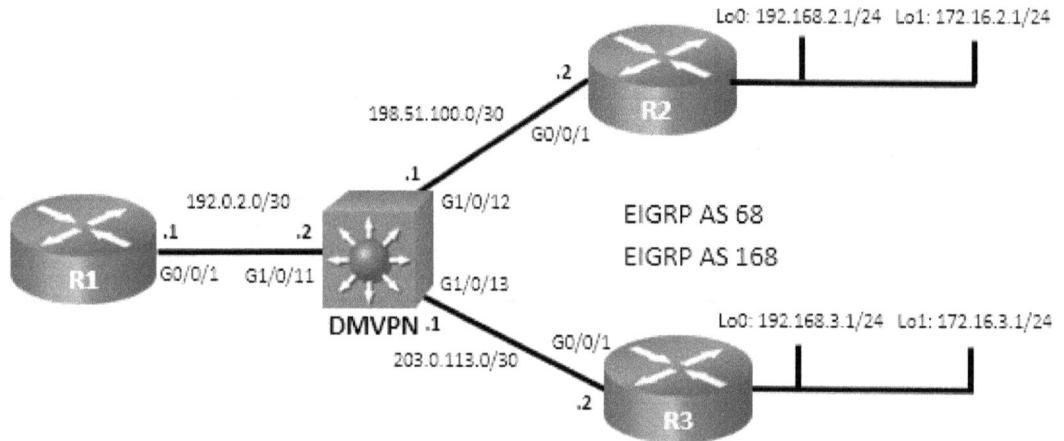

Addressing Table

Device	Interface	IPv4 Address
R1	G0/0/1	192.0.2.1/30
	Tunnel 1	100.100.100.1/29
R2	G0/0/1	198.51.100.2/30
	Loopback 0	192.168.1.1/24
	Loopback 1	172.16.1.1/24
	Tunnel 1	100.100.100.2/29
R3	G0/0/1	203.0.113.2/30
	Loopback 0	192.168.3.1/24
	Loopback 1	172.16.3.1/24
	Tunnel 1	100.100.100.3/29

Objectives

Part 1: Build the Network and Configure Basic Device Settings

Part 2: Configure and Verify DMVPN Phase 1

Part 3: Configure EIGRP Routing for the Tunnel Networks

Background/Scenario

In this lab you will create a Dynamic Multipoint Virtual Private Network (DMVPN) that consists of a hub router with two spokes. You will implement a DMVPN Phase 1 hub-to-spoke topology.

DMVPN is a Cisco IOS Software solution for building scalable IPsec Virtual Private Networks (VPNs). Cisco DMVPN uses a centralized architecture to provide easier implementation and management for deployments that require granular access controls for diverse user communities, including mobile workers, telecommuters, and extranet users. The centralized architecture involves designating one or more routers as multipoint GRE hub routers that are used to connect spoke, or branch, routers to VPN services.

Cisco DMVPN allows branch locations to communicate directly with each other over the public WAN or internet, such as when using voice over IP (VoIP) between two branch offices, But it does not require a permanent VPN connection between sites. It enables dynamic deployment of IPsec VPNs and improves network performance by reducing latency and jitter, while optimizing head office bandwidth utilization.

DMVPN combines GRE tunnels, IPsec encryption, and Next Hop Resolution Protocol (NHRP) routing. The use of multipoint GRE (mGRE) enables a single physical interface on the hub router to support tunnel connections from multiple spoke routers, thus providing the scalability required to connect many locations over VPN to a single hub, such as at a headquarters site. Securing the VPN with crypto profiles makes it unnecessary to define static crypto maps. This enables dynamic discovery of tunnel endpoints. In addition, DMVPN enables the dynamic establishment and teardown of spoke-to-spoke tunnels when they are required. This feature will be covered in a subsequent lab.

DMVPN was released in phases that provide different capabilities to the DMVPN network.

Phase 1: This phase provides mechanisms (NHRP, mGRE) for creating the spoke-to-hub DMVPN topology. Spoke routers that are configured with NHRP register with the hub router which then dynamically creates spoke-to-hub tunnels over a single physical interface. In this phase, mGRE is only configured on the hub router, with the spoke router establishing GRE tunnels to connect to the hub. Phase 1 supported only spoke-to-hub communication, meaning that all inter-spoke tunnel traffic had to pass through the hub router and essentially traverse two tunnels.

Phase 2: This phase enabled dynamic establishment and teardown of spoke-to-spoke tunnels as required by network traffic. Spoke-to-spoke communication requires multiple tunnels to be connected on a single spoke router physical interface. To accomplish this, the spoke routers are configured with mGRE like the hub router. The hub router orchestrates the dynamic establishment of the spoke-to-spoke tunnels.

Phase 3: This phase improved upon Phase 2 by using a different mechanism for the establishment of spoke-to-spoke tunnels. In this case, instead of the hub forwarding NHRP resolution messages between the spoke endpoints, the hub creates a NHRP redirection message that is sent to the spoke that is initiating the tunnel. The spoke uses this information to establish the tunnel by forwarding the NHRP resolution request message directly to the spoke router, rather than relying on the hub to do so. This allows the spoke routers to create routing table entries for the spoke-to-spoke networks and also enables the distribution of route summaries from the hub router to the spokes. Finally, Phase 3 enables a hierarchical tree-based VPN architecture in which central hub routers connect other, regional hubs and their spokes. This allows for the establishment of tunnels between routers that are not connected to the same regional hub.

Save your configurations from this lab. You will use them as a starting point for the DMVPN Phase 3 lab that follows.

In this lab, you will configure DMVPN Phase 1.

Note: This lab does not include the configuration of IPSec to secure the tunnels. This essential procedure will be covered in a later lab.

Note: This lab is an exercise in configuring and verifying various implementations of DMVPN topologies and does not reflect networking best practices.

Note: The routers used with CCNP hands-on labs are Cisco 4221s with Cisco IOS XE Release 16.9.4 (universalk9 image). The Layer 3 switch is a Cisco Catalyst 3650 with Cisco IOS XE Release 16.9.4 (universalk9 image). Other routers, Layer 3 switches, and Cisco IOS versions can be used. Depending on the model and Cisco IOS version, the commands available and the output produced might vary from what is shown in the labs.

Note: Make sure that the switches have been erased and have no startup configurations. If you are unsure, please contact your instructor.

Required Resources

- 3 Routers (Cisco 4221 with Cisco IOS XE Release 16.9.4 universal image or comparable)

- 1 Layer 3 switch (Cisco 3650 with Cisco IOS Release 16.9.4 universal image or comparable)

- 1 PC (Choice of operating system with a terminal emulation program installed)

- Console cables to configure the Cisco IOS devices via the console ports

- Ethernet cables as shown in the topology

Instructions

Part 1: Build the Network and Configure Basic Device Settings

In Part 1, you will set up the network topology and configure basic settings.

Step 1. Cable the network as shown in the topology.

Attach the devices as shown in the topology diagram, and cable as necessary.

Step 2. Configure initial settings for each router and the Layer 3 switch.

    a. Console into each device, enter global configuration mode, and apply the initial settings for the lab. Initial configurations for each device are provided below.

    Hub Router R1

```
hostname R1
no ip domain lookup
banner motd # R1, Implement a DMVPN hub #
line con 0
 exec-timeout 0 0
 logging synchronous
 exit
line vty 0 4
 privilege level 15
 password cisco123
 exec-timeout 0 0
 logging synchronous
```

```
 login
 exit
ip route 0.0.0.0 0.0.0.0 g0/0/1
interface g0/0/1
 ip address 192.0.2.1 255.255.255.252
 no shutdown
 exit
end
```

Spoke Router R2

```
hostname R2
no ip domain lookup
banner motd # R2, Implement DMVPN Spoke 1 #
line con 0
 exec-timeout 0 0
 logging synchronous
 exit
line vty 0 4
 privilege level 15
 password cisco123
 exec-timeout 0 0
 logging synchronous
 login
 exit
ip route 0.0.0.0 0.0.0.0 g0/0/1
interface g0/0/1
 ip address 198.51.100.2 255.255.255.252
 no shutdown
 exit
interface loopback 0
 ip address 192.168.2.1 255.255.255.0
 no shutdown
 exit
interface loopback 1
 ip address 172.16.2.1 255.255.255.0
 no shutdown
 exit
end
```

Spoke Router R3

```
hostname R3
no ip domain lookup
banner motd # R3, Implement DMVPN Spoke 2 #
line con 0
 exec-timeout 0 0
 logging synchronous
 exit
line vty 0 4
 privilege level 15
 password cisco123
 exec-timeout 0 0
 logging synchronous
```

```
 login
 exit
ip route 0.0.0.0 0.0.0.0 g0/0/1
interface g0/0/1
 ip address 203.0.113.2 255.255.255.252
 no shutdown
 exit
interface loopback 0
 ip address 192.168.3.1 255.255.255.0
 no shutdown
 exit
interface loopback 1
 ip address 172.16.3.1 255.255.255.0
 no shutdown
 exit
end
```

DMVPN Layer 3 Switch

```
hostname DMVPN
no ip domain lookup
ip routing
banner motd # DMVPN, DMVPN cloud switch #
line con 0
 exec-timeout 0 0
 logging synchronous
 exit
line vty 0 4
 privilege level 15
 password cisco123
 exec-timeout 0 0
 logging synchronous
 login
 exit
interface g1/0/11
 no switchport
 ip address 192.0.2.2 255.255.255.252
 no shutdown
 exit
interface g1/0/12
 no switchport
 ip address 198.51.100.1 255.255.255.252
 no shutdown
 exit
interface g1/0/13
 no switchport
 ip address 203.0.113.1 255.255.255.252
 no shutdown
 exit
ip route 192.168.2.0 255.255.255.0 g1/0/12
ip route 172.16.2.0 255.255.255.0 g1/0/12
```

```
ip route 192.168.3.0 255.255.255.0 g1/0/13
ip route 172.16.3.0 255.255.255.0 g1/0/13
end
```

   b. Set the clock on each device to UTC time.

   c. Save the running configuration to the startup configuration.

Part 2: Configure and Verify DMVPN Phase 1

In this part of the lab, you will configure DMVPN Phase 1 to create DMVPN tunnels between the spoke routers R2 and R3, and the hub router, R1. DMVPN is very flexible and there are many options for implementation beyond what is being done in this lab.

In Phase 1 DMVPN, all spoke router traffic must pass through the hub router as shown in the topology diagram.

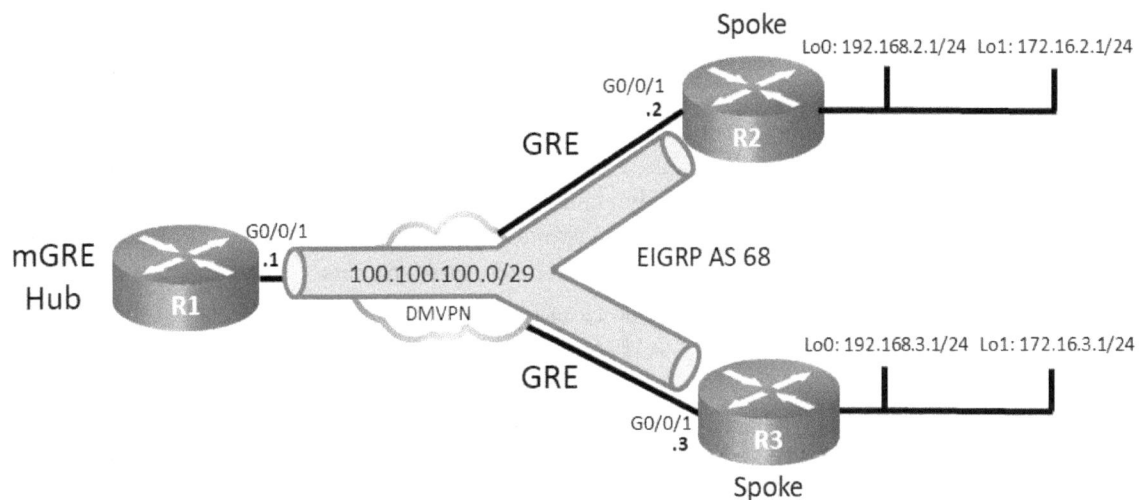

Note: In this lab, you will need to change the configuration of the DMVPN Layer 3 switch. Normally, you would not need to configure this device. The DMVPN switch is simulating the ISP transport network.

Step 1.   Verify connectivity in the underlay network.

          From R1, ping the Loopback 0 interfaces of R2 and R3.

```
R1# ping 192.168.2.1
Type escape sequence to abort.
Sending 5, 100-byte ICMP Echos to 192.168.3.1, timeout is 2 seconds:
!!!!!
Success rate is 100 percent (5/5), round-trip min/avg/max = 1/1/1 ms


R1# ping 192.168.3.1
Type escape sequence to abort.
Sending 5, 100-byte ICMP Echos to 192.168.3.1, timeout is 2 seconds:
!!!!!
Success rate is 100 percent (5/5), round-trip min/avg/max = 1/1/1 ms
```

Step 2.   Configure the tunnel interface on the hub router.

DMVPN requires configuration of tunnel interfaces like GRE. In fact, DMVPN Phase 1 uses GRE and mGRE mode tunnels. The hub router tunnel interface will use mGRE. This will enable it to create multiple tunnels on a single interface.

a.   On R1, create the tunnel interface, set the tunnel mode to mGRE, and establish the tunnel source as Loopback 0. For DMVPN Phase 1, a tunnel key is also required when multiple tunnels will be established from a single interface. Finally, address the interface on the 100.100.100.0/29 network. The overlay network will use this subnet for all members of the DMVPN network. The hub router interface does not require a tunnel destination because it is a multipoint interface.

```
R1(config)# interface tunnel 1
R1(config-if)# tunnel mode gre multipoint
R1(config-if)# tunnel source GigabitEthernet0/0/1
R1(config-if)# tunnel key 999
R1(config-if)# ip address 100.100.100.1 255.255.255.248
```

b.   Configure the hub router as an NHRP server (NHS). NHRP enables DMVPN to dynamically learn the NBMA physical addresses of devices in the network. The NHRP network ID must be consistent between the hub and spokes in the DMVPN network. You configure authentication to add a layer of security. Finally, configure the interface as multicast dynamic, which enables the hub to dynamically add spoke routers to the NHRP table when spokes initiate a tunnel. This enables the use of dynamic routing protocols between the hub and spoke routers.

```
R1(config-if)# ip nhrp network-id 1
R1(config-if)# ip nhrp authentication NHRPauth
R1(config-if)# ip nhrp map multicast dynamic
```

c.   Because DMVPN networks add information to packet headers, the interface should be fine-tuned to participate in the DMVPN network. In addition, configure the interface bandwidth so that routing protocols that use bandwidth values will function properly.

```
R1(config-if)# bandwidth 4000
R1(config-if)# ip mtu 1400
R1(config-if)# ip tcp adjust-mss 1360
R1(config-if)# end
```

d.   Verify the tunnel interface configuration with the **show interface tunnel 1** command.

```
R1# show interface tunnel 1
Tunnel1 is up, line protocol is up
  Hardware is Tunnel
  Internet address is 100.100.100.1/29
  MTU 9972 bytes, BW 4000 Kbit/sec, DLY 50000 usec,
     reliability 255/255, txload 1/255, rxload 1/255
  Encapsulation TUNNEL, loopback not set
  Keepalive not set
  Tunnel linestate evaluation up
  Tunnel source 192.0.2.1 (GigabitEthernet0/0/1)
   Tunnel Subblocks:
      src-track:
         Tunnel1 source tracking subblock associated with GigabitEthernet0/0/1
          Set of tunnels with source GigabitEthernet0/0/1, 1 member (includes
iterators), on interface <OK>
```

```
Tunnel protocol/transport multi-GRE/IP
  Key 0x3E7, sequencing disabled
  Checksumming of packets disabled
Tunnel TTL 255, Fast tunneling enabled
Tunnel transport MTU 1472 bytes
Tunnel transmit bandwidth 8000 (kbps)
Tunnel receive bandwidth 8000 (kbps)
Last input never, output never, output hang never
Last clearing of "show interface" counters 00:02:06
Input queue: 0/375/0/0 (size/max/drops/flushes); Total output drops: 0
Queueing strategy: fifo
Output queue: 0/0 (size/max)
5 minute input rate 0 bits/sec, 0 packets/sec
5 minute output rate 0 bits/sec, 0 packets/sec
   0 packets input, 0 bytes, 0 no buffer
   Received 0 broadcasts (0 IP multicasts)
   0 runts, 0 giants, 0 throttles
   0 input errors, 0 CRC, 0 frame, 0 overrun, 0 ignored, 0 abort
   0 packets output, 0 bytes, 0 underruns
   0 output errors, 0 collisions, 0 interface resets
   0 unknown protocol drops
   0 output buffer failures, 0 output buffers swapped out
```

Step 3. Configure the R2 and R3 spoke router tunnel interfaces.

In contrast to NHS interfaces, configuration of the NHRP client (NHC) tunnel interfaces uses regular GRE and also requires the configuration of a tunnel destination address.

a. On R2, create the tunnel interface and configure the GRE tunnel parameters. Set the tunnel destination as the address of the GigabitEthernet 0/0/1 interface of the hub router. Configuring the tunnel source as the Loopback 0 interface provides a stable source for the tunnel. The tunnel key must match the key that is configured on the hub router. Finally, configure the overlay network IP address for the tunnel interface.

```
R2(config)# interface tunnel 1
R2(config-if)# tunnel mode gre ip
R2(config-if)# tunnel source loopback 0
R2(config-it)# tunnel destination 192.0.2.1
R2(config-if)# tunnel key 999
R2(config-if)# ip address 100.100.100.2 255.255.255.248
```

b. Configure the tunnel interface as an NHRP client. You will need to designate the underlay address of the NHRP server and map the NHRP server underlay address to its overlay address.

```
R2(config-if)# ip nhrp network-id 1
R2(config-if)# ip nhrp authentication NHRPauth
R2(config-if)# ip nhrp nhs 100.100.100.1
R2(config-if)# ip nhrp map multicast 192.0.2.1
R2(config-if)# ip nhrp map 100.100.100.1 192.0.2.1
```

c. Adjust settings on the interface to accommodate the GRE packet overhead.

```
R2(config-if)# ip mtu 1400
R2(config-if)# ip tcp adjust-mss 1360
```

d. Repeat this configuration on router R3 using the commands above and information from the addressing table.

_____

_____

_____

_____

_____

e. Verify your configurations with the **show interface** command. If your configurations are correct, you should be able to successfully ping the interface addresses of the over-lay network (100.100.100.0/29) from each router.

f. View the status of DMVPN with the **show dmvpn** and **show dmvpn detail** commands. Become familiar with the output of each. The output from the hub router is shown.

```
R1# show dmvpn detail
Legend: Attrb --> S - Static, D - Dynamic, I - Incomplete
        N - NATed, L - Local, X - No Socket
        T1 - Route Installed, T2 - Nexthop-override
        C - CTS Capable, I2 - Temporary
        # Ent --> Number of NHRP entries with same NBMA peer
        NHS Status: E --> Expecting Replies, R --> Responding, W --> Waiting
        UpDn Time --> Up or Down Time for a Tunnel
==========================================================================

Interface Tunnel1 is up/up, Addr. is 100.100.100.1, VRF ""
   Tunnel Src./Dest. addr: 192.0.2.1/Multipoint, Tunnel VRF ""
   Protocol/Transport: "multi-GRE/IP", Protect ""
   Interface State Control: Disabled
   nhrp event-publisher : Disabled
Type:Hub, Total NBMA Peers (v4/v6): 2

# Ent   Peer NBMA Addr Peer Tunnel Add State  UpDn Tm Attrb    Target Network
-----   -------------- --------------- ----- -------- ----- ----------------
    1   192.168.2.1      100.100.100.2   UP 00:05:45    D    100.100.100.2/32
    1   192.168.3.1      100.100.100.3   UP 00:06:15    D    100.100.100.3/32


Crypto Session Details:
-------------------------------------------------------------------------


Pending DMVPN Sessions:
```

The output shows the status of the tunnel, the tunnel address and the tunnel source address. The table shows the NBMA (underlay) addresses of the known DMVPN peers. That address is derived from the tunnel source addresses that you configured on R2 and R3. The tunnel source is the Loopback 0 interface address. The overlay network peer tunnel interface addresses are also shown. The state of the entries in the table must be UP in order for data to travel on the tunnels. If configured, the crypto settings for the tunnel would be shown. You will secure the tunnels in a later lab.

Repeat this command on the spoke routers so that you become familiar with the command output.

g.   Verify the status of NHRP by viewing the contents of the NHRP with the **show ip nhrp detail** command. Output is shown for the hub router. Note that it displays information for both of the dynamic tunnels between the spoke routers and the hub.

```
R1# show ip nhrp detail
100.100.100.2/32 via 100.100.100.2
    Tunnel1 created 00:25:06, expire 00:08:13
    Type: dynamic, Flags: registered nhop
    NBMA address: 192.168.2.1
    Preference: 255
100.100.100.3/32 via 100.100.100.3
    Tunnel1 created 00:25:36, expire 00:08:18
    Type: dynamic, Flags: registered nhop
    NBMA address: 192.168.3.1
    Preference: 255
```

Part 3: Configure EIGRP Routing for the Tunnel Networks

Cisco recommends that EIGRP be used to route DMVPN networks. Because the hub router uses a single interface to reach multiple networks, the split horizon rule will affect connectivity. For that reason, split horizon should be disabled on the hub router. Finally, improve network performance by configuring the spoke routers as stubs.

You will configure two separate EIGRP routing processes to route the overlay and underlay networks. Each network will use a different autonomous system number. Care should be taken not to route the underlay interface network using the same routing process as the overlay network. This can disrupt routing protocol operation on the hub router, severely impact performance, and possibly cause the router to crash. This condition is called recursive routing. Cisco IOS should detect recursive routing and provide syslog information regarding this error. In addition, IOS will temporarily disable the tunnel interface until recursive routing has stopped.

Step 1.   Configure dynamic routing for the overlay network.

a.   Remove the static routes from the three routers and the Layer 3 switch by pasting the commands below into the console of the appropriate devices.

R1, R2, and R3

```
no ip route 0.0.0.0 0.0.0.0 g0/0/1
```

DMVPN switch

```
no ip route 192.168.2.0 255.255.255.0 g1/0/12
no ip route 172.16.2.0 255.255.255.0 g1/0/12
no ip route 192.168.3.0 255.255.255.0 g1/0/13
no ip route 172.16.3.0 255.255.255.0 g1/0/13
```

b.   Create a named EIGRP process with the name **DMVPN_TUNNEL_NET**. This process and AS will route the overlay network. Note that the DMVPN switch must also be configured. Add the tunnel interface and Loopback 1 interface networks to the routing process. Loopback 1 simulates a LAN that will be sending traffic through the tunnel. Note that split horizon is disabled on the hub router tunnel interface. Also note that the two hub routers are configured as stub routers. Configure the three routers as follows:

```
R1(config)# router eigrp DMVPN_TUNNEL_NET
R1(config-router)# address-family ipv4 unicast autonomous-system 68
R1(config-router-af)# eigrp router-id 1.1.1.1
R1(config-router-af)# network 100.100.100.0 255.255.255.248
R1(config-router-af)# af-interface tunnel 1
R1(config-router-af-interface)# no split-horizon
R2(config)# router eigrp DMVPN_TUNNEL_NET
R2(config-router)# address-family ipv4 unicast autonomous-system 68
R2(config-router-af)# eigrp router-id 2.2.2.2
R2(config-router-af)# network 100.100.100.0 255.255.255.248
R2(config-router-af)# network 172.16.2.0 255.255.255.0
R2(config-router-af)# eigrp stub connected
R3(config)# router eigrp DMVPN_TUNNEL_NET
R3(config-router)# address-family ipv4 unicast autonomous-system 68
R3(config-router-af)# eigrp router-id 3.3.3.3
R3(config-router-af)# network 100.100.100.0 255.255.255.248
R3(config-router-af)# network 172.16.3.0 255.255.255.0
R3(config-router-af)# eigrp stub connected
```

Step 2. Configure dynamic routing for the underlay network.

a. Create a new named EIGRP process with the name **DMVPN_TRANS_NET** with AS **168**. This process and AS will route the underlay, or transport, network. Note that the DMVPN Layer 3 switch did not need to have routing knowledge for the tunnel network. However, it does need to be configured to route between the point-to-point underlay networks. After configuring the DMVPN Layer 3 switch with EIGRP, you will see EIGRP new adjacency syslog messages on the DMVPN switch for EIGRP AS 168 and on R1, R2, and R3 for EIGRP AS 68 and EIGRP 168.

```
R1(config)# router eigrp DMVPN_TRANS_NET
R1(config-router)# address-family ipv4 unicast autonomous-system 168
R1(config-router-af)# eigrp router-id 10.1.1.1
R1(config-router-af)# network 192.0.2.0 255.255.255.252

R2(config)# router eigrp DMVPN_TRANS_NET
R2(config-router)# address-family ipv4 unicast autonomous-system 168
R2(config-router-af)# eigrp router-id 20.2.2.2
R2(config-router-af)# network 198.51.100.0 255.255.255.252
R1(config-router-af)# network 192.168.2.0 255.255.255.0
R2(config-router-af)# eigrp stub connected

R3(config)# router eigrp DMVPN_TRANS_NET
R3(config-router)# address-family ipv4 unicast autonomous-system 168
R3(config-router-af)# eigrp router-id 30.3.3.3
R3(config-router-af)# network 203.0.113.0 255.255.255.252
R3(config-router-af)# network 192.168.3.0 255.255.255.0
R3(config-router-af)# eigrp stub connected

DMVPN(config)# router eigrp DMVPN_TRANS_NET
DMVPN(config-router)# address-family ipv4 unicast autonomous-system 168
```

```
DMVPN(config-router-af)# eigrp router-id 40.4.4.4
DMVPN(config-router-af)# network 203.0.113.0 255.255.255.252
DMVPN(config-router-af)# network 192.0.2.0 255.255.255.252
DMVPN(config-router-af)# network 198.51.100.0 255.255.255.252
```

b. Verify dynamic routing on all three routers by using the **show ip route eigrp | begin Gateway** and **show ip route eigrp** commands followed by the AS number. Do this on all three routers to verify that the underlay and overlay networks appear in the routing tables for the correct AS. Output is shown for the hub router.

```
R1# show ip route eigrp | begin Gateway
Gateway of last resort is not set


      172.16.0.0/24 is subnetted, 2 subnets
D        172.16.2.0 [90/26880640] via 100.100.100.2, 00:04:09, Tunnel1
D        172.16.3.0 [90/26880640] via 100.100.100.3, 00:04:04, Tunnel1
D     192.168.2.0/24 [90/16000] via 192.0.2.2, 00:04:14, GigabitEthernet0/0/1
D     192.168.3.0/24 [90/16000] via 192.0.2.2, 00:04:14, GigabitEthernet0/0/1
      198.51.100.0/30 is subnetted, 1 subnets
D        198.51.100.0 [90/15360] via 192.0.2.2, 00:04:14, GigabitEthernet0/0/1
      203.0.113.0/30 is subnetted, 1 subnets
D        203.0.113.0 [90/15360] via 192.0.2.2, 00:04:14, GigabitEthernet0/0/1


R1# show ip route eigrp 68 | begin Gateway
Gateway of last resort is not set


Gateway of last resort is not set


      172.16.0.0/24 is subnetted, 2 subnets
D        172.16.2.0 [90/26880640] via 100.100.100.2, 06:04:37, Tunnel1
D        172.16.3.0 [90/26880640] via 100.100.100.3, 06:04:41, Tunnel1


R1# show ip route eigrp 168 | begin Gateway
Gateway of last resort is not set


D     192.168.2.0/24 [90/16000] via 192.0.2.2, 00:16:08, GigabitEthernet0/0/1
D     192.168.3.0/24 [90/16000] via 192.0.2.2, 00:14:55, GigabitEthernet0/0/1
      198.51.100.0/30 is subnetted, 1 subnets
D        198.51.100.0 [90/15360] via 192.0.2.2, 00:17:48, GigabitEthernet0/0/1
      203.0.113.0/30 is subnetted, 1 subnets
D        203.0.113.0 [90/15360] via 192.0.2.2, 00:17:48, GigabitEthernet0/0/1
```

Step 3. Verify DMVPN Phase 1 operation.

You have completed the configuration of DMVPN Phase 1. Verify communication as follows:

a. On R1 execute a **traceroute** to the Loopback 1 interface IP addresses on R2 and R3. You should see the path use the tunnel network.

```
R1# traceroute 172.16.2.1
Type escape sequence to abort.
Tracing the route to 172.16.2.1
VRF info: (vrf in name/id, vrf out name/id)
  1 100.100.100.2 1 msec *  1 msec
```

```
R1# traceroute 172.16.3.1
Type escape sequence to abort.
Tracing the route to 172.16.3.1
VRF info: (vrf in name/id, vrf out name/id)
  1 100.100.100.3 1 msec *  1 msec
```

b. On R1, execute a **traceroute** to the Loopback 0 interface IP addresses on R2 and R3. You should see the path use the physical point-to-point networks of the underlay transport network.

```
R1# traceroute 192.168.2.1
Type escape sequence to abort.
Tracing the route to 192.168.2.1
VRF info: (vrf in name/id, vrf out name/id)
  1 192.0.2.2 2 msec 4 msec 1 msec
  2 198.51.100.2 1 msec *  1 msec
```

```
R1# traceroute 192.168.3.1
Type escape sequence to abort.
Tracing the route to 192.168.3.1
VRF info: (vrf in name/id, vrf out name/id)
  1 192.0.2.2 2 msec 1 msec 1 msec
  2 203.0.113.2 1 msec *  1 msec
```

c. Repeat the **traceroute** commands on R2 and R3.

d. You have successfully configured a DMVPN Phase 1 network. Save your configurations. You will use them as the starting point for the DMVPN Phase 3 lab.

Router Interface Summary Table

Router Model	Ethernet Interface #1	Ethernet Interface #2	Serial Interface #1	Serial Interface #2
1800	Fast Ethernet 0/0 (F0/0)	Fast Ethernet 0/1 (F0/1)	Serial 0/0/0 (S0/0/0)	Serial 0/0/1 (S0/0/1)
1900	Gigabit Ethernet 0/0 (G0/0)	Gigabit Ethernet 0/1 (G0/1)	Serial 0/0/0 (S0/0/0)	Serial 0/0/1 (S0/0/1)
2801	Fast Ethernet 0/0 (F0/0)	Fast Ethernet 0/1 (F0/1)	Serial 0/1/0 (S0/1/0)	Serial 0/1/1 (S0/1/1)
2811	Fast Ethernet 0/0 (F0/0)	Fast Ethernet 0/1 (F0/1)	Serial 0/0/0 (S0/0/0)	Serial 0/0/1 (S0/0/1)
2900	Gigabit Ethernet 0/0 (G0/0)	Gigabit Ethernet 0/1 (G0/1)	Serial 0/0/0 (S0/0/0)	Serial 0/0/1 (S0/0/1)
4221	Gigabit Ethernet 0/0/0 (G0/0/0)	Gigabit Ethernet 0/0/1 (G0/0/1)	Serial 0/1/0 (S0/1/0)	Serial 0/1/1 (S0/1/1)
4300	Gigabit Ethernet 0/0/0 (G0/0/0)	Gigabit Ethernet 0/0/1 (G0/0/1)	Serial 0/1/0 (S0/1/0)	Serial 0/1/1 (S0/1/1)

Note: To find out how the router is configured, look at the interfaces to identify the type of router and how many interfaces the router has. There is no way to effectively list all the combinations of configurations for each router class. This table includes identifiers for the possible combinations of Ethernet and Serial interfaces in the device. The table does not include any other type of interface, even though a specific router may contain one. An example of this might be an ISDN BRI interface. The string in parenthesis is the legal abbreviation that can be used in Cisco IOS commands to represent the interface.

Device Config Final – Notes

19.1.4 Lab - Implement a DMVPN Phase 3 Spoke-to-Spoke Topology

Topology

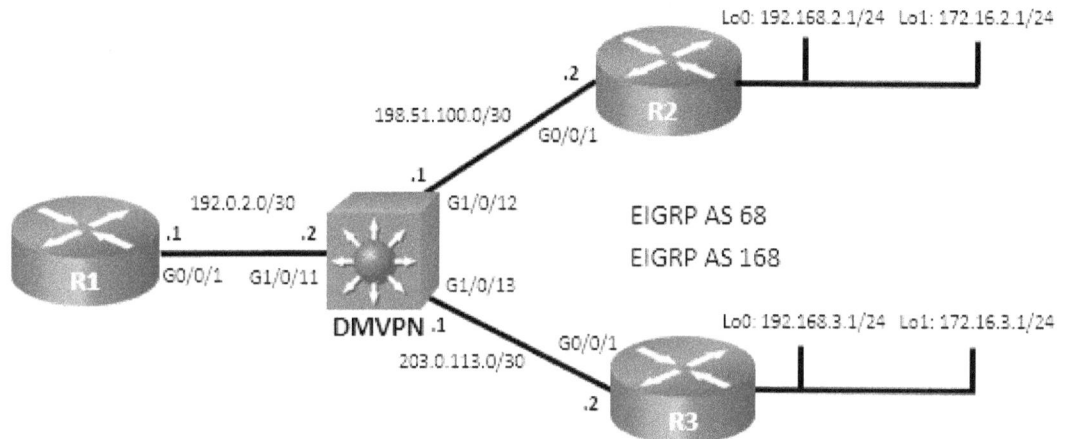

Addressing Table

Device	Interface	IPv4 Address
R1	G0/0/1	192.0.2.1/30
	Tunnel 1	100.100.100.1/29
R2	G0/0/1	198.51.100.2/30
	Loopback 0	192.168.2.1/24
	Loopback 1	172.16.2.1/24
	Tunnel 1	100.100.100.2/29
R3	G0/0/1	203.0.113.2/30
	Loopback 0	192.168.3.1/24
	Loopback 1	172.16.3.1/24
	Tunnel 1	100.100.100.3/29

Objectives

In this lab, you will create a Dynamic Multipoint Virtual Private Network (DMVPN) that consists of a hub router with two spokes. You will implement a DMVPN Phase 3 spoke-to-spoke topology.

Part 1: Build the Network and Configure Basic Device Settings

Part 2: Configure DMVPN Phase 3

Part 3: Verify DMVPN Phase 3

Background/Scenario

In this lab, you will modify a DMVPN Phase 1 network to create a Phase 3 spoke-to-spoke network. The configurations from the previous Phase 1 lab are required as the starting point for this lab.

Phase 3 improved upon Phase 2 by using a different mechanism for routing between destinations in spoke-to-spoke tunnels. Because of this, Phase 3 can be seen as superseding Phase 2. For that reason, there is no Phase 2 lab.

In DMVPN Phase 3, the spoke routers create routing table entries in which NHRP manipulates the routing information for tunnel destinations by overriding the next hop that was determined by the routing protocol. NHRP provides the actual destination network as the next hop. This enables the distribution of route summaries from the hub router to the spokes. Finally, Phase 3 enables a hierarchical tree-based VPN architecture in which central hub routers connect other, regional hubs and their spokes. This allows for the establishment of tunnels between routers that are not connected to the same regional hub.

In Phase 1 DMVPN, all spoke router traffic must pass through the hub router, as shown in the topology diagram.

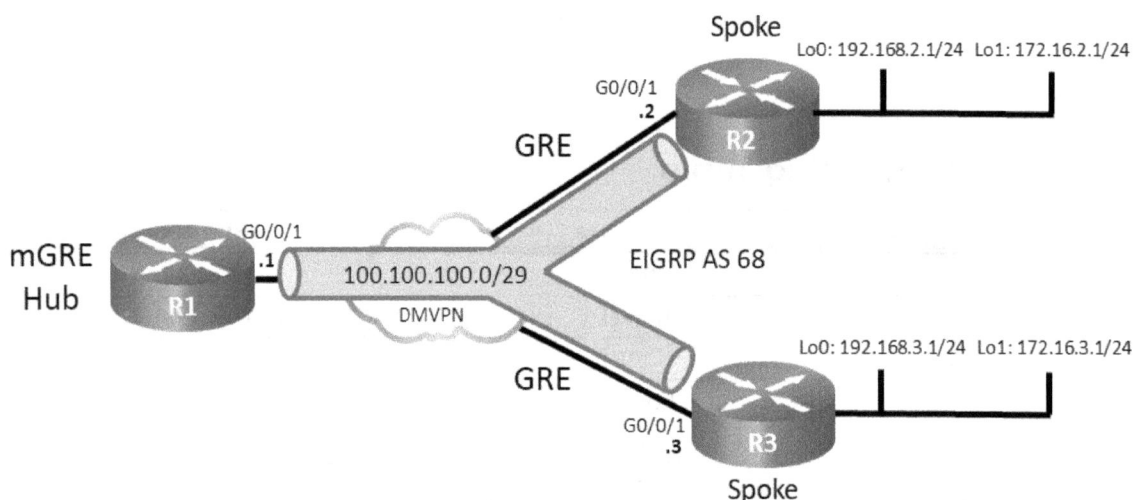

In a Phase 3 spoke-to-spoke DMVPN, traffic initially travels to the hub. The traffic then travels directly between the spoke routers over the spoke-to-spoke tunnel.

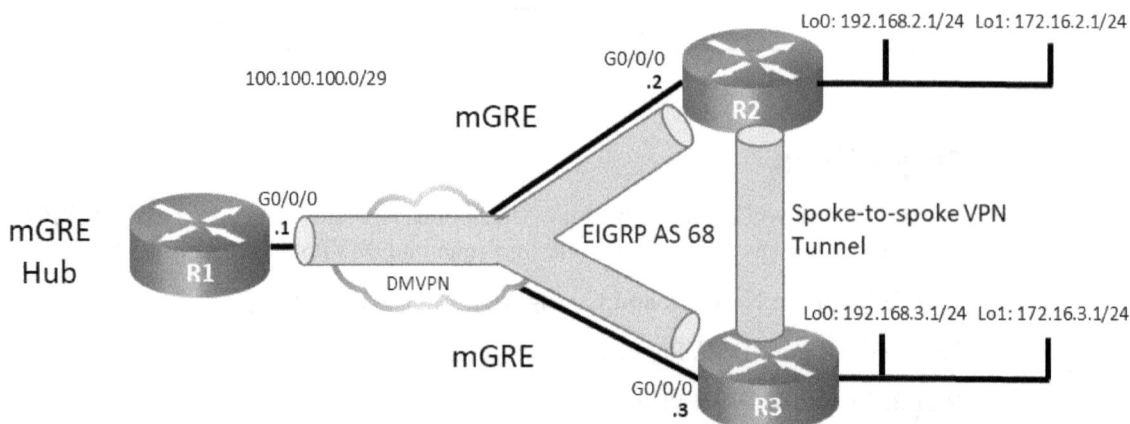

Note: This lab does not include the configuration of IPSec to secure the tunnels. This essential procedure will be covered in a later lab.

Note: This lab is an exercise in configuring and verifying various implementations of DMVPN topologies and does not reflect networking best practices.

Note: The routers used with CCNP hands-on labs are the Cisco 4221 with Cisco IOS XE Release 16.9.4 (universalk9 image). The Layer 3 switch is a Cisco Catalyst 3650 with Cisco IOS XE Release 16.9.4 (universalk9 image). Other routers, Layer 3 switches, and Cisco IOS versions can be used. Depending on the model and Cisco IOS version, the commands available and the output produced might vary from what is shown in the labs.

Note: Make sure that the devices have been erased and have no startup configurations. If you are unsure, please contact your instructor.

Required Resources

- 3 Routers (Cisco 4221 with Cisco IOS XE Release 16.9.4 universal image or comparable)
- 1 Layer 3 switch (Cisco 3650 with Cisco IOS Release 16.9.4 universal image or comparable)
- 1 PC (Choice of operating system with a terminal emulation program installed)
- Console cables to configure the Cisco IOS devices via the console ports
- Ethernet cables as shown in the topology

Instructions

Part 1: Build the Network and Configure Basic Device Settings

In Part 1, you will set up the network topology and configure basic settings if the network is not already configured. This lab uses the same topology and final configurations from the **Lab - Implement a DMVPN Phase 1 Hub-to-Spoke Topology.**

Step 1.   Cable the network as shown in the topology.

Attach the devices as shown in the topology diagram, and cable as necessary.

Step 2.   Configure initial settings for each router and the Layer 3 switch.

Console into each device, enter global configuration mode, and apply the initial settings for the lab if the devices are not already configured.

Step 3.   Verify connectivity in the network.

    a.   From R1, ping the loopback interfaces of R2 and R3. All pings should be successful. This verifies that full connectivity exists in the underlay, or transport, network.

```
R1# ping 192.168.2.1
Type escape sequence to abort.
Sending 5, 100-byte ICMP Echos to 192.168.3.1, timeout is 2 seconds:
!!!!!
Success rate is 100 percent (5/5), round-trip min/avg/max = 1/1/1 ms

R1# ping 192.168.3.1
Type escape sequence to abort.
Sending 5, 100-byte ICMP Echos to 192.168.3.1, timeout is 2 seconds:
!!!!!
Success rate is 100 percent (5/5), round-trip min/avg/max = 1/1/1 ms
```

b.   From R2 traceroute to the Loopback 1 interface of R3. You should see the following output. Note that traffic from the R2 spoke router travels through the R1 hub router to reach R3. This is the DMVPN Phase 1 traffic flow. Repeat the traceroute to verify this result.

```
R2# traceroute 172.16.3.1
Type escape sequence to abort.
Tracing the route to 172.16.3.1
VRF info: (vrf in name/id, vrf out name/id)
  1 100.100.100.1 1 msec 1 msec 1 msec
  2 100.100.100.3 1 msec *  2 msec
```

c.   On the spoke routers, issue the **show ip route** command. Note that EIGRP shows the next hop for the Loopback 1 networks to be the R1 tunnel interface address. That is, both spoke routers will use R1 to send traffic on the overlay network to the LANs.

```
R2# show ip route eigrp | begin Gateway
Gateway of last resort is not set

      172.16.0.0/16 is variably subnetted, 3 subnets, 2 masks
D        172.16.3.0/24 [90/102400640] via 100.100.100.1, 00:21:14, Tunnel1
      192.0.2.0/30 is subnetted, 1 subnets
D        192.0.2.0 [90/15360] via 198.51.100.1, 00:21:21, GigabitEthernet0/0/1
D     192.168.3.0/24
         [90/16000] via 198.51.100.1, 00:21:21, GigabitEthernet0/0/1
      203.0.113.0/30 is subnetted, 1 subnets
D        203.0.113.0
         [90/15360] via 198.51.100.1, 00:21:21, GigabitEthernet0/0/1
R3# show ip route eigrp | begin Gateway
Gateway of last resort is not set

      172.16.0.0/16 is variably subnetted, 3 subnets, 2 masks
D        172.16.2.0/24 [90/102400640] via 100.100.100.1, 00:22:49, Tunnel1
      192.0.2.0/30 is subnetted, 1 subnets
D        192.0.2.0 [90/15360] via 203.0.113.1, 00:22:59, GigabitEthernet0/0/1
D     192.168.2.0/24
         [90/16000] via 203.0.113.1, 00:22:59, GigabitEthernet0/0/1
      198.51.100.0/30 is subnetted, 1 subnets
D        198.51.100.0
         [90/15360] via 203.0.113.1, 00:22:59, GigabitEthernet0/0/1
```

Part 2: Configure DMVPN Phase 3

In this part of the lab, you will configure DMVPN Phase 3 to create DMVPN tunnels between the spoke routers. Initially, the services of the hub are required to begin data transfer over the VPN. However, the hub multipoint tunnel interface detects that traffic is "hairpinning," or exiting the interface on which it was received. The hub router sends redirect messages to the two spoke routers that are communicating. The spoke routers use the redirect messages to dynamically create a tunnel to connect to each other. In this way, the hub router does not need to forward spoke-to-spoke traffic after the setup of the spoke-to-spoke tunnel.

Step 1. Modify the tunnel interfaces on the spoke routers.

This lab builds on the Phase 1 lab that was previously completed. On the spoke routers R2 and R3, modify the configurations of the Tunnel 1 interfaces. The network will essentially be configured for DMVPN Phase 2 at the conclusion of this step.

a. On R2, set the tunnel mode to mGRE. You will see an error message that states that point-to-multipoint tunnels cannot have static tunnel destinations configured.

```
R2(config)# interface tunnel 1
R2(config-if)# tunnel mode gre multipoint
  Tunnel set mode failed. p2mp tunnels cannot have a tunnel destination.
```

b. Remove the tunnel destination configuration. You will see the interface go down and the EIGRP adjacency with the hub router be lost. Reconfigure the tunnel mode as GRE multipoint. You will see the interface come back up and the EIGRP adjacency will be restored.

```
R2(config-if)# no tunnel destination
*Mar 27 12:46:31.496: %LINEPROTO-5-UPDOWN: Line protocol on Interface Tunnel1,
changed state to down
*Mar 27 12:46:31.499: %DUAL-5-NBRCHANGE: EIGRP-IPv4 68: Neighbor 100.100.100.1
(Tunnel1) is down: interface down
R2(config-if)# tunnel mode gre multipoint
*Mar 27 12:46:46.629: %LINEPROTO-5-UPDOWN: Line protocol on Interface Tunnel1,
changed state to up
*Mar 27 12:46:48.008: %DUAL-5-NBRCHANGE: EIGRP-IPv4 68: Neighbor 100.100.100.1
(Tunnel1) is up: new adjacency
```

c. Repeat this process for the R3 tunnel interface. At this point, you have essentially configured DMVPN Phase 2. You may want to verify the R1 can ping R2 and R3 loopbacks again.

d. Verify the tunnel interface configuration is now using mGRE with the **show interface tunnel 1** command.

```
R3# show interface tunnel 1 | include Tunnel protocol
   Tunnel protocol/transport multi-GRE/IP
```

Step 2. Modify the configuration on the spoke routers to enable NHRP routing shortcuts.

Configure the spoke router interfaces on R2 and R3 to enable NHRP to add spoke networks as next hops in the EIGRP routing table. The hub router will send NHRP redirect messages to the spokes. NHRP uses information in the redirect messages to manipulate the EIGRP routing table and create the next hop shortcuts.

```
R2(config)# interface tunnel 1
R2(config-if)# ip nhrp shortcut
R3(config)# interface tunnel 1
R3(config-if)# ip nhrp shortcut
```

Step 3. Modify the configuration of the hub router to send NHRP redirect messages.

The Tunnel 1 interface of the hub router needs to be configured to send NHRP redirect messages to the spokes when it detects spoke-to-spoke traffic. Issue the **ip nhrp redirect** command on R1.

```
R1(config)# interface tunnel 1
R1(config-if)# ip nhrp redirect
```

Part 3: Verify DMVPN Phase 3

Now that DMVPN Phase 3 is complete, you can test dynamic spoke-to-spoke tunnel creation.

Step 1. Observe dynamic tunnel creation.

    a. Return to R2. Initiate a **traceroute** to the simulated LAN interface on R3. The path will pass through R1 as it does in DMVPN Phase 1.

```
R2# traceroute 172.16.3.1
Type escape sequence to abort.
Tracing the route to 172.16.3.1
VRF info: (vrf in name/id, vrf out name/id)
  1 100.100.100.1 1 msec 1 msec 1 msec
  2 100.100.100.3 1 msec *  2 msec
```

    b. Issue the **traceroute** command again. You will now see that R1 has enabled direct spoke-to-spoke communication between R2 and R3. This tunnel will expire after ten minutes by default. The tunnel dynamically reopens after data is sent to the spoke router again.

```
R2# traceroute 172.16.3.1
Type escape sequence to abort.
Tracing the route to 172.16.3.1
VRF info: (vrf in name/id, vrf out name/id)
  1 100.100.100.3 1 msec *  1 msec
```

    c. Repeat the **traceroute** commands on R3 to 172.16.2.1.

Step 2. View the routing table.

    a. On R2, issue the command to view the EIGRP routes that are in the routing table. Compare this routing table with the DMVPN Phase 1 routing from earlier in this lab. You should see that EIGRP shows the next hop interface to the 172.16.3.0 network to be the unchanged. However, the route is flagged with % indicating that NHRP is has overridden the next hop entry with its own value. Also notice that an NHRP route now appears in the routing table. This route indicates that the tunnel interface for R3 is considered to be directly connected by NHRP.

```
R2# show ip route
Codes: L - local, C - connected, S - static, R - RIP, M - mobile, B - BGP
       D - EIGRP, EX - EIGRP external, O - OSPF, IA - OSPF inter area
       N1 - OSPF NSSA external type 1, N2 - OSPF NSSA external type 2
       E1 - OSPF external type 1, E2 - OSPF external type 2
       i - IS-IS, su - IS-IS summary, L1 - IS-IS level-1, L2 - IS-IS level-2
       ia - IS-IS inter area, * - candidate default, U - per-user static route
       o - ODR, P - periodic downloaded static route, H - NHRP, l - LISP
       a - application route
       + - replicated route, % - next hop override, p - overrides from PfR

Gateway of last resort is not set

      100.0.0.0/8 is variably subnetted, 3 subnets, 2 masks
C        100.100.100.0/29 is directly connected, Tunnel1
L        100.100.100.2/32 is directly connected, Tunnel1
H        100.100.100.3/32 is directly connected, 00:03:53, Tunnel1
      172.16.0.0/16 is variably subnetted, 3 subnets, 2 masks
```

```
C       172.16.2.0/24 is directly connected, Loopback1
L       172.16.2.1/32 is directly connected, Loopback1
D   %   172.16.3.0/24 [90/102400640] via 100.100.100.1, 02:53:00, Tunnel1
        192.0.2.0/30 is subnetted, 1 subnets
D       192.0.2.0 [90/15360] via 198.51.100.1, 03:38:15, GigabitEthernet0/0/1
        192.168.2.0/24 is variably subnetted, 2 subnets, 2 masks
C       192.168.2.0/24 is directly connected, Loopback0
L       192.168.2.1/32 is directly connected, Loopback0
D     192.168.3.0/24
          [90/16000] via 198.51.100.1, 03:38:15, GigabitEthernet0/0/1
        198.51.100.0/24 is variably subnetted, 2 subnets, 2 masks
C       198.51.100.0/30 is directly connected, GigabitEthernet0/0/1
L       198.51.100.2/32 is directly connected, GigabitEthernet0/0/1
        203.0.113.0/30 is subnetted, 1 subnets
D       203.0.113.0
          [90/15360] via 198.51.100.1, 03:38:15, GigabitEthernet0/0/1
```

b. View the value that is actually used for the next hop address. It is marked with NHO for next hop override. This indicates that the next hop to 172.16.3.0 is R3 through the spoke-to-spoke tunnel.

```
R2# show ip route next-hop-override | begin Gateway
Gateway of last resort is not set

        100.0.0.0/8 is variably subnetted, 3 subnets, 2 masks
C       100.100.100.0/29 is directly connected, Tunnel1
L       100.100.100.2/32 is directly connected, Tunnel1
H       100.100.100.3/32 is directly connected, 00:17:39, Tunnel1
        172.16.0.0/16 is variably subnetted, 3 subnets, 2 masks
C       172.16.2.0/24 is directly connected, Loopback1
L       172.16.2.1/32 is directly connected, Loopback1
D   %   172.16.3.0/24 [90/102400640] via 100.100.100.1, 00:17:41, Tunnel1
                      [NHO] [90/255] via 100.100.100.3, 00:17:39, Tunnel1
        192.0.2.0/30 is subnetted, 1 subnets
D       192.0.2.0 [90/15360] via 198.51.100.1, 01:02:56, GigabitEthernet0/0/1
        192.168.2.0/24 is variably subnetted, 2 subnets, 2 masks
C       192.168.2.0/24 is directly connected, Loopback0
L       192.168.2.1/32 is directly connected, Loopback0
D     192.168.3.0/24
          [90/16000] via 198.51.100.1, 01:02:56, GigabitEthernet0/0/1
        198.51.100.0/24 is variably subnetted, 2 subnets, 2 masks
C       198.51.100.0/30 is directly connected, GigabitEthernet0/0/1
L       198.51.100.2/32 is directly connected, GigabitEthernet0/0/1
        203.0.113.0/30 is subnetted, 1 subnets
D       203.0.113.0
          [90/15360] via 198.51.100.1, 01:02:56, GigabitEthernet0/0/1
```

Step 3. Verify the DMVPN

a. View the DMVPN information with the spoke-to-spoke tunnel open. The first entry in the table is the static tunnel between R2 and the hub router. The DLX flag indicates that the entry is the local network. The DT1 entry is the route through the overlay network, and the DT2 entry the next hop override to the target network. If the spoke-to-spoke tunnel has closed, generate traffic to reopen it with traceroute.

```
R2# show dmvpn detail
Legend: Attrb --> S - Static, D - Dynamic, I - Incomplete
        N - NATed, L - Local, X - No Socket
        T1 - Route Installed, T2 - Nexthop-override
        C - CTS Capable, I2 - Temporary
        # Ent --> Number of NHRP entries with same NBMA peer
        NHS Status: E --> Expecting Replies, R --> Responding, W --> Waiting
        UpDn Time --> Up or Down Time for a Tunnel
==========================================================================
Interface Tunnel1 is up/up, Addr. is 100.100.100.2, VRF ""
   Tunnel Src./Dest. addr: 192.168.2.1/Multipoint, Tunnel VRF ""
   Protocol/Transport: "multi-GRE/IP", Protect ""
   Interface State Control: Disabled
   nhrp event-publisher : Disabled


IPv4 NHS:
100.100.100.1  RE priority = 0 cluster = 0
Type:Spoke, Total NBMA Peers (v4/v6): 3


# Ent   Peer NBMA Addr Peer Tunnel Add State  UpDn Tm Attrb    Target Network
-----   -------------- --------------- ----- -------- -----    --------------
    1 192.0.2.1          100.100.100.1   UP 03:22:47    S  100.100.100.1/32
    1 192.168.2.1        100.100.100.2   UP 00:01:36  DLX  100.100.100.2/32
    2 192.168.3.1        100.100.100.3   UP 00:01:36  DT1  100.100.100.3/32
      192.168.3.1        100.100.100.3   UP 00:01:36  DT2    172.16.3.0/24



Crypto Session Details:
------------------------------------------------------------------------------
--


Pending DMVPN Sessions:
```

b. Repeat this command on R3.

Step 4.   View NHRP Mappings

Note: If it has been more than 10 minutes since you completed your traceroutes, repeat them now to re-establish the spoke-to-spoke tunnels.

---

a. On R2 and R3, issue the **show ip nhrp detail** command to view the NHRP information for the routers. Note the correspondence between the entries in the output of this command with the output from the **show dmvpn detail** command. The first entry is the static entry to the hub router. The second entry is the local route, corresponding to the DTX entry in output above. The third entry corresponds to the DT1 entry and is flagged with **router nhop rib**. This indicates that the router has an explicit method to reach the tunnel IP address using an NBMA address and has an associated route installed in the routing table.

The flag **rib nho** indicates that the router has found an identical route in the routing table that belongs to a different protocol (EIGRP in this case.). NHRP has overridden the other protocol's next-hop entry for the network by installing a next-hop shortcut in the routing table. This corresponds to the DT2 entry in the **show dmvpn detail** output.

```
R2# show ip nhrp detail
100.100.100.1/32 via 100.100.100.1
    Tunnel1 created 04:03:01, never expire
    Type: static, Flags:
    NBMA address: 192.0.2.1
    Preference: 255
100.100.100.2/32 via 100.100.100.2
    Tunnel1 created 00:00:21, expire 00:09:38
    Type: dynamic, Flags: router unique local
    NBMA address: 192.168.2.1
    Preference: 255
     (no-socket)
   Requester: 100.100.100.3 Request ID: 9
100.100.100.3/32 via 100.100.100.3
    Tunnel1 created 00:00:21, expire 00:09:38
    Type: dynamic, Flags: router nhop rib
    NBMA address: 192.168.3.1
    Preference: 255
172.16.3.0/24 via 100.100.100.3
    Tunnel1 created 00:00:21, expire 00:09:38
    Type: dynamic, Flags: router rib nho
    NBMA address: 192.168.3.1
    Preference: 255
```

b.   You have successfully configured a DMVPN Phase 3 network.

Router Interface Summary Table

Router Model	Ethernet Interface #1	Ethernet Interface #2	Serial Interface #1	Serial Interface #2
1800	Fast Ethernet 0/0 (F0/0)	Fast Ethernet 0/1 (F0/1)	Serial 0/0/0 (S0/0/0)	Serial 0/0/1 (S0/0/1)
1900	Gigabit Ethernet 0/0 (G0/0)	Gigabit Ethernet 0/1 (G0/1)	Serial 0/0/0 (S0/0/0)	Serial 0/0/1 (S0/0/1)
2801	Fast Ethernet 0/0 (F0/0)	Fast Ethernet 0/1 (F0/1)	Serial 0/1/0 (S0/1/0)	Serial 0/1/1 (S0/1/1)
2811	Fast Ethernet 0/0 (F0/0)	Fast Ethernet 0/1 (F0/1)	Serial 0/0/0 (S0/0/0)	Serial 0/0/1 (S0/0/1)
2900	Gigabit Ethernet 0/0 (G0/0)	Gigabit Ethernet 0/1 (G0/1)	Serial 0/0/0 (S0/0/0)	Serial 0/0/1 (S0/0/1)
4221	Gigabit Ethernet 0/0/0 (G0/0/0)	Gigabit Ethernet 0/0/1 (G0/0/1)	Serial 0/1/0 (S0/1/0)	Serial 0/1/1 (S0/1/1)
4300	Gigabit Ethernet 0/0/0 (G0/0/0)	Gigabit Ethernet 0/0/1 (G0/0/1)	Serial 0/1/0 (S0/1/0)	Serial 0/1/1 (S0/1/1)

Note: To find out how the router is configured, look at the interfaces to identify the type of router and how many interfaces the router has. There is no way to effectively list all the combinations of configurations for each router class. This table includes identifiers for the possible combinations of Ethernet and Serial interfaces in the device. The table does not include any other type of interface, even though a specific router may contain one. An example of this might be an ISDN BRI interface. The string in parenthesis is the legal abbreviation that can be used in Cisco IOS commands to represent the interface.

Device Config Final – Notes

19.1.5 Lab - Implement an IPv6 DMVPN Phase 3 Spoke-to-Spoke Topology

Topology

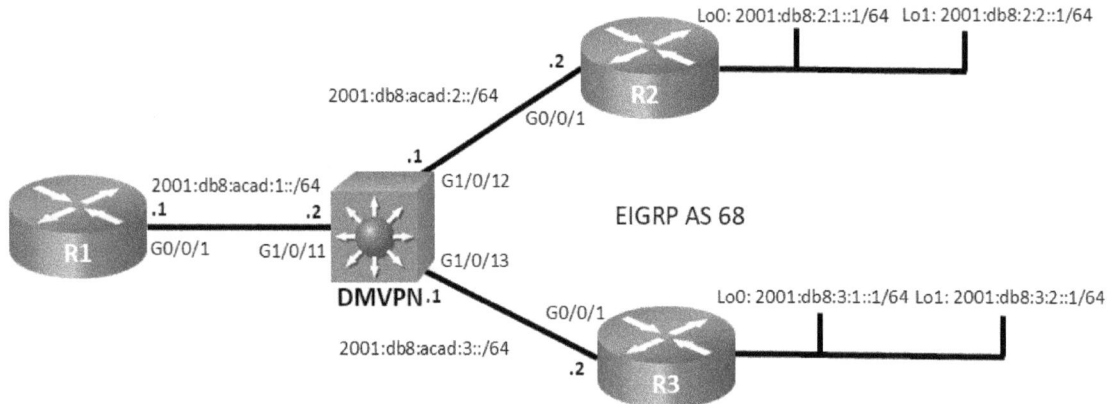

Addressing Table

Device	Interface	IPv6 Address	Link Local
R1	G0/0/0	2001.db8:acad:1::1/64	fe80::1
	Tunnel 1	2001:db8:cafe:100::1/64	fe80::2001
R2	G0/0/0	2001:db8:acad:2::2/64	fe80::2
	Loopback 0	2001:db8:2:1::1/64	fe80::2
	Loopback 1	2001:db8:2:2::1/64	fe80::2
	Tunnel 1	2001:db8:cafe:100::2/64	fe80::2002
R3	G0/0/0	2001:db8:acad:3::2/64	fe80::3
	Loopback 0	2001:db8:3:1::1/64	fe80::3
	Loopback 1	2001:db8:3:2::1/64	fe80::3
	Tunnel 1	2001:db8:cafe:100::3/64	fe80::2003

Objectives

In this lab, you will create a Dynamic Multipoint Virtual Private Network (DMVPN) that consists of a hub router with two spoke routers. You will implement a DMVPN Phase 3 spoke-to-spoke topology using IPv6.

Part 1: Build the Network and Configure Basic Device Settings

Part 2: Implement IPv6 DMVPN Phase 3

Part 3: Configure EIGRP for IPv6

Background/Scenario

In this lab you will configure IPv6 DMVPN Phase 3, which is very similar to the configuration with IPv4. Most of the tunnel and NHRP commands have direct parallels in IPv6. In addition, the configuration process and the differences between hub and spoke configuration is also similar. You will dynamically route overlay and transport networks over EIGRP for IPv6.

IPv6 DMVPN can be implemented in three different address type scenarios:

- **IPv4 over IPv6** - IPv4 is the protocol that is used on the tunnel and IPv6 is used in the physical transport network.

- **IPv6 over IPv4** - IPv6 is the tunnel protocol and IPv4 is the protocol that is used in the physical transport network.

- **IPv6 over IPv6** - Both the transport and tunnel networks use IPv6.

In this lab, you will configure the IPv6 over IPv6 scenario.

Note: This lab does not include the configuration of IPsec to secure the tunnels. This essential procedure will be covered in a later lab.

Note: This lab is an exercise in configuring and verifying various implementations of DMVPN topologies and does not reflect networking best practices.

Note: The routers used with CCNP hands-on labs are Cisco 4221s with Cisco IOS XE Release 16.9.4 (universalk9 image). The Layer 3 switch is a Cisco Catalyst 3650 with Cisco IOS XE Release 16.9.4 (universalk9 image). Other routers, Layer 3 switches, and Cisco IOS versions can be used. Depending on the model and Cisco IOS version, the commands available and the output produced might vary from what is shown in the labs.

Note: Make sure that the routers and switches have been erased and have no startup configurations. If you are unsure, please contact your instructor.

Required Resources

- 3 Routers (Cisco 4221 with Cisco IOS XE Release 16.9.4 universal image or comparable)

- 1 Layer 3 switch (Cisco 3650 with Cisco IOS Release 16.9.4 universal image or comparable)

- 1 PC (Choice of operating system with a terminal emulation program installed)

- Console cables to configure the Cisco IOS devices via the console ports

- Ethernet cables as shown in the topology

Instructions

Part 1: Build the Network and Configure Basic Device Settings

In Part 1, you will set up the network topology and configure basic settings.

Step 1. Cable the network as shown in the topology.

Connect the devices as shown in the topology diagram.

Step 2.   Configure initial settings for each router and the Layer 3 switch.

a.   Console into each device, enter global configuration mode, and apply the initial settings for the lab. Initial configurations for each device are provided below:

Hub Router R1

```
hostname R1
ipv6 unicast-routing
no ip domain lookup
banner motd # R1, Implement a DMVPN hub #
line con 0
 exec-timeout 0 0
 logging synchronous
 exit
line vty 0 4
 privilege level 15
 password cisco123
 exec-timeout 0 0
 logging synchronous
 login
 exit
ipv6 route ::/0 2001:db8:acad:1::2
interface g0/0/1
 ipv6 address 2001:db8:acad:1::1/64
 ipv6 address fe80::1 link-local
 no shutdown
end
```

Spoke Router R2

```
hostname R2
ipv6 unicast-routing
no ip domain lookup
banner motd # R2, Implement DMVPN Spoke 1 #
line con 0
 exec-timeout 0 0
 logging synchronous
 exit
line vty 0 4
 privilege level 15
 password cisco123
 exec-timeout 0 0
 logging synchronous
 login
 exit
ipv6 route ::/0 2001:db8:acad:2::1
interface g0/0/1
 ipv6 address 2001:db8:acad:2::2/64
 ipv6 address fe80::2 link-local
 no shutdown
 exit
interface loopback 0
 ipv6 address 2001:db8:2:1::1/64
```

```
 ipv6 address fe80::2 link-local
 no shutdown
 exit
interface loopback 1
 ipv6 address 2001:db8:2:2::1/64
 ipv6 address fe80::2 link-local
 no shutdown
 exit
```

Spoke Router R3

```
hostname R3
ipv6 unicast-routing
no ip domain lookup
banner motd # R3, Implement DMVPN Spoke 2 #
line con 0
 exec-timeout 0 0
 logging synchronous
 exit
line vty 0 4
 privilege level 15
 password cisco123
 exec-timeout 0 0
 logging synchronous
 login
 exit
ipv6 route ::/0 2001:db8:acad:3::1
interface g0/0/1
 ipv6 address 2001:db8:acad:3::2/64
 ipv6 address fe80::3 link-local
 no shutdown
 exit
interface loopback 0
 ipv6 address 2001:db8:3:1::1/64
 ipv6 address fe80::3 link-local
 exit
interface loopback 1
 ipv6 address 2001:db8:3:2::1/64
 ipv6 address fe80::3 link-local
 exit
end
```

DMVPN Layer 3 Switch

```
hostname DMVPN
ipv6 unicast-routing
ip routing
no ip domain lookup
banner motd # DMVPN, DMVPN cloud switch #
line con 0
 exec-timeout 0 0
 logging synchronous
 exit
line vty 0 4
```

```
                      privilege level 15
                      password cisco123
                      exec-timeout 0 0
                      logging synchronous
                      login
                  interface g1/0/11
                      no switchport
                      ipv6 address 2001:db8:acad:1::2/64
                      ipv6 address fe80::4 link-local
                      no shutdown
                      exit
                  interface g1/0/12
                      no switchport
                      ipv6 address 2001:db8:acad:2::1/64
                      ipv6 address fe80::4 link-local
                      no shutdown
                      exit
                  interface g1/0/13
                      no switchport
                      ipv6 address 2001:db8:acad:3::1/64
                      ipv6 address fe80::4 link-local
                      no shutdown
                      exit
                  ipv6 route 2001:db8:2:1::/64 2001:db8:acad:2::2
                  ipv6 route 2001:db8:2:2::/64 2001:db8:acad:2::2
                  ipv6 route 2001:db8:3:1::/64 2001:db8:acad:3::2
                  ipv6 route 2001:db8:3:2::/64 2001:db8:acad:3::2
                  end
```

b. Set the clock on each device to UTC time.

c. Save the running configuration to the startup configuration.

Note: In this lab, you will need to preconfigure the DMVPN Layer 3 switch. Normally, you would not need to configure this device. The DMVPN switch is simulating the ISP transport network.

Part 2: Implement IPv6 DMVPN Phase 3

In this part of the lab, you will configure IPv6 DMVPN Phase 3 to create DMVPN tunnels between the spoke routers R2 and R3, and the hub router, R1. DMVPN is very flexible and there are many options for implementation beyond what is being done in this lab.

In Phase 3 DMVPN, dynamic IPv6 spoke-to-spoke tunnels will be created between spoke routers after the initiating spoke router sends initial traffic to the hub.

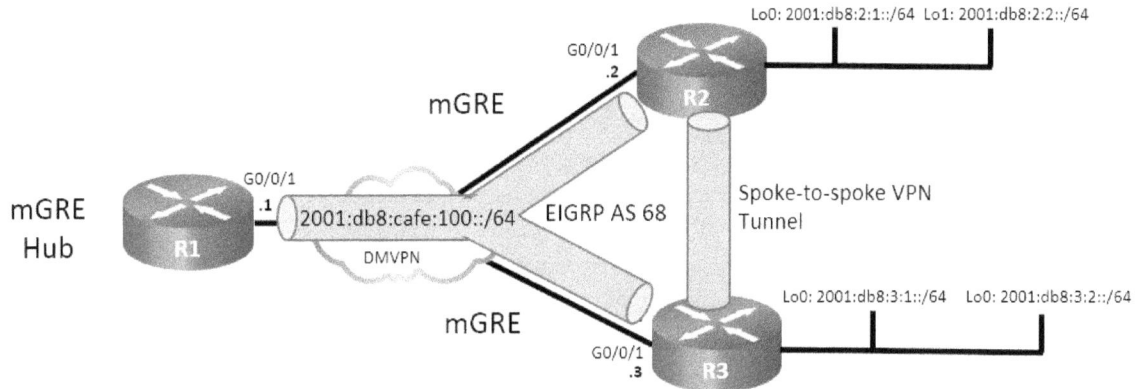

Step 1. Verify connectivity in the underlay network.

From R1, ping the Loopback 0 interfaces of R2 and R3.

```
R1# ping 2001:db8:2:1::1
Type escape sequence to abort.
Sending 5, 100-byte ICMP Echos to 2001:DB8:2:1::1, timeout is 2 seconds:
!!!!!
Success rate is 100 percent (5/5), round-trip min/avg/max = 1/1/1 ms
R1# ping 2001:db8:3:1::1
Type escape sequence to abort.
Sending 5, 100-byte ICMP Echos to 2001:DB8:3:1::1, timeout is 2 seconds:
!!!!!
Success rate is 100 percent (5/5), round-trip min/avg/max = 1/1/1 ms
```

Step 2. Configure the tunnel interface on the hub router.

As you know, DMVPN requires configuration of tunnel interfaces like GRE. Unlike GRE tunnels, DMVPN Phase 3 uses multipoint GRE (mGRE) mode tunnels. When configuring tunnel interfaces, care must be taken to use unique IPv6 link local addresses on all tunnel interfaces. The tunnel interfaces do not require a tunnel destination because the tunnel interfaces are multipoint.

  a. On R1, create the tunnel interface, set the tunnel mode to mGRE, and establish the tunnel source as Loopback 0. A tunnel key is also required when multiple tunnels will be established from a single interface. Finally, address the interface. The overlay network will use the same IPv6 network for all tunnel interfaces of the DMVPN.

```
R1(config)# interface tunnel 1
R1(config-if)# tunnel mode gre multipoint ipv6
R1(config-if)# tunnel source GigabitEthernet0/0/1
R1(config-if)# tunnel key 999
R1(config-if)# ipv6 address 2001:db8:cafe:100::1/64
R1(config-if)# ipv6 address fe80::2001 link-local
```

  b. Configure the hub router as an NHRP server (NHS). Spoke routers require the services of the NHS to establish dynamic tunnels.

NHRP enables DMVPN to dynamically learn the NBMA physical addresses of devices in the network. The NHRP network ID must be consistent between the hub and spokes in the DMVPN network. You configure authentication to add a layer of security. Finally, configure the interface as multicast dynamic, which enables the NHS to dynamically add spoke routers to the NHRP table when spokes initiate a tunnel. This enables the use of dynamic routing protocols between the hub and spoke routers.

The **ipv6 nhrp redirect** command is required to enable the hub router to support DMVPN Phase 3.

```
R1(config-if)# ipv6 nhrp network-id 1
R1(config-if)# ipv6 nhrp authentication NHRPauth
R1(config-if)# ipv6 nhrp map multicast dynamic
R1(config-if)# ipv6 nhrp redirect
```

c. Because DMVPN networks add information to packet headers, the interface should be fine-tuned to participate in the DMVPN network. In addition, configure the interface bandwidth so that routing protocols that use bandwidth values will function properly.

```
R1(config-if)# bandwidth 4000
R1(config-if)# ipv6 mtu 1380
R1(config-if)# ipv6 tcp adjust-mss 1360
```

d. Verify the tunnel interface configuration with the **show interface tunnel 1** and **show ipv6 interface tunnel 1** commands.

```
R1# show interface tunnel 1
Tunnel1 is up, line protocol is up
  Hardware is Tunnel
  MTU 1452 bytes, BW 4000 Kbit/sec, DLY 50000 usec,
     reliability 255/255, txload 1/255, rxload 1/255
  Encapsulation TUNNEL, loopback not set
  Keepalive not set
  Tunnel linestate evaluation up
  Tunnel source 2001:DB8:ACAD:1::1 (GigabitEthernet0/0/1)
   Tunnel Subblocks:
      src-track:
         Tunnel1 source tracking subblock associated with GigabitEthernet0/0/1
          Set of tunnels with source GigabitEthernet0/0/1, 1 member (includes
iterators), on interface <OK>
  Tunnel protocol/transport multi-GRE/IPv6
    Key 0x3E7, sequencing disabled
    Checksumming of packets disabled
  Tunnel TTL 255
  Path MTU Discovery, ager 10 mins, min MTU 1280
  Tunnel transport MTU 1452 bytes
  Tunnel transmit bandwidth 8000 (kbps)
  Tunnel receive bandwidth 8000 (kbps)
  Last input never, output never, output hang never
  Last clearing of "show interface" counters 00:02:45
  Input queue: 0/375/0/0 (size/max/drops/flushes); Total output drops: 20
  Queueing strategy: fifo
  Output queue: 0/0 (size/max)
  5 minute input rate 0 bits/sec, 0 packets/sec
  5 minute output rate 0 bits/sec, 0 packets/sec
     0 packets input, 0 bytes, 0 no buffer
     Received 0 broadcasts (0 IP multicasts)
     0 runts, 0 giants, 0 throttles
     0 input errors, 0 CRC, 0 frame, 0 overrun, 0 ignored, 0 abort
```

```
            0 packets output, 0 bytes, 0 underruns
            0 output errors, 0 collisions, 0 interface resets
            0 unknown protocol drops
            0 output buffer failures, 0 output buffers swapped out

R1# show ipv6 interface tunnel 1
Tunnel1 is up, line protocol is up
  IPv6 is enabled, link-local address is FE80::2001
  No Virtual link-local address(es):
  Global unicast address(es):
    2001:DB8:CAFE:100::1, subnet is 2001:DB8:CAFE:100::/64
  Joined group address(es):
    FF02::1
    FF02::2
    FF02::1:FF00:1
    FF02::1:FF00:2001
  MTU is 1380 bytes
  ICMP error messages limited to one every 100 milliseconds
  ICMP redirects are enabled
  ICMP unreachables are sent
  Input features: IPv6 TCP Adjust MSS
  Output features: IPv6 TCP Adjust MSS
  ND DAD is not supported
  ND reachable time is 30000 milliseconds (using 30000)
  ND advertised reachable time is 0 (unspecified)
  ND advertised retransmit interval is 0 (unspecified)
  ND router advertisements live for 1800 seconds
  ND advertised default router preference is Medium
  ND RAs are suppressed (periodic)
  Hosts use stateless autoconfig for addresses.
```

Step 3. Configure the R2 and R3 spoke router tunnel interfaces.

In DMVPN Phase 3, the NHRP client (NHC) tunnel interfaces use mGRE as does the NHS hub router. Much of the interface configuration is the same as for the NHS tunnel interface. However, instead of the interfaces being configured to send NHRP redirect messages, the interfaces are configured to create shortcuts, or spoke-to-spoke tunnels.

   a. On R2, create the tunnel interface and configure the mGRE tunnel parameters. Configuring the tunnel source as the Loopback 0 interface provides a stable source for the tunnel. The tunnel key must match the key that is configured on the hub router. Configure the overlay network IPv6 addresses for the tunnel interface. Note that no static tunnel destination is configured, because these are multipoint interfaces.

```
R2(config)# interface tunnel 1
R2(config-if)# tunnel mode gre multipoint ipv6
R2(config-if)# tunnel source loopback 0
R2(config-if)# tunnel key 999
R2(config-if)# ipv6 address 2001:db8:cafe:100::2/64
R2(config-if)# ipv6 address fe80::2002 link-local
```

b. Configure the tunnel interface as an NHRP client. You will need to designate the underlay address of the NHRP server and map the NHRP server underlay address to its overlay address.

```
R2(config-if)# ipv6 nhrp network-id 1
R2(config-if)# ipv6 nhrp authentication NHRPauth
R2(config-if)# ipv6 nhrp nhs 2001:db8:cafe:100::1 nbma 2001:db8:acad:1::1 multicast
R2(config-if)# ipv6 nhrp map multicast dynamic
R2(config-if)# ipv6 nhrp shortcut
```

c. Adjust settings on the interface to accommodate the GRE packet overhead.

```
R2(config-if)# bandwidth 4000
R2(config-if)# ipv6 mtu 1380
R2(config-if)# ipv6 tcp adjust-mss 1360
```

d. Repeat this configuration on router R3 using the commands above and information from the addressing table.

_____

_____

_____

_____

_____

e. Verify your configurations with the **show interface** command. If your configurations are correct, you should be able to successfully ping the interface addresses of the overlay network from each router.

f. Go to R1 and view the status of DMVPN with the **show dmvpn** and **show dmvpn detail** commands. Become familiar with the output of each.

```
R1# show dmvpn detail
Legend: Attrb --> S - Static, D - Dynamic, I - Incomplete
        N - NATed, L - Local, X - No Socket
        T1 - Route Installed, T2 - Nexthop-override
        C - CTS Capable, I2 - Temporary
        # Ent --> Number of NHRP entries with same NBMA peer
        NHS Status: E --> Expecting Replies, R --> Responding, W --> Waiting
        UpDn Time --> Up or Down Time for a Tunnel
========================================================================

Interface Tunnel1 is up/up, Addr. is 2001:DB8:CAFE:100::1, VRF ""
    Tunnel Src./Dest. addr: 2001:DB8:ACAD:1::1/Multipoint, Tunnel VRF ""
    Protocol/Transport: "multi-GRE/IPv6", Protect ""
    Interface State Control: Disabled
    nhrp event-publisher : Disabled
Type:Hub, Total NBMA Peers (v4/v6): 2
    1.Peer NBMA Address: 2001:DB8:2:1::1
        Tunnel IPv6 Address: 2001:DB8:CAFE:100::2
        IPv6 Target Network: 2001:DB8:CAFE:100::2/128
        # Ent: 2, Status: UP, UpDn Time: 00:12:54, Cache Attrib: D
```

```
    2.Peer NBMA Address: 2001:DB8:2:1::1
        Tunnel IPv6 Address: 2001:DB8:CAFE:100::2
        IPv6 Target Network: FE80::2002/128
        # Ent: 0, Status: UP, UpDn Time: 00:12:54, Cache Attrib: D
    3.Peer NBMA Address: 2001:DB8:3:1::1
        Tunnel IPv6 Address: 2001:DB8:CAFE:100::3
        IPv6 Target Network: 2001:DB8:CAFE:100::3/128
        # Ent: 2, Status: UP, UpDn Time: 00:06:32, Cache Attrib: D
    4.Peer NBMA Address: 2001:DB8:3:1::1
        Tunnel IPv6 Address: 2001:DB8:CAFE:100::3
        IPv6 Target Network: FE80::2003/128
        # Ent: 0, Status: UP, UpDn Time: 00:06:32, Cache Attrib: D


Crypto Session Details:
-------------------------------------------------------------------------


Pending DMVPN Sessions:
```

The output shows the status of the tunnel, the tunnel address and the tunnel source address. The list of peers shows the NBMA (underlay) addresses of the DMVPN peers that were learned by NHRP. These addresses come from the tunnel source Loopback 0 addresses. Although there are only two peers known (R2 and R3), there are two entries for each. The first entry shows the tunnel target network interface address, and the second gives the link local address. The status of the entries in the table must be UP for data to travel on the tunnels. The attribute D indicates the tunnels are dynamic. If configured, the crypto settings for the tunnel would be shown. You will secure the tunnels in a later lab.

Repeat this command on the spoke routers so that you become familiar with the command output.

g. Verify the status of NHRP by viewing the contents of the NHRP cache with the show ipv6 nhrp detail command. Output is shown for the hub router. Note that it displays information for both of the dynamic tunnels between the spoke routers and the hub.

```
R1# show ipv6 nhrp detail
2001:DB8:CAFE:100::2/128 via 2001:DB8:CAFE:100::2
    Tunnel1 created 00:27:29, expire 00:07:21
    Type: dynamic, Flags: registered nhop
    NBMA address: 2001:DB8:2:1::1
    Preference: 255
2001:DB8:CAFE:100::3/128 via 2001:DB8:CAFE:100::3
    Tunnel1 created 00:21:07, expire 00:08:52
    Type: dynamic, Flags: registered nhop
    NBMA address: 2001:DB8:3:1::1
    Preference: 255
FE80::2002/128 via 2001:DB8:CAFE:100::2
    Tunnel1 created 00:27:29, expire 00:07:21
    Type: dynamic, Flags: registered
    NBMA address: 2001:DB8:2:1::1
    Preference: 255
```

```
FE80::2003/128 via 2001:DB8:CAFE:100::3
    Tunnel1 created 00:21:07, expire 00:08:52
    Type: dynamic, Flags: registered
    NBMA address: 2001:DB8:3:1::1
    Preference: 255
```

This output provides details about the tunnel endpoints that are known to NHRP. This incudes the overlay and transport interface addresses for the known peers.

Part 3: Configure EIGRP for IPv6

In this scenario, you will create two EIGRP for IPv6 routing processes for two different ASs. AS 68 will route the tunnel network and the LANs to be accessed across the tunnels. AS 168 will route the transport network in order to ensure connectivity between the underlay networks that the tunnel network relies upon.

Initially, static routes were configured in the topology to enable initial testing of network connectivity after the topology was set up for the lab. You no longer need these static routes and will replace them with EIGRPv6.

Step 1.   Remove static routes.

    a.   Remove the preconfigured static routes from the three routers by pasting the commands below into the console of the appropriate devices.

R1

```
no ipv6 route ::/0 2001:db8:acad:1::2
```

R2

```
no ipv6 route ::/0 2001:db8:acad:2::1
```

R3

```
no ipv6 route ::/0 2001:db8:acad:3::1
```

DMVPN

```
no ipv6 route 2001:db8:2:1::/64 2001:db8:acad:2::2
no ipv6 route 2001:db8:2:2::/64 2001:db8:acad:2::2
no ipv6 route 2001:db8:3:1::/64 2001:db8:acad:3::2
no ipv6 route 2001:db8:3:2::/64 2001:db8:acad:3::2
```

Note: Normally devices in the DMVPN cloud would require no intervention from enterprise networking staff. However, for the purposes of this lab, some configuration of the DMVPN Layer 3 switch is required.

    b.   Create classic mode IPv6 EIGRP processes with AS **68**. This process and AS will route the overlay network. Add the tunnel interface and Loopback 1 interface networks to the routing process. Loopback 1 simulates a LAN that will be sending traffic through the tunnel. Note that split horizon is disabled on the hub and spoke router tunnel interfaces. Also note that the two spoke routers are configured as stub routers. Configure the three routers as follows:

```
R1(config)# ipv6 router eigrp 68
R1(config-router)# eigrp router-id 1.1.1.1
R1(config-router)# interface tunnel 1
R1(config-if)# ipv6 eigrp 68
R1(config-if)# no ipv6 split-horizon eigrp 68
```

```
R2(config)# ipv6 router eigrp 68
R2(config-router)# eigrp router-id 2.2.2.2
R2(config-router)# interface tunnel 1
R2(config-if)# ipv6 eigrp 68
R2(config-if)# no ipv6 split-horizon eigrp 68
R2(config-if)# interface loopback 1
R2(config-if)# ipv6 eigrp 68
R3(config)# ipv6 router eigrp 68
R3(config-router)# eigrp router-id 3.3.3.3
R3(config-router)# interface tunnel 1
R3(config-if)# ipv6 eigrp 68
R3(config-if)# no ipv6 split-horizon eigrp 68
R3(config-if)# interface loopback 1
R3(config-if)# ipv6 eigrp 68
```

Step 2.   Configure dynamic routing for the underlay network.

   a.   Create new classic mode EIGRP processes for AS **168**. This process and AS will route the underlay, or transport, network. Note that the DMVPN Layer 3 switch did not need to have routing knowledge for the tunnel network. However, it does need to be configured to route between the point-to-point underlay networks. Split horizon does not need to be disabled for this AS because the underlay network is a point-to-point network.

```
R1(config)# ipv6 router eigrp 168
R1(config-router)# eigrp router-id 10.1.1.1
R1(config-router)# interface GigabitEthernet 0/0/1
R1(config-if)# ipv6 eigrp 168
R2(config)# ipv6 router eigrp 168
R2(config-router)# eigrp router-id 20.2.2.2
R2(config-router)# interface GigabitEthernet 0/0/1
R2(config-if)# ipv6 eigrp 168
R2(config-if)# interface loopback 0
R2(config-if)# ipv6 eigrp 168
R3(config)# ipv6 router eigrp 168
R3(config-router)# eigrp router-id 30.3.3.3
R3(config-router)# interface GigabitEthernet 0/0/1
R3(config-if)# ipv6 eigrp 168
R3(config-if)# interface loopback 0
R3(config-if)# ipv6 eigrp 168
DMVPN(config)# ipv6 router eigrp 168
DMVPN(config-router)# eigrp router-id 40.4.4.4
DMVPN(config-router)# interface GigabitEthernet 1/0/11
DMVPN(config-if)# ipv6 eigrp 168
DMVPN(config-router)# interface GigabitEthernet 1/0/12
DMVPN(config-if)# ipv6 eigrp 168
DMVPN(config-router)# interface GigabitEthernet 1/0/13
DMVPN(config-if)# ipv6 eigrp 168
```

Note: Normally devices in the DMVPN cloud would require no intervention from enterprise networking staff. However, for the purposes of this lab, some configuration of the DMVPN Layer 3 switch is required.

b. Verify dynamic routing on all three routers by using the **show ipv6 route eigrp** command. Do this on all three routers to verify that the underlay and overlay networks appear in the routing tables. Output is shown for the hub router.

```
R1# show ipv6 route eigrp
<output omitted>
D   2001:DB8:2:1::/64 [90/131072]
      via FE80::4, GigabitEthernet0/0/1
D   2001:DB8:2:2::/64 [90/2048000]
     via FE80::2002, Tunnel1
D   2001:DB8:3:1::/64 [90/131072]
      via FE80::4, GigabitEthernet0/0/1
D   2001:DB8:3:2::/64 [90/2048000]
     via FE80::2003, Tunnel1
D   2001:DB8:ACAD:2::/64 [90/3072]
      via FE80::4, GigabitEthernet0/0/1
D   2001:DB8:ACAD:3::/64 [90/3072]
      via FE80::4, GigabitEthernet0/0/1
```

Step 3.   Verify DMVPN Phase 3 operation.

You have completed configuration of DMVPN Phase 3. Verify communication as follows:

a. On R1, execute a **traceroute** to the Loopback 1 interface IP addresses on R2 and R3. You should see the path use the tunnel network.

```
R1# traceroute 2001:db8:2:2::1
Type escape sequence to abort.
Tracing the route to 2001:DB8:2:2::1

  1 2001:DB8:CAFE:100::2 2 msec 1 msec 1 msec


R1# traceroute 2001:db8:3:2::1
Type escape sequence to abort.
Tracing the route to 2001:DB8:3:2::1

  1 2001:DB8:CAFE:100::3 1 msec 1 msec 1 msec
```

b. On R1, execute a **traceroute** to the Loopback 0 interface IP addresses on R2 and R3. You should see the path use the physical point-to-point networks of the underlay transport network.

```
R1# traceroute 2001:db8:2:1::1
Type escape sequence to abort.
Tracing the route to 2001:DB8:2:1::1

  1 2001:DB8:ACAD:1::2 2 msec 1 msec 2 msec
  2 2001:DB8:ACAD:2::2 1 msec 0 msec 0 msec


R1# traceroute 2001:db8:3:1::1
Type escape sequence to abort.
Tracing the route to 2001:DB8:3:1::1
```

```
    1 2001:DB8:ACAD:1::2 2 msec 2 msec 1 msec
    2 2001:DB8:ACAD:3::2 1 msec 1 msec 1 msec
```

   c. Repeat the traceroute commands between R2 and R3.

Step 4. Observe dynamic tunnel creation.

   a. Return to R2. Initiate a **traceroute** to the simulated LAN interface (Loopback 1) on R3. The path will pass through R1 as it does in DMVPN Phase 1.

```
R2# traceroute 2001:db8:3:2::1
Type escape sequence to abort.
Tracing the route to 2001:DB8:3:2::1


    1 2001:DB8:CAFE:100::1 1 msec 1 msec 1 msec
    2 2001:DB8:CAFE:100::3 2 msec 1 msec
```

   b. Issue the **traceroute** command again. You will now see that DMVPN hub, R1, has enabled direct spoke-to-spoke communication between R2 and R3. R1 is no longer in the path, instead, the path is directly to R3. This tunnel will expire after ten minutes by default. The tunnel dynamically reopens after data is sent to the spoke router again.

```
R2# traceroute 2001:db8:3:2::1
Type escape sequence to abort.
Tracing the route to 2001:DB8:3:2::1


    1 2001:DB8:CAFE:100::3 1 msec 1 msec 1 msec
```

   c. You have successfully configured a DMVPN Phase 3 network. Feel free to explore the IPv6 versions of the DMVPN Phase 3 verification commands you used for IPv4 DMVPN.

Router Interface Summary Table

Router Model	Ethernet Interface #1	Ethernet Interface #2	Serial Interface #1	Serial Interface #2
1800	Fast Ethernet 0/0 (F0/0)	Fast Ethernet 0/1 (F0/1)	Serial 0/0/0 (S0/0/0)	Serial 0/0/1 (S0/0/1)
1900	Gigabit Ethernet 0/0 (G0/0)	Gigabit Ethernet 0/1 (G0/1)	Serial 0/0/0 (S0/0/0)	Serial 0/0/1 (S0/0/1)
2801	Fast Ethernet 0/0 (F0/0)	Fast Ethernet 0/1 (F0/1)	Serial 0/1/0 (S0/1/0)	Serial 0/1/1 (S0/1/1)
2811	Fast Ethernet 0/0 (F0/0)	Fast Ethernet 0/1 (F0/1)	Serial 0/0/0 (S0/0/0)	Serial 0/0/1 (S0/0/1)
2900	Gigabit Ethernet 0/0 (G0/0)	Gigabit Ethernet 0/1 (G0/1)	Serial 0/0/0 (S0/0/0)	Serial 0/0/1 (S0/0/1)
4221	Gigabit Ethernet 0/0/0 (G0/0/0)	Gigabit Ethernet 0/0/1 (G0/0/1)	Serial 0/1/0 (S0/1/0)	Serial 0/1/1 (S0/1/1)
4300	Gigabit Ethernet 0/0/0 (G0/0/0)	Gigabit Ethernet 0/0/1 (G0/0/1)	Serial 0/1/0 (S0/1/0)	Serial 0/1/1 (S0/1/1)

Note: To find out how the router is configured, look at the interfaces to identify the type of router and how many interfaces the router has. There is no way to effectively list all the combinations of configurations for each router class. This table includes identifiers for the possible combinations of Ethernet and Serial interfaces in the device. The table does not include any other type of interface, even though a specific router may contain one. An example of this might be an ISDN BRI interface. The string in parenthesis is the legal abbreviation that can be used in Cisco IOS commands to represent the interface.

Device Config Final – Notes

CHAPTER 20

Securing DMVPN Tunnels

20.1.2 Lab - Configure Secure DMVPN Tunnels

Topology

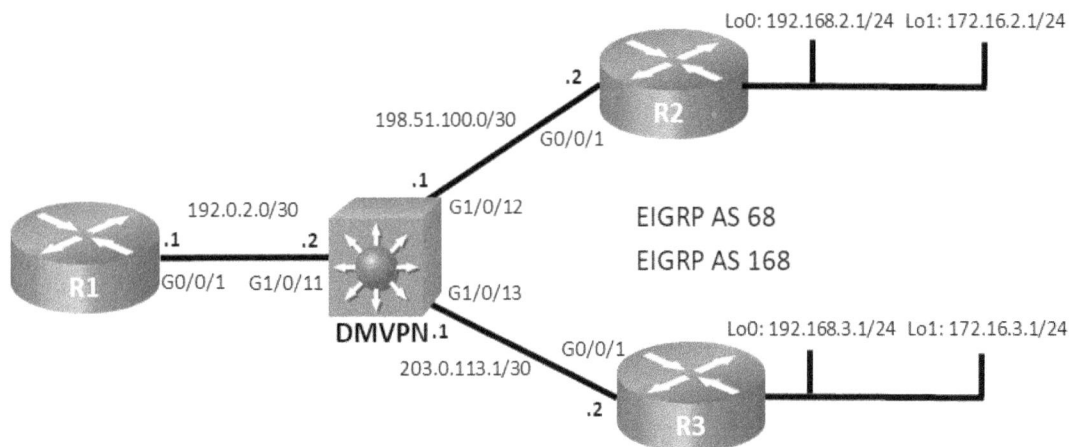

Addressing Table

Device	Interface	IPv4 Address
R1	G0/0/1	192.0.2.1/24
	Tunnel 1	100.100.100.1/29
R2	G0/0/1	198.51.100.2/24
	Loopback 0	192.168.1.1/24
	Loopback 1	172.16.1.1/24
	Tunnel 1	100.100.100.2/29
R3	G0/0/1	203.0.113.2/24
	Loopback 0	192.168.3.1/24
	Loopback 1	172.16.3.1/24
	Tunnel 1	100.100.100.3/29

Objectives

Part 1: Build the Network and Verify DMVPN Phase 3 Operation

Part 2: Secure DMVPN Phase 3 Tunnels

Background/Scenario

In previous labs, you have configured DMVPN Phase 1 and Phase 3 networks, including configuration of DMVPN Phase 3 with IPv6. However, in those labs, IPsec was not used to encrypt and protect data travelling on the tunnels. IPsec functionality is essential to DMVPN implementation. In this lab, you will work with the DMVPN Phase 3 implementation from the Implement a DMVPN Phase 3 Spoke-to-Spoke Topology lab. You will start with a working configuration and then apply IPsec to the spoke-to-hub and spoke-to-spoke tunnels. Finally, you will verify the operation of the secured tunnels.

Note: The routers used with CCNA hands-on labs are Cisco 4221 with Cisco IOS XE Release 16.9.4 (universalk9 image). The switch used is a Cisco Catalyst 3650 with Cisco IOS XE Release 16.9.4 (universalk9 image). Other routers, Layer 3 switches, and Cisco IOS versions can be used. Depending on the model and Cisco IOS version, the commands available and the output produced might vary from what is shown in the labs. Refer to the Router Interface Summary Table at the end of the lab for the correct interface identifiers.

Required Resources

- 3 Routers (Cisco 4221 with Cisco IOS XE Release 16.9.4 universal image or comparable)

- 1 Switch (Cisco 3560 with Cisco IOS XE Release 16.9.4 universal image or comparable)

- 1 PC (Choice of operating system with a terminal emulation program installed)

- Console cables to configure the Cisco IOS devices via the console ports

- Ethernet cables as shown in the topology

Instructions

Part 1: Build the Network and Verify DMVPN Phase 3 Operation

In Part 1, you will set up the network topology and configure basic settings if the network is not already configured. This lab uses the same topology and final configurations from the **Implement a DMVPN Phase 3 Spoke-to-Spoke Topology** lab.

Step 1.   Cable the network as shown in the topology.

Connect the devices as shown in the topology diagram.

Step 2.   Configure initial settings for each router and the Layer 3 switch.

Console into each device, enter global configuration mode, and apply the initial settings for the lab if the devices are not already configured.

Step 3.   Verify connectivity in the network.

a.   From R1, **ping** the loopback interfaces of R2 and R3. All pings should be successful. This verifies that full connectivity exists in the underlay, or transport, network.

```
R1# ping 192.168.2.1
Type escape sequence to abort.
Sending 5, 100-byte ICMP Echos to 192.168.3.1, timeout is 2 seconds:
!!!!!
Success rate is 100 percent (5/5), round-trip min/avg/max = 1/1/1 ms
```

```
R1# ping 192.168.3.1
Type escape sequence to abort.
Sending 5, 100-byte ICMP Echos to 192.168.3.1, timeout is 2 seconds:
!!!!!
Success rate is 100 percent (5/5), round-trip min/avg/max = 1/1/1 ms
```

Step 4. Verify DMVPN Phase 3 operation.

 a. Return to R2. Initiate a **traceroute** to the simulated LAN interface on R3. The path will pass through R1 as it does in a DMVPN Phase 1 network.

Note: The first trace may fail if the DMVPN switch CAM table is empty.

```
R2# traceroute 172.16.3.1
Type escape sequence to abort.
Tracing the route to 172.16.3.1
VRF info: (vrf in name/id, vrf out name/id)
  1 100.100.100.1 1 msec 1 msec 1 msec
  2 100.100.100.3 1 msec *  2 msec
```

 b. Issue the **traceroute** command again. You will now see that R1 has enabled direct spoke-to-spoke communication between R2 and R3. This tunnel will expire and close dynamically. The tunnel reopens after data for the spoke router is sent again.

```
R2# traceroute 172.16.3.1
Type escape sequence to abort.
Tracing the route to 172.16.3.1
VRF info: (vrf in name/id, vrf out name/id)
  1 100.100.100.3 1 msec *  1 msec
```

Part 2: Secure DMVPN Phase 3 Tunnels

Now that the tunnels have been configured and DMVPN connectivity has been verified, the tunnels can be secured with IPsec.

Step 1. Create the IKE policy.

Create an IKE policy that defines the hash algorithm, encryption type, key exchange method, Diffie-Hellman group, and the authentication method.

```
R1(config)# crypto isakmp policy 99
R1(config-isakmp)# hash sha384
R1(config-isakmp)# encryption aes 256
R1(config-isakmp)# group 14
R1(config-isakmp)# authentication pre-share
R1(config-isakmp)# exit
```

Step 2. Configure the ISAKMP key.

Configure the pre-shared key and peer address. Use 0.0.0.0 to match multiple peer addresses. Use a key of **DMVPN@key#**.

```
R1(config)# crypto isakmp key DMVPN@key# address 0.0.0.0
```

Step 3.   Create and configure the IPsec transform set.

Configure the IPsec transform set. Use **DMVPN_TRANS** as the transform set name. Specify **esp-aes** with a 256-bit key as the encryption transform and **esp-sha384-hmac** as the authentication transform. Configure the transform set to use IPsec **transport** mode for the tunnels.

```
R1(config)# crypto ipsec transform-set DMVPN_TRANS esp-aes 256 esp-sha384-hmac
R1(cfg-crypto-trans)# mode transport
R1(cfg-crypto-trans)# exit
```

Step 4.   Create the IPsec profile.

Create an IPsec profile with the name **DMVPN_PROFILE**. Associate the **DMVPN_TRANS** transform set with the profile.

```
R1(config)# crypto ipsec profile DMVPN_PROFILE
R1(ipsec-profile)# set transform-set DMVPN_TRANS
R1(ipsec-profile)# exit
```

Step 5.   Apply the IPsec profile to the tunnel interface.

Finally, apply the IPsec profile to the tunnel interface. After you apply the profile, you will see the that IPsec is now active and you will lose adjacency with R2 and R3 until their respective ends of the tunnel are configured.

```
R1(config)# interface tunnel 1
R1(config-if)# tunnel protection ipsec profile DMVPN_PROFILE
R1(config-if)# exit
*Mar 30 07:39:32.398: %CRYPTO-6-ISAKMP_ON_OFF: ISAKMP is ON
R1(config)#
*Mar 30 07:39:32.963: %IOSXE-3-PLATFORM: R0/0: cpp_cp: QFP:0.0 Thread:001
TS:00000000594132950499 %IPSEC-3-RECVD_PKT_NOT_IPSEC: Rec'd packet not an IPSEC
packet, dest_addr= 192.0.2.1, src_addr= 192.168.2.1, prot= 47
*Mar 30 07:39:43.664: %DUAL-5-NBRCHANGE: EIGRP-IPv4 68: Neighbor 100.100.100.2
(Tunnel1) is down: holding time expired
*Mar 30 07:39:44.235: %DUAL-5-NBRCHANGE: EIGRP-IPv4 68: Neighbor 100.100.100.3
(Tunnel1) is down: holding time expired
R1(config)#
```

Step 6.   Configure R2 and R3 with IPsec.

Repeat this configuration on the R2 and R3 routers.

_____

_____

_____

_____

_____

Step 7.   Verify DMVPN Phase 3 operation.

   a.   As was done previously, test the operation of the spoke-to-spoke DMVPN. Return to R2. Initiate a **traceroute** to the simulated LAN interface on R3. The path will pass through R1 as it does in a DMVPN Phase 1 network.

```
R2# traceroute 172.16.3.1
Type escape sequence to abort.
Tracing the route to 172.16.3.1
VRF info: (vrf in name/id, vrf out name/id)
```

```
  1 100.100.100.1 1 msec 1 msec 1 msec
  2 100.100.100.3 1 msec *  2 msec
```

b. Issue the **traceroute** command again. You will now see that R1 has enabled direct spoke-to-spoke communication between R2 and R3. This tunnel will expire and close dynamically. The tunnel reopens after data for the spoke router is sent again.

```
R2# traceroute 172.16.3.1
Type escape sequence to abort.
Tracing the route to 172.16.3.1
VRF info: (vrf in name/id, vrf out name/id)
  1 100.100.100.3 1 msec *  1 msec
```

Step 8. Verify IPsec configuration.

Note: Shut down a tunnel interface to clear its IPsec socket if you wish to explore the outputs before and after spoke-to-spoke tunnel establishment.

---

a. To show information about the IPsec profiles that are configured on a device, issue the **show crypto ipsec profile** command. Note that the profile that was previously configured is shown along with a default profile.

```
R2# show crypto ipsec profile
IPSEC profile DMVPN_PROFILE
        Security association lifetime: 4608000 kilobytes/3600 seconds
        Responder-Only (Y/N): N
        PFS (Y/N): N
        Mixed-mode : Disabled
        Transform sets={
                DMVPN_TRANS:  { esp-256-aes esp-sha384-hmac  } ,
        }

IPSEC profile default
        Security association lifetime: 4608000 kilobytes/3600 seconds
        Responder-Only (Y/N): N
        PFS (Y/N): N
        Mixed-mode : Disabled
        Transform sets={
                default:  { esp-aes esp-sha-hmac  } ,
        }
```

b. It is very important to verify that tunnel traffic will be encrypted. On R1, issue the **show dmvpn detail** command. As the hub router, R1 should see the spoke peers. The first part of the output shows the tunnel interface status and the peer table. Both peers should be shown with their transport and overlay interface addresses, as you have seen previously.

The Crypto Session Details portion of the output should contain information about the status of the encrypted tunnels. Both of the spoke routers should appear in this output also. Note that the transform set that you configured is also displayed in the Crypto Session output.

```
R1# show dmvpn detail
<output omitted>
Interface Tunnel1 is up/up, Addr. is 100.100.100.1, VRF ""
   Tunnel Src./Dest. addr: 192.0.2.1/Multipoint, Tunnel VRF ""
```

```
        Protocol/Transport: "multi-GRE/IP", Protect "DMVPN_PROFILE"
        Interface State Control: Disabled
        nhrp event-publisher : Disabled
Type:Hub, Total NBMA Peers (v4/v6): 2

# Ent  Peer NBMA Addr Peer Tunnel Add State  UpDn Tm Attrb    Target Network
-----  -------------- --------------- ----- -------- ----- ----------------
    1  192.168.2.1      100.100.100.2    UP 00:04:25     D  100.100.100.2/32
    1  192.168.3.1      100.100.100.3    UP 00:04:59     D  100.100.100.3/32


Crypto Session Details:
-----------------------------------------------------------------------------
--


Interface: Tunnel1
Session: [0x7F6E17B867D0]
  Session ID: 0
  IKEv1 SA: local 192.0.2.1/500 remote 192.168.2.1/500 Active
        Capabilities:(none) connid:1001 lifetime:23:59:19
  Session ID: 0
  IKEv1 SA: local 192.0.2.1/500 remote 192.168.2.1/500 Active
        Capabilities:(none) connid:1002 lifetime:23:59:28
  Crypto Session Status: UP-ACTIVE
  fvrf: (none), Phase1_id: 192.168.2.1
  IPSEC FLOW: permit 47 host 192.0.2.1 host 192.168.2.1
        Active SAs: 4, origin: crypto map
        Inbound:  #pkts dec'ed 17 drop 0 life (KB/Sec) 4607998/3568
        Outbound: #pkts enc'ed 16 drop 0 life (KB/Sec) 4607999/3568
    Outbound SPI : 0xD2E76488, transform : esp-256-aes esp-sha384-hmac
    Socket State: Open

Interface: Tunnel1
Session: [0x7F6E17B86950]
  Session ID: 0
  IKEv1 SA: local 192.0.2.1/500 remote 192.168.3.1/500 Active
        Capabilities:(none) connid:1004 lifetime:23:59:48
  Session ID: 0
  IKEv1 SA: local 192.0.2.1/500 remote 192.168.3.1/500 Active
        Capabilities:(none) connid:1003 lifetime:23:59:40
  Crypto Session Status: UP-ACTIVE
  fvrf: (none), Phase1_id: 192.168.3.1
  IPSEC FLOW: permit 47 host 192.0.2.1 host 192.168.3.1
        Active SAs: 6, origin: crypto map
        Inbound:  #pkts dec'ed 11 drop 0 life (KB/Sec) 4607999/3588
        Outbound: #pkts enc'ed 10 drop 0 life (KB/Sec) 4607999/3588
    Outbound SPI : 0xCB3D3313, transform : esp-256-aes esp-sha384-hmac
    Socket State: Open


Pending DMVPN Sessions:
```

c. Issue the **show crypto ipsec sa** command on R2 to display the security associations (sa) that have been made by R2. This output is for the spoke-to-hub tunnel between R1 and R2 prior to the establishment of the spoke-to-spoke tunnel. This command provides additional details regarding the IPsec status of the tunnel, encrypted and decrypted packet statistics, and other details regarding characteristics of the encrypted tunnel.

```
R2# show crypto ipsec sa

interface: Tunnel1
    Crypto map tag: Tunnel1-head-0, local addr 192.168.2.1

   protected vrf: (none)
   local  ident (addr/mask/prot/port): (192.168.2.1/255.255.255.255/47/0)
   remote ident (addr/mask/prot/port): (192.0.2.1/255.255.255.255/47/0)
   current_peer 192.0.2.1 port 500
     PERMIT, flags={origin_is_acl,}
    #pkts encaps: 125, #pkts encrypt: 125, #pkts digest: 125
    #pkts decaps: 126, #pkts decrypt: 126, #pkts verify: 126
    #pkts compressed: 0, #pkts decompressed: 0
    #pkts not compressed: 0, #pkts compr. failed: 0
    #pkts not decompressed: 0, #pkts decompress failed: 0
    #send errors 0, #recv errors 0

     local crypto endpt.: 192.168.2.1, remote crypto endpt.: 192.0.2.1
     plaintext mtu 1458, path mtu 1514, ip mtu 1514, ip mtu idb Loopback0
     current outbound spi: 0x97C1D18A(2546061706)
     PFS (Y/N): N, DH group: none

     inbound esp sas:
      spi: 0xD2E76488(3538379912)
        transform: esp-256-aes esp-sha384-hmac ,
        in use settings ={Transport, }
        conn id: 2003, flow_id: ESG:3, sibling_flags FFFFFFFF80000008, crypto
map: Tunnel1-head-0
        sa timing: remaining key lifetime (k/sec): (4607984/3047)
        IV size: 16 bytes
        replay detection support: Y
        Status: ACTIVE(ACTIVE)

     inbound ah sas:

     inbound pcp sas:

     outbound esp sas:
      spi: 0x97C1D18A(2546061706)
        transform: esp-256-aes esp-sha384-hmac ,
        in use settings ={Transport, }
        conn id: 2004, flow_id: ESG:4, sibling_flags FFFFFFFF80000008, crypto
map: Tunnel1-head-0
        sa timing: remaining key lifetime (k/sec): (4607990/3047)
        IV size: 16 bytes
        replay detection support: Y
```

```
        Status: ACTIVE(ACTIVE)

      outbound ah sas:

      outbound pcp sas:
```

The output below is for the same command after the spoke-to-spoke tunnel is open. Entries exist for both the tunnel to R1 and the spoke-to-spoke tunnel between R2 and R3.

R2# **show crypto ipsec sa**

```
interface: Tunnel1
    Crypto map tag: Tunnel1-head-0, local addr 192.168.2.1

   protected vrf: (none)
   local  ident (addr/mask/prot/port): (192.168.2.1/255.255.255.255/47/0)
   remote ident (addr/mask/prot/port): (192.168.3.1/255.255.255.255/47/0)
   current_peer 192.168.3.1 port 500
     PERMIT, flags={origin_is_acl,}
    #pkts encaps: 0, #pkts encrypt: 0, #pkts digest: 0
    #pkts decaps: 0, #pkts decrypt: 0, #pkts verify: 0
    #pkts compressed: 0, #pkts decompressed: 0
    #pkts not compressed: 0, #pkts compr. failed: 0
    #pkts not decompressed: 0, #pkts decompress failed: 0
    #send errors 0, #recv errors 0

     local crypto endpt.: 192.168.2.1, remote crypto endpt.: 192.168.3.1
     plaintext mtu 1458, path mtu 1514, ip mtu 1514, ip mtu idb Loopback0
     current outbound spi: 0x658E8CF5(1703841013)
     PFS (Y/N): N, DH group: none

     inbound esp sas:
      spi: 0xFA8FC9F2(4203727346)
        transform: esp-256-aes esp-sha384-hmac ,
        in use settings ={Transport, }
        conn id: 2005, flow_id: ESG:5, sibling_flags FFFFFFFF80000008, crypto
map: Tunnel1-head-0
        sa timing: remaining key lifetime (k/sec): (4608000/3316)
        IV size: 16 bytes
        replay detection support: Y
        Status: ACTIVE(ACTIVE)
      spi: 0x59C41A42(1506024002)
        transform: esp-256-aes esp-sha384-hmac ,
        in use settings ={Transport, }
        conn id: 2007, flow_id: ESG:7, sibling_flags FFFFFFFF80004008, crypto
map: Tunnel1-head-0
        sa timing: remaining key lifetime (k/sec): (4608000/3326)
        IV size: 16 bytes
        replay detection support: Y
        Status: ACTIVE(ACTIVE)
```

```
      inbound ah sas:

      inbound pcp sas:

      outbound esp sas:
       spi: 0x60CC6F77(1624010615)
          transform: esp-256-aes esp-sha384-hmac ,
          in use settings ={Transport, }
          conn id: 2006, flow_id: ESG:6, sibling_flags FFFFFFFF80000008, crypto
map: Tunnel1-head-0
          sa timing: remaining key lifetime (k/sec): (4608000/3316)
          IV size: 16 bytes
          replay detection support: Y
          Status: ACTIVE(ACTIVE)
        spi: 0x658E8CF5(1703841013)
          transform: esp-256-aes esp-sha384-hmac ,
          in use settings ={Transport, }
          conn id: 2008, flow_id: ESG:8, sibling_flags FFFFFFFF80004008, crypto
map: Tunnel1-head-0
          sa timing: remaining key lifetime (k/sec): (4608000/3326)
          IV size: 16 bytes
          replay detection support: Y
          Status: ACTIVE(ACTIVE)

      outbound ah sas:

      outbound pcp sas:

   protected vrf: (none)
   local  ident (addr/mask/prot/port): (192.168.2.1/255.255.255.255/47/0)
   remote ident (addr/mask/prot/port): (192.0.2.1/255.255.255.255/47/0)
   current_peer 192.0.2.1 port 500
     PERMIT, flags={origin_is_acl,}
    #pkts encaps: 67, #pkts encrypt: 67, #pkts digest: 67
    #pkts decaps: 67, #pkts decrypt: 67, #pkts verify: 67
    #pkts compressed: 0, #pkts decompressed: 0
    #pkts not compressed: 0, #pkts compr. failed: 0
    #pkts not decompressed: 0, #pkts decompress failed: 0
    #send errors 0, #recv errors 0

     local crypto endpt.: 192.168.2.1, remote crypto endpt.: 192.0.2.1
     plaintext mtu 1458, path mtu 1514, ip mtu 1514, ip mtu idb Loopback0
     current outbound spi: 0x97C1D18A(2546061706)
     PFS (Y/N): N, DH group: none

      inbound esp sas:
       spi: 0xD2E76488(3538379912)
          transform: esp-256-aes esp-sha384-hmac ,
          in use settings ={Transport, }
          conn id: 2003, flow_id: ESG:3, sibling_flags FFFFFFFF80000008, crypto
```

```
map: Tunnel1-head-0
        sa timing: remaining key lifetime (k/sec): (4607991/3305)
        IV size: 16 bytes
        replay detection support: Y
        Status: ACTIVE(ACTIVE)


    inbound ah sas:


    inbound pcp sas:


    outbound esp sas:
     spi: 0x97C1D18A(2546061706)
        transform: esp-256-aes esp-sha384-hmac ,
        in use settings ={Transport, }
        conn id: 2004, flow_id: ESG:4, sibling_flags FFFFFFFF80000008, crypto
map: Tunnel1-head-0
        sa timing: remaining key lifetime (k/sec): (4607995/3305)
        IV size: 16 bytes
        replay detection support: Y
        Status: ACTIVE(ACTIVE)


    outbound ah sas:


    outbound pcp sas:
```

d. On R2 issue the **show crypto isakmp sa** command to view the Internet Security Association Management Protocol (ISAKMP) SAs between the peers. Before the formation of the spoke-to-spoke tunnel, SAs have been made between R2 and R3, but no further negotiations have occurred, as indicated by the MM_NO_STATE state of the two SAs between the routers.

```
R2# show crypto isakmp sa
IPv4 Crypto ISAKMP SA
dst             src             state           conn-id status
192.0.2.1       192.168.2.1     QM_IDLE         1001 ACTIVE
192.168.2.1     192.0.2.1       QM_IDLE         1002 ACTIVE
192.168.3.1     192.168.2.1     MM_NO_STATE     1004 ACTIVE (deleted)
192.168.2.1     192.168.3.1     MM_NO_STATE     1003 ACTIVE (deleted)

IPv6 Crypto ISAKMP SA
```

After traffic has established the spoke-to-spoke tunnel, the SAs all show the QM_IDLE state. The SAs have been fully negotiated and are available for further ISAKMP quick mode exchanges.

Note: ISAKMP modes are outside the scope of this course.

```
R2# show crypto isakmp sa
IPv4 Crypto ISAKMP SA
dst             src             state           conn-id status
192.0.2.1       192.168.2.1     QM_IDLE         1001 ACTIVE
192.168.2.1     192.0.2.1       QM_IDLE         1002 ACTIVE
```

```
        192.168.3.1    192.168.2.1    QM_IDLE         1004 ACTIVE
        192.168.2.1    192.168.3.1    QM_IDLE         1003 ACTIVE

    IPv6 Crypto ISAKMP SA
```

e. You have successfully configured and verified IPsec on DMVPN Phase 3 tunnels.

Router Interface Summary Table

Router Model	Ethernet Interface #1	Ethernet Interface #2	Serial Interface #1	Serial Interface #2
1800	Fast Ethernet 0/0 (F0/0)	Fast Ethernet 0/1 (F0/1)	Serial 0/0/0 (S0/0/0)	Serial 0/0/1 (S0/0/1)
1900	Gigabit Ethernet 0/0 (G0/0)	Gigabit Ethernet 0/1 (G0/1)	Serial 0/0/0 (S0/0/0)	Serial 0/0/1 (S0/0/1)
2801	Fast Ethernet 0/0 (F0/0)	Fast Ethernet 0/1 (F0/1)	Serial 0/1/0 (S0/1/0)	Serial 0/1/1 (S0/1/1)
2811	Fast Ethernet 0/0 (F0/0)	Fast Ethernet 0/1 (F0/1)	Serial 0/0/0 (S0/0/0)	Serial 0/0/1 (S0/0/1)
2900	Gigabit Ethernet 0/0 (G0/0)	Gigabit Ethernet 0/1 (G0/1)	Serial 0/0/0 (S0/0/0)	Serial 0/0/1 (S0/0/1)
4221	Gigabit Ethernet 0/0/0 (G0/0/0)	Gigabit Ethernet 0/0/1 (G0/0/1)	Serial 0/1/0 (S0/1/0)	Serial 0/1/1 (S0/1/1)
4300	Gigabit Ethernet 0/0/0 (G0/0/0)	Gigabit Ethernet 0/0/1 (G0/0/1)	Serial 0/1/0 (S0/1/0)	Serial 0/1/1 (S0/1/1)

Note: To find out how the router is configured, look at the interfaces to identify the type of router and how many interfaces the router has. There is no way to effectively list all the combinations of configurations for each router class. This table includes identifiers for the possible combinations of Ethernet and Serial interfaces in the device. The table does not include any other type of interface, even though a specific router may contain one. An example of this might be an ISDN BRI interface. The string in parenthesis is the legal abbreviation that can be used in Cisco IOS commands to represent the interface.

Device Config Final – Notes

CHAPTER 21

Troubleshooting ACLs and Prefix Lists

21.1.2 Lab - Troubleshoot IPv4 ACLs

Topology

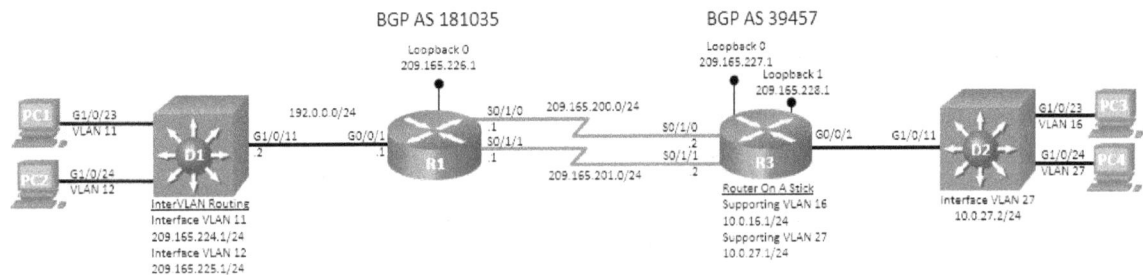

Addressing Table

Device	Interface	IP Address	Subnet Mask
R1	G0/0/1	192.0.0.1	255.255.255.0
	S0/1/0	209.165.200.1	255.255.255.0
	S0/1/1	209.165.201.1	255.255.255.0
	Loopback0	209.165.226.1	255.255.255.0
R3	G0/0/1.16	10.0.16.1	255.255.255.0
	G0/0/1.27	10.0.27.1	255.255.255.0
	S0/1/0	209.165.200.2	255.255.255.0
	S0/1/1	209.165.201.2	255.255.255.0
	Loopback0	209.165.227.1	255.255.255.0
	Loopback1	209.165.228.1	255.255.255.0
D1	G1/0/11	192.0.0.2	255.255.255.0
	VLAN 11	209.165.224.1	255.255.255.0
	VLAN 12	209.165.225.1	255.255.255.0
PC1	NIC	DHCP	
PC2	NIC	DHCP	
PC3	NIC	DHCP	
PC4	NIC	DHCP	

Objectives

Troubleshoot network issues related to the configuration and operation of ACLs for IPv4.

Background/Scenario

In this topology, R1 and D1 are OSPF neighbors, while R1 and R3 are BGP neighbors. Switch D1 provides inter-VLAN routing for two subnets. R3 provides inter-VLAN routing for two subnets, and switch D2 provides connectivity for the two VLANs supporting those subnets. The BGP relationship between R1 and R3 is established using EBGP multihop between the router's respective Loopback 0 interfaces. You will be loading configurations with intentional errors onto the network. Your tasks are to FIND the error(s), document your findings and the command(s) or method(s) used to fix them, FIX the issue(s) presented here and then test the network to ensure both of the following conditions are met:

1. the complaint received in the ticket is resolved

2. full reachability is restored

Note: The routers used with CCNP hands-on labs are Cisco 4221 with Cisco IOS XE Release 16.9.4 (universalk9 image). The switches used in the labs are Cisco Catalyst 3650 with Cisco IOS XE Release 16.9.4 (universalk9 image). Other routers, switches, and Cisco IOS versions can be used. Depending on the model and Cisco IOS version, the commands available and the output produced might vary from what is shown in the labs. Refer to the Router Interface Summary Table at the end of the lab for the correct interface identifiers.

Note: Make sure that the devices have been erased and have no startup configurations. If you are unsure, contact your instructor.

Required Resources

- 2 Routers (Cisco 4221 with Cisco IOS XE Release 16.9.4 universal image or comparable)
- 2 Switches (Cisco 3560 with Cisco IOS XE Release 16.9.4 universal image or comparable)
- 4 PCs (Choice of operating system with terminal emulation program installed)
- Console cables to configure the Cisco IOS devices via the console ports
- Ethernet and serial cables as shown in the topology

Instructions

Part 1: Trouble Ticket 21.1.2.1

Scenario:

A security consultant worked overnight making R1 and R3 compliant with RFC 1918. After the consultant finished the task, a business-critical connection between PC1 and PC3 is no longer operational. The task of finding and fixing the error(s) is now your job.

Use the commands listed below to load the configuration files for this trouble ticket:

Device	Command
R1	copy flash:/enarsi/21.1.2.1-r1-config.txt run
R3	copy flash:/enarsi/21.1.2.1-r3-config.txt run
D1	copy flash:/enarsi/21.1.2.1-d1-config.txt run
D2	copy flash:/enarsi/21.1.2.1-d2-config.txt run

- PCs 1, 2, 3, and 4 receive their addressing via DHCP for IPv4.

- Passwords on all devices are **cisco12345**. If a username is required, use **admin**.

- When you have fixed the ticket, change the MOTD on EACH DEVICE using the following command:

  banner motd # This is $(hostname) FIXED from ticket <ticket number> #

- Then save the configuration by issuing the **wri** command (on each device).

- Inform your instructor that you are ready for the next ticket.

- After the instructor approves your solution for this ticket, issue the **reset.now** privileged EXEC command. This script will clear your configurations and reload the devices.

Part 2: Trouble Ticket 21.1.2.2

Scenario:

A junior network administrator has attempted to tune access control lists to improve security. After doing so, PC2 is no longer able to communicate with devices with the IPv4 addresses 209.165.227.1 or 209.165.228.1. This problem needs to be solved to allow for business operations to continue.

Use the commands listed below to load the configuration files for this trouble ticket:

Device	Command
R1	`copy flash:/enarsi/21.1.2.2-r1-config.txt run`
R3	`copy flash:/enarsi/21.1.2.2-r3-config.txt run`
D1	`copy flash:/enarsi/21.1.2.2-d1-config.txt run`
D2	`copy flash:/enarsi/21.1.2.2-d2-config.txt run`

- PCs 1, 2, 3, and 4 receive their addressing via DHCP for IPv4.

- Passwords on all devices are **cisco12345**. If a username is required, use **admin**.

- When you have fixed the ticket, change the MOTD on EACH DEVICE using the following command:

  banner motd # This is $(hostname) FIXED from ticket <ticket number> #

- Then save the configuration by issuing the **wri** command (on each device).

- Inform your instructor that you are ready for the next ticket.

- After the instructor approves your solution for this ticket, issue the **reset.now** privileged EXEC command. This script will clear your configurations and reload the devices.

Part 3: Trouble Ticket 21.1.2.3

Scenario:

Security is an important consideration in your network. Over the weekend, a junior network administrator was working to improve remote access security with BGP AS 181035. It is 8:00 Monday morning, and router R1 and switch D1 are refusing Telnet connections. You need to find and fix this error as soon as possible.

Use the commands listed below to load the configuration files for this trouble ticket:

Device	Command
R1	`copy flash:/enarsi/21.1.2.3-r1-config.txt run`
R3	`copy flash:/enarsi/21.1.2.3-r3-config.txt run`
D1	`copy flash:/enarsi/21.1.2.3-d1-config.txt run`
D2	`copy flash:/enarsi/21.1.2.3-d2-config.txt run`

- PCs 1, 2, 3, and 4 receive their addressing via DHCP for IPv4.

- Passwords on all devices are **cisco12345**. If a username is required, use **admin**.

- When you have fixed the ticket, change the MOTD on EACH DEVICE using the following command:

  banner motd # This is $(hostname) FIXED from ticket <ticket number> #

- Then save the configuration by issuing the **wri** command (on each device).

- Inform your instructor that you are ready for the next ticket.

- After the instructor approves your solution for this ticket, issue the **reset.now** privileged EXEC command. This script will clear your configurations and reload the devices.

Router Interface Summary Table

Router Model	Ethernet Interface #1	Ethernet Interface #2	Serial Interface #1	Serial Interface #2
1800	Fast Ethernet 0/0 (F0/0)	Fast Ethernet 0/1 (F0/1)	Serial 0/0/0 (S0/0/0)	Serial 0/0/1 (S0/0/1)
1900	Gigabit Ethernet 0/0 (G0/0)	Gigabit Ethernet 0/1 (G0/1)	Serial 0/0/0 (S0/0/0)	Serial 0/0/1 (S0/0/1)
2801	Fast Ethernet 0/0 (F0/0)	Fast Ethernet 0/1 (F0/1)	Serial 0/1/0 (S0/1/0)	Serial 0/1/1 (S0/1/1)
2811	Fast Ethernet 0/0 (F0/0)	Fast Ethernet 0/1 (F0/1)	Serial 0/0/0 (S0/0/0)	Serial 0/0/1 (S0/0/1)
2900	Gigabit Ethernet 0/0 (G0/0)	Gigabit Ethernet 0/1 (G0/1)	Serial 0/0/0 (S0/0/0)	Serial 0/0/1 (S0/0/1)
4221	Gigabit Ethernet 0/0/0 (G0/0/0)	Gigabit Ethernet 0/0/1 (G0/0/1)	Serial 0/1/0 (S0/1/0)	Serial 0/1/1 (S0/1/1)
4300	Gigabit Ethernet 0/0/0 (G0/0/0)	Gigabit Ethernet 0/0/1 (G0/0/1)	Serial 0/1/0 (S0/1/0)	Serial 0/1/1 (S0/1/1)

Note: To find out how the router is configured, look at the interfaces to identify the type of router and how many interfaces the router has. There is no way to effectively list all the combinations of configurations for each router class. This table includes identifiers for the possible combinations of Ethernet and Serial interfaces in the device. The table does not include any other type of interface, even though a specific router may contain one. An example of this might be an ISDN BRI interface. The string in parenthesis is the legal abbreviation that can be used in Cisco IOS commands to represent the interface.

Device Config Final – Notes

21.1.3 Lab - Troubleshoot IPv6 ACLs

Topology

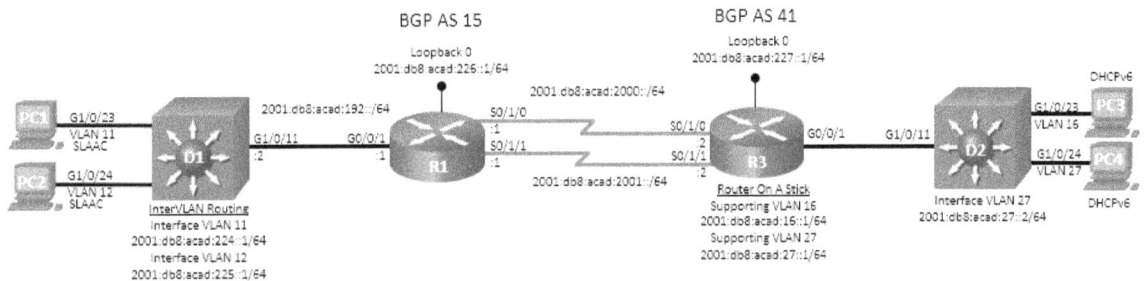

Addressing Table

Device	Interface	IPv6 Address/Prefix Length	Link-Local Address
R1	G0/0/1	2001:db8:acad:192::1/64	fe80::1:1
	S0/1/0	2001:db8:acad:2000::1/64	fe80::1:2
	S0/1/1	2001:db8:acad:2001::/1/64	fe80::1:3
	Loopback 0	2001:db8:acad:226::1/64	fe80::1:4
R3	G0/0/1.16	2001:db8:acad:16::1/64	fe80::3:1
	G0/0/1.27	2001:db8:acad:27::1/64	fe80::3:2
	Loopback 1	2001:db8:acad:227::1/64	fe80::3:3
D1	G1/0/11	2001:db8:acad:192::2/64	fe80::d1:1
	VLAN 11	2001:db8:acad:224::1/64	fe80::d1:2
	VLAN 12	2001:db8:acad:225::1/64	fe80::d1:3
D2	VLAN 27	2001:db8:acad:27::2/64	fe80::d2:1
	G1/0/11	2001:db8:1d1::2/64	fe80::d1:1
	Loopback 0	2001:db8:acad:1000::1/64	fe80::d1:2
	Loopback 1	2001:db8:acad:1001::1/64	fe80::d1:3
PC1	NIC	SLAAC	EUI-64/CGA
PC2	NIC	SLAAC	EUI-64/CGA
PC3	NIC	DHCPv6	EUI-64/CGA
PC4	NIC	DHCPv6	EUI-64/CGA

Objectives

Troubleshoot network issues related to the configuration and operation of IPv6 ACLs.

Background/Scenario

In this topology, R1 and R3 are BGP neighbors. R1 speaks for BGP ASN 15, while R3 speaks for BGP ASN 41. They are peered via their respective Loopback 0 interface using BGP Multi-hop across the serial interfaces that connect them. R1 and D1 have an OSPFv3 adjacency, with R1 providing a default route. R3 is performing Router-On-A-Stick for VLANs 16 and 27. The host connected to D1 is using SLAAC to determine their IPv6 Global Unicast Address (GUA), while the host connected to D2 is using DHCPv6 to determine their IPv6 GUA. You will be loading configurations with intentional errors onto the network. Your tasks are to FIND the error(s), document your findings and the command(s) or method(s) used to fix them, FIX the issue(s) presented here, and then test the network to ensure both of the following conditions are met:

1. the complaint received in the ticket is resolved

2. full reachability is restored

Note: The routers used with CCNP hands-on labs are Cisco 4221 with Cisco IOS XE Release 16.9.4 (universalk9 image). The switches used in the labs are Cisco Catalyst 3650 with Cisco IOS XE Release 16.9.4 (universalk9 image). Other routers, switches, and Cisco IOS versions can be used. Depending on the model and Cisco IOS version, the commands available and the output produced might vary from what is shown in the labs. Refer to the Router Interface Summary Table at the end of the lab for the correct interface identifiers.

Note: Make sure that the devices have been erased and have no startup configurations. If you are unsure, contact your instructor.

Required Resources

- 2 Routers (Cisco 4221 with Cisco IOS XE Release 16.9.4 universal image or comparable)

- 2 Switches (Cisco 3560 with Cisco IOS XE Release 16.9.4 universal image or comparable)

- 4 PCs (Choice of operating system with terminal emulation program installed)

- Console cables to configure the Cisco IOS devices via the console ports

- Ethernet and serial cables as shown in the topology

Part 1: Trouble Ticket 21.1.3.1

Scenario:

The night shift completed work in an attempt to secure the network. Network hosts, represented by PC1 and PC2 in this topology, are now unable to generate an IPv6 GUA.

Use the commands listed below to load the configuration files for this trouble ticket:

Device	Command
R1	`copy flash:/enarsi/21.1.3.1-r1-config.txt run`
R3	`copy flash:/enarsi/21.1.3.1-r3-config.txt run`
D1	`copy flash:/enarsi/21.1.3.1-d1-config.txt run`
D2	`copy flash:/enarsi/21.1.3.1-d2-config.txt run`

- PCs 1, 2, 3, and 4 should be configured for dynamic acquisition of an IPv6 address.

- Passwords on all devices are **cisco12345**. If a username is required, use **admin**.

- After you have corrected the ticket, change the MOTD on EACH DEVICE using the following command:

  banner motd # This is $(hostname) FIXED from ticket <ticket number> #

- Save the configuration by issuing the **wri** command (on each device).

- Inform your instructor that you are ready for the next ticket.

- After the instructor approves your solution for this ticket, issue the privileged EXEC command **reset.now**. This script will clear your configurations and reload the devices.

Part 2: Trouble Ticket 21.1.3.2

Scenario:

The night shift completed work in an attempt to secure the network. This morning it was discovered that PC3 and PC4 are no longer able to reach D1 interfaces VLAN 11 and VLAN 12 using the **ping** command. This must be fixed to allow for normal business operations.

Use the commands listed below to load the configuration files for this trouble ticket:

Device	Command
R1	copy flash:/enarsi/21.1.3.2-r1-config.txt run
R3	copy flash:/enarsi/21.1.3.2-r3-config.txt run
D1	copy flash:/enarsi/21.1.3.2-d1-config.txt run
D2	copy flash:/enarsi/21.1.3.2-d2-config.txt run

- PCs 1, 2, 3, and 4 should be configured for dynamic acquisition of an IPv6 address.

- Passwords on all devices are **cisco12345**. If a username is required, use **admin**.

- Once you have fixed the ticket, change the MOTD on EACH DEVICE using the following command:

  banner motd # This is $(hostname) FIXED from ticket <ticket number> #

- Then save the configuration by issuing the **wri** command (on each device).

- Inform your instructor that you are ready for the next ticket.

- After the instructor approves your solution for this ticket, issue the privileged EXEC command **reset.now**. This script will clear your configurations and reload the devices.

Part 3: Trouble Ticket 21.1.3.3

Scenario:

The night shift completed work in an attempt to secure the network. It was discovered this morning that PC3 and PC4 are no longer able to obtain DHCPv6 addresses. This must be fixed to allow for normal business operations.

Use the commands listed below to load the configuration files for this trouble ticket:

Device	Command
R1	`copy flash:/enarsi/21.1.3.3-r1-config.txt run`
R3	`copy flash:/enarsi/21.1.3.3-r3-config.txt run`
D1	`copy flash:/enarsi/21.1.3.3-d1-config.txt run`
D2	`copy flash:/enarsi/21.1.3.3-d2-config.txt run`

- PCs 1, 2, 3, and 4 should be configured for dynamic acquisition of an IPv6 address.

- Passwords on all devices are **cisco12345**. If a username is required, use **admin**.

- Once you have fixed the ticket, change the MOTD on EACH DEVICE using the following command:

  banner motd # This is $(hostname) FIXED from ticket <ticket number> #

- Then save the configuration by issuing the **wri** command (on each device).

- Inform your instructor that you are ready for the next ticket.

- After the instructor approves your solution for this ticket, issue the privileged EXEC command **reset.now**. This script will clear your configurations and reload the devices.

Router Interface Summary Table

Router Model	Ethernet Interface #1	Ethernet Interface #2	Serial Interface #1	Serial Interface #2
1800	Fast Ethernet 0/0 (F0/0)	Fast Ethernet 0/1 (F0/1)	Serial 0/0/0 (S0/0/0)	Serial 0/0/1 (S0/0/1)
1900	Gigabit Ethernet 0/0 (G0/0)	Gigabit Ethernet 0/1 (G0/1)	Serial 0/0/0 (S0/0/0)	Serial 0/0/1 (S0/0/1)
2801	Fast Ethernet 0/0 (F0/0)	Fast Ethernet 0/1 (F0/1)	Serial 0/1/0 (S0/1/0)	Serial 0/1/1 (S0/1/1)
2811	Fast Ethernet 0/0 (F0/0)	Fast Ethernet 0/1 (F0/1)	Serial 0/0/0 (S0/0/0)	Serial 0/0/1 (S0/0/1)
2900	Gigabit Ethernet 0/0 (G0/0)	Gigabit Ethernet 0/1 (G0/1)	Serial 0/0/0 (S0/0/0)	Serial 0/0/1 (S0/0/1)
4221	Gigabit Ethernet 0/0/0 (G0/0/0)	Gigabit Ethernet 0/0/1 (G0/0/1)	Serial 0/1/0 (S0/1/0)	Serial 0/1/1 (S0/1/1)
4300	Gigabit Ethernet 0/0/0 (G0/0/0)	Gigabit Ethernet 0/0/1 (G0/0/1)	Serial 0/1/0 (S0/1/0)	Serial 0/1/1 (S0/1/1)

Note: To find out how the router is configured, look at the interfaces to identify the type of router and how many interfaces the router has. There is no way to effectively list all the combinations of configurations for each router class. This table includes identifiers for the possible combinations of Ethernet and Serial interfaces in the device. The table does not include any other type of interface, even though a specific router may contain one. An example of this might be an ISDN BRI interface. The string in parenthesis is the legal abbreviation that can be used in Cisco IOS commands to represent the interface.

Device Config Final – Notes

21.1.4 Lab - Troubleshoot Prefix Lists

Topology

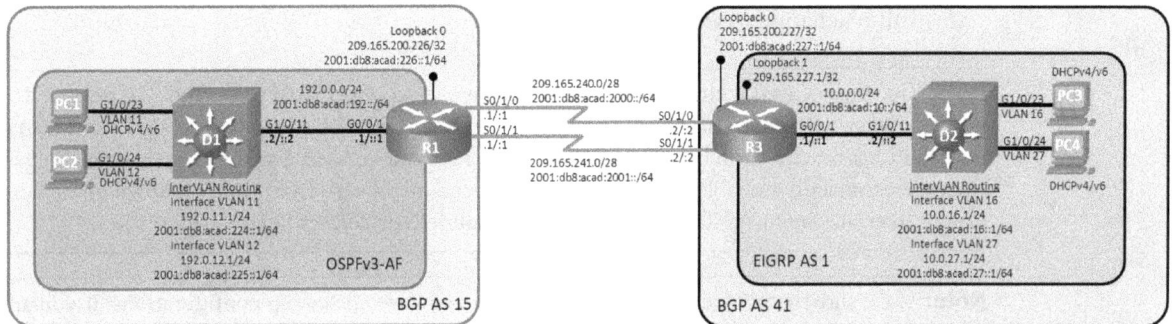

Addressing Table

Device	Interface	IPv4 Address/Prefix Length	IPv6 Address/Prefix Length	Link-Local Address
R1	G0/0/1	192.0.0.1/24	2001:db8:acad:192::1/64	fe80::1:1
	S0/1/0	209.165.240.1/28	2001:db8:acad:2000::1/64	fe80::1:2
	S0/1/1	209.165.241.1/28	2001:db8:acad:2001::1/64	fe80::1:3
	Loopback 0	209.165.200.226/32	2001:db8:acad:226::1/64	fe80::1:4
R3	G0/0/1	10.0.0.1/24	2001:db8:acad:10::1/64	fe80::3:1
	S0/1/0	209.165.240.2/28	2001:db8:acad:2000::2/64	fe80::3:2
	S0/1/1	209.165.241.2/28	2001:db8:acad:2001::/2/64	fe80::3:3
	Loopback 0	209.165.200.227/32	2001:db8:acad:227::1/65	fe80::3:4
	Loopback 1	209.165.227.1/32	N/A	N/A
D1	G1/0/11	192.0.0.2/24	2001:db8:acad:192::2/64	fe80::d1:1
	VLAN 11	192.0.11.1/24	2001:db8:acad:224::1/64	fe80::d1:2
	VLAN 12	192.0.12.1/24	2001:db8:acad:225::1/64	fe80::d1:3
D2	G1/0/11	10.0.0.2/24	2001:db8:acad:10::2/64	fe80::d2:1
	VLAN 16	10.0.16.1/24	2001:db8:acad:16::1/64	fe80::d2:2
	VLAN 27	10.0.27.1/24	2001:db8:acad:27::1/64	fe80::d2:3

Objectives

Troubleshoot network issues related to the configuration and operation of prefix lists.

Background/Scenario

In this topology, R1 and R3 are multi-hop MP-BGP neighbors, with an established adjacency using both IPv4 and IPv6. R1 and D1 have an OSPFv3 adjacency, and D1 is providing interVLAN routing. R3 and D2 have a Named-EIGPR adjacency, and D1 is providing InterVLAN routing. R3 is performing NAT for all networks in BGP AS 41, overloading onto the IPv4 address of Loopback 0. You will be loading configurations with intentional errors onto the network. Your tasks are to FIND the error(s), document your findings and the command(s) or method(s) used to fix them, FIX the issue(s) presented here, and then test the network to ensure both of the following conditions are met:

1. the complaint received in the ticket is resolved

2. full reachability is restored

Note: The routers used with CCNP hands-on labs are Cisco 4221 with Cisco IOS XE Release 16.9.4 (universalk9 image). The switches used in the labs are Cisco Catalyst 3650 with Cisco IOS XE Release 16.9.4 (universalk9 image). Other routers, switches, and Cisco IOS versions can be used. Depending on the model and Cisco IOS version, the commands available and the output produced might vary from what is shown in the labs. Refer to the Router Interface Summary Table at the end of the lab for the correct interface identifiers.

Note: Make sure that the devices have been erased and have no startup configurations. If you are unsure, contact your instructor.

Required Resources

- 2 Routers (Cisco 4221 with Cisco IOS XE Release 16.9.4 universal image or comparable)
- 2 Switches (Cisco 3560 with Cisco IOS XE Release 16.9.4 universal image or comparable)
- 4 PCs (With terminal emulation program, such as Tera Term)
- Console cables to configure the Cisco IOS devices via the console ports
- Ethernet and serial cables as shown in the topology

Part 1: Trouble Ticket 21.1.4.1

Scenario:

You are the senior network engineer for BGP AS 41. Budget cuts have taken a toll on your network. Therefore, Switch D2 is not the most robust system available, even though it performs a critical function in the network. You tasked the night shift to reduce the amount of information switch D2 has to deal with. You have come in to work to find that, although the D2 routing table is now very small, your network is unable to communicate with networks in BGP AS 15. You have to get this fixed!

Use the commands listed below to load the configuration files for this trouble ticket:

Device	Command
R1	`copy flash:/enarsi/21.1.4.1-r1-config.txt run`
R3	`copy flash:/enarsi/21.1.4.1-r3-config.txt run`
D1	`copy flash:/enarsi/21.1.4.1-d1-config.txt run`
D2	`copy flash:/enarsi/21.1.4.1-d2-config.txt run`

- PCs 1, 2, 3, and 4 should be configured to receive dynamically assigned addresses (both IPv4 and IPv6).
- Passwords on all devices are **cisco12345**. If a username is required, use **admin**.

- After you have fixed the ticket, change the MOTD on EACH DEVICE using the following command:

  banner motd # This is $(hostname) FIXED from ticket <ticket number> #

- Then save the configuration by issuing the **wri** command (on each device).

- Inform your instructor that you are ready for the next ticket.

- After the instructor approves your solution for this ticket, issue the privileged EXEC command **reset.now**. This script will clear your configurations and reload the devices.

Part 2: Trouble Ticket 21.1.4.2

Scenario:

You are the senior network engineer for BGP AS 41. After careful review of bandwidth utilization on the two links between AS 15 and AS 41, you have suggested to management that some adjustments be put in place to equalize the utilization of the two circuits. Specifically, you want to cause IPv6 traffic to use the S0/1/1 link. Management was so pleased with the suggestion that you were told to forgo the normal change control procedures and get this implemented as soon as possible. Just as you were starting to plan the changes, you were called away to an urgent budget meeting. You left your second-in-charge with the task to develop and implement the solution. When you returned, you found that things were not working in the way you had stipulated.

Use the commands listed below to load the configuration files for this trouble ticket:

Device	Command
R1	`copy flash:/enarsi/21.1.4.2-r1-config.txt run`
R3	`copy flash:/enarsi/21.1.4.2-r3-config.txt run`
D1	`copy flash:/enarsi/21.1.4.2-d1-config.txt run`
D2	`copy flash:/enarsi/21.1.4.2-d2-config.txt run`

- PCs 1, 2, 3, and 4 should be configured to receive dynamically assigned addresses (both IPv4 and IPv6.)

- Passwords on all devices are **cisco12345**. If a username is required, use **admin**.

- After you have fixed the ticket, change the MOTD on EACH DEVICE using the following command:

  banner motd # This is $(hostname) FIXED from ticket <ticket number> #

- Then save the configuration by issuing the **wri** command (on each device).

- Inform your instructor that you are ready for the next ticket.

- After the instructor approves your solution for this ticket, issue the privileged EXEC command **reset.now**. This script will clear your configurations and reload the devices.

Router Interface Summary Table

Router Model	Ethernet Interface #1	Ethernet Interface #2	Serial Interface #1	Serial Interface #2
1800	Fast Ethernet 0/0 (F0/0)	Fast Ethernet 0/1 (F0/1)	Serial 0/0/0 (S0/0/0)	Serial 0/0/1 (S0/0/1)
1900	Gigabit Ethernet 0/0 (G0/0)	Gigabit Ethernet 0/1 (G0/1)	Serial 0/0/0 (S0/0/0)	Serial 0/0/1 (S0/0/1)
2801	Fast Ethernet 0/0 (F0/0)	Fast Ethernet 0/1 (F0/1)	Serial 0/1/0 (S0/1/0)	Serial 0/1/1 (S0/1/1)
2811	Fast Ethernet 0/0 (F0/0)	Fast Ethernet 0/1 (F0/1)	Serial 0/0/0 (S0/0/0)	Serial 0/0/1 (S0/0/1)
2900	Gigabit Ethernet 0/0 (G0/0)	Gigabit Ethernet 0/1 (G0/1)	Serial 0/0/0 (S0/0/0)	Serial 0/0/1 (S0/0/1)
4221	Gigabit Ethernet 0/0/0 (G0/0/0)	Gigabit Ethernet 0/0/1 (G0/0/1)	Serial 0/1/0 (S0/1/0)	Serial 0/1/1 (S0/1/1)
4300	Gigabit Ethernet 0/0/0 (G0/0/0)	Gigabit Ethernet 0/0/1 (G0/0/1)	Serial 0/1/0 (S0/1/0)	Serial 0/1/1 (S0/1/1)

Note: To find out how the router is configured, look at the interfaces to identify the type of router and how many interfaces the router has. There is no way to effectively list all the combinations of configurations for each router class. This table includes identifiers for the possible combinations of Ethernet and Serial interfaces in the device. The table does not include any other type of interface, even though a specific router may contain one. An example of this might be an ISDN BRI interface. The string in parenthesis is the legal abbreviation that can be used in Cisco IOS commands to represent the interface.

Device Config Final – Notes

CHAPTER 22

Infrastructure Security

22.1.2 Lab - Troubleshoot IOS AAA Authentication

Topology

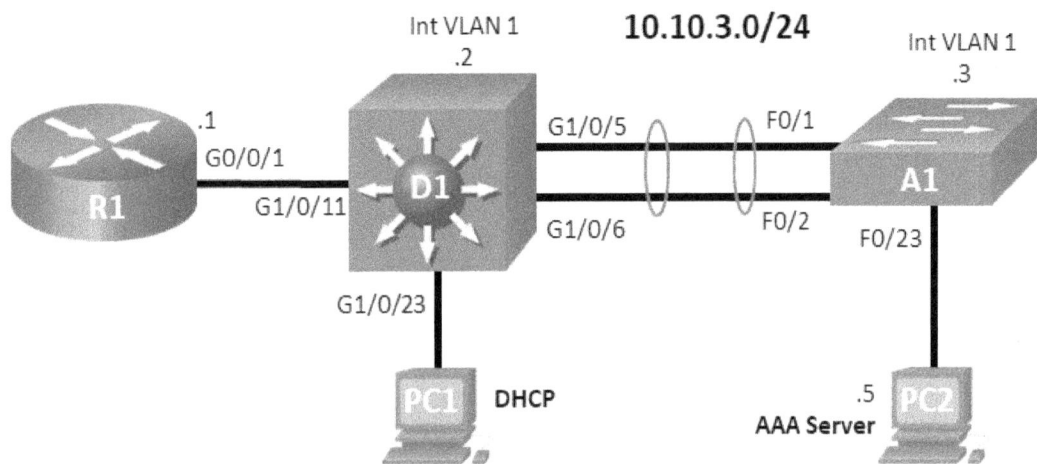

Addressing Table

Device	Interface	IP Address	Subnet Mask
R1	G0/0/1	10.10.3.1	255.255.255.0
D1	VLAN 1	10.10.3.2	255.255.255.0
A1	VLAN 1	10.10.3.3	255.255.255.0
PC1	NIC	DHCP	
PC2	NIC	10.10.3.5	255.255.255.0

Objectives

Troubleshoot authentication issues related to the configuration and operation of AAA. Router R1 is configured for inter-VLAN routing and DHCP to provide support for PC1. You will be loading configurations with intentional errors onto the network. Your tasks are to FIND the error(s), document your findings and the command(s) or method(s) used to fix them, FIX the issue(s) presented here, and then test the network to ensure both of the following conditions are met:

1. the complaint received in the ticket is resolved

2. the AAA process occurs as specified

Background/Scenario

Using AAA-based services allows for more granular control of access to your devices. In this lab, you will troubleshoot issues arising from the operation of local and server-based AAA.

Note: The routers used with CCNP hands-on labs are Cisco 4221 with Cisco IOS XE Release 16.9.4 (universalk9 image). The switches used in the labs are Cisco Catalyst 3650 with Cisco IOS XE Release 16.9.4 (universalk9 image) and Cisco Catalyst 2960s with Cisco IOS Release 15.2(2) (lanbasek9 image). Other routers, switches, and Cisco IOS versions can be used. Depending on the model and Cisco IOS version, the commands available and the output produced might vary from what is shown in the labs. Refer to the Router Interface Summary Table at the end of the lab for the correct interface identifiers.

Note: Make sure that the routers and switches have been erased and have no startup configurations. If you are unsure, contact your instructor.

Required Resources

- 1 Router (Cisco 4221 with Cisco IOS XE Release 16.9.4 universal image or comparable)
- 1 Switch (Cisco 3650 with Cisco IOS XE Release 16.9.4 universal image or comparable)
- 1 Switch (Cisco 2960 with Cisco IOS Release 15.2(2) lanbasek9 image or comparable)
- 1 PC (Choice of operating system with a terminal emulation program installed)
- 1 PC (Cisco Network Academy CCNP VM running in a virtual machine client or a server with TACACS+ and RADIUS servers installed, configured and running)
- Console cables to configure the Cisco IOS devices via the console ports
- Ethernet cables as shown in the topology

Instructions

Part 1: Trouble Ticket 22.1.2.1

Scenario:

As the result of a network security audit, a policy change was implemented on the routers and switches at the branch office to force stronger access control for management of the devices. All logins were to be authenticated using AAA Method Lists. Everything appeared to go well with the change, and remote access to the devices functions as expected. About a month later, the local branch IT tech attempted to use the console connection to upgrade the IOS on **Switch A1** and was unable to gain access to the device with the local username and password combination provided (username **admin**, password **cisco1234**).

The privileged EXEC password is **cisco12345cisco**.

Use the commands listed below to load the configuration files for this trouble ticket:

Device	Command
R1	`copy flash:/enarsi/22.1.2.1-r1-config.txt run`
D1	`copy flash:/enarsi/22.1.2.1-d1-config.txt run`
A1	`copy flash:/enarsi/22.1.2.1-a1-config.txt run`

- PC1 should be configured for and receive an address from an IPv4 DHCP server. PC2 must be statically configured with the IP address in the addressing table.

- Passwords on all devices are **cisco1234**. If a username is required, use **admin**.

- After you have fixed the ticket, change the MOTD on EACH DEVICE using the following command:

  banner motd # This is $(hostname) FIXED from ticket <ticket number> #

- Save the configuration by issuing the **wri** command (on each device).

- Inform your instructor that you are ready for the next ticket.

- After the instructor approves your solution for this ticket, issue the privileged EXEC command **reset.now**. This script will clear your configurations and reload the devices.

Part 2: Trouble Ticket 22.1.2.2

Scenario:

Recently, the RADIUS server at the main office was replaced. The previous server, which is running a standard RADIUS server on Linux, was shipped out to the branch office to be used to authenticate access to the VTY ports on the switches and routers. The server was plugged into **switch A1**. The main office technician logged into **switch D1** remotely and reconfigured it to use RADIUS authentication. As soon as the main office technician logged out of D1 to test the RADIUS authentication, the tech was no longer able to login via Telnet. The local branch office technician now needs to connect to D1 via a console connection and fix the RADIUS authentication issue. The console connection is configured to use local login (username **admin**, password **cisco1234**). The remote access username and password is **raduser** and **upass123**.

The privileged EXEC password is **cisco12345cisco**.

Use the commands listed below to load the configuration files for this trouble ticket:

Device	Command
R1	`copy flash:/enarsi/22.1.2.2-r1-config.txt run`
D1	`copy flash:/enarsi/22.1.2.2-d1-config.txt run`
A1	`copy flash:/enarsi/22.1.2.2-a1-config.txt run`

- PC1 should be configured for and receive an address from an IPv4 DHCP server. PC2 must be statically configured with the IP address in the addressing table.

- Passwords on all devices are **cisco1234**. If a username is required, use **admin**.

- The username and password configured on the RADIUS server is **raduser** and **upass123**.

- After you have fixed the ticket, change the MOTD on EACH DEVICE using the following command:

  banner motd # This is $(hostname) FIXED from ticket <ticket number> #

- Save the configuration by issuing the **wri** command (on each device).

- Inform your instructor that you are ready for the next ticket.

- After the instructor approves your solution for this ticket, issue the privileged EXEC command **reset.now**. This script will clear your configurations and reload the devices.

Router Interface Summary Table

Router Model	Ethernet Interface #1	Ethernet Interface #2	Serial Interface #1	Serial Interface #2
1800	Fast Ethernet 0/0 (F0/0)	Fast Ethernet 0/1 (F0/1)	Serial 0/0/0 (S0/0/0)	Serial 0/0/1 (S0/0/1)
1900	Gigabit Ethernet 0/0 (G0/0)	Gigabit Ethernet 0/1 (G0/1)	Serial 0/0/0 (S0/0/0)	Serial 0/0/1 (S0/0/1)
2801	Fast Ethernet 0/0 (F0/0)	Fast Ethernet 0/1 (F0/1)	Serial 0/1/0 (S0/1/0)	Serial 0/1/1 (S0/1/1)
2811	Fast Ethernet 0/0 (F0/0)	Fast Ethernet 0/1 (F0/1)	Serial 0/0/0 (S0/0/0)	Serial 0/0/1 (S0/0/1)
2900	Gigabit Ethernet 0/0 (G0/0)	Gigabit Ethernet 0/1 (G0/1)	Serial 0/0/0 (S0/0/0)	Serial 0/0/1 (S0/0/1)
4221	Gigabit Ethernet 0/0/0 (G0/0/0)	Gigabit Ethernet 0/0/1 (G0/0/1)	Serial 0/1/0 (S0/1/0)	Serial 0/1/1 (S0/1/1)
4300	Gigabit Ethernet 0/0/0 (G0/0/0)	Gigabit Ethernet 0/0/1 (G0/0/1)	Serial 0/1/0 (S0/1/0)	Serial 0/1/1 (S0/1/1)

Note: To find out how the router is configured, look at the interfaces to identify the type of router and how many interfaces the router has. There is no way to effectively list all the combinations of configurations for each router class. This table includes identifiers for the possible combinations of Ethernet and Serial interfaces in the device. The table does not include any other type of interface, even though a specific router may contain one. An example of this might be an ISDN BRI interface. The string in parenthesis is the legal abbreviation that can be used in Cisco IOS commands to represent the interface.

Device Config Final – Notes

_____

_____

_____

_____

_____

_____

_____

_____

_____

_____

_____

_____

_____

_____

_____

_____

22.1.3 Lab - Troubleshoot uRPF

Topology

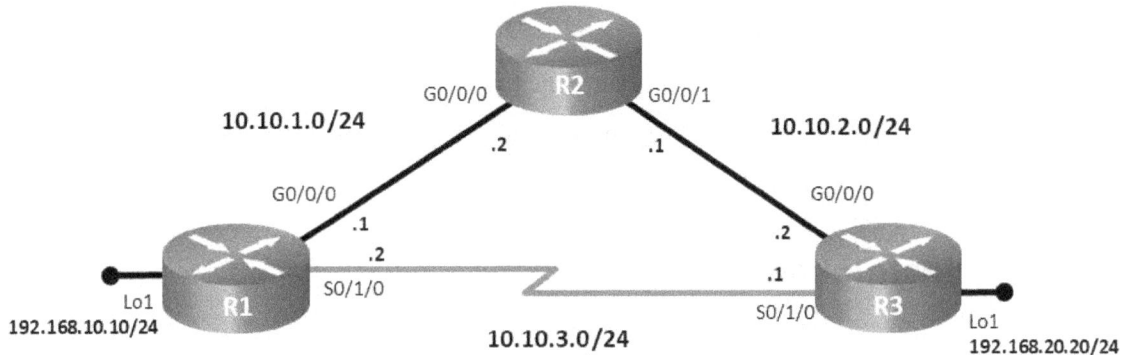

Addressing Table

Device	Interface	IP Address	Subnet Mask
R1	G0/0/0	10.10.1.1	255.255.255.0
	S0/1/0	10.10.3.2	255.255.255.0
	Lo1	192.168.10.10	255.255.255.0
R2	G0/0/0	10.10.1.2	255.255.255.0
	G0/0/1	10.10.2.1	255.255.255.0
R3	G0/0/0	10.10.2.2	255.255.255.0
	S0/1/0	10.10.3.1	255.255.255.0
	Lo1	192.168.20.20	255.255.255.0

Objectives

Troubleshoot issues related to the configuration and operation of uRPF.

Background/Scenario

uRPF is a security feature that helps limit or even eliminate spoofed IP packets on a network. In this lab, you will be loading configurations with intentional errors onto the network. Your tasks are to FIND the error(s), document your findings and the command(s) or method(s) used to fix them, FIX the issue(s) presented here, and then test the network to ensure both of the following conditions are met:

1. The trouble ticket has been resolved

2. The network is fully functioning

Note: The routers used with CCNP hands-on labs are Cisco 4221 with Cisco IOS XE Release 16.9.4 (universalk9 image). Other routers and Cisco IOS versions can be used. Depending on the model and Cisco IOS version, the commands available and the output produced might vary from what is shown in the labs. Refer to the Router Interface Summary Table at the end of the lab for the correct interface identifiers.

Note: Make sure that the devices have been erased and have no startup configurations. If you are unsure, contact your instructor.

Required Resources

- 3 Routers (Cisco 4221 with Cisco IOS XE Release 16.9.4 universal image or comparable)
- Console cables to configure the Cisco IOS devices via the console ports
- 1 PC (Choice of operating system with a terminal emulation program installed)
- Ethernet and serial cables as shown in the topology

Instructions

Part 1: Trouble Ticket 22.1.3.1

Scenario:

As a security measure, uRPF was implemented on router R1 to ensure a malicious actor could not circumvent access control restrictions using a spoofed IP address. A fellow colleague was tasked with configuring uRPF on R1 to ensure that any spoofed IP packets received are dropped. However, after the implementation, R3's loopback address has lost connectivity to the 192.168.10.0/24 network.

Step 1.   Cable the network as shown in the topology.

   a.   Attach the devices as shown in the topology diagram, and cable as necessary.

   b.   Use the commands listed below to load the configuration files for this trouble ticket:

Device	Command
R1	`copy flash:/enarsi/22.1.3.1-r1-config.txt run`
R2	`copy flash:/enarsi/22.1.3.1-r2-config.txt run`
R3	`copy flash:/enarsi/22.1.3.1-r3-config.txt run`

Note: Passwords on all devices are **cisco12345**.

Step 2.   Troubleshoot Ticket.

Troubleshoot and repair the issue. All devices, including loopback addresses, should be able to ping each other.

Step 3.   Complete the Ticket.

   a.   After you have fixed the ticket, change the MOTD on Router R1 using the following command:

   banner motd # This is $(*hostname*) FIXED from ticket <*ticket number*> #

   b.   Verify that uRPF is enabled, configured correctly and all devices, including loopback addresses, can ping each other. Then save the configuration by issuing the **wri** command.

   c.   Inform your instructor that you have completed the ticket.

d. After the instructor approves your solution for this ticket, issue the **reset.now** privileged EXEC command on each device. This script will clear your configurations and reload the devices.

Router Interface Summary Table

Router Model	Ethernet Interface #1	Ethernet Interface #2	Serial Interface #1	Serial Interface #2
1800	Fast Ethernet 0/0 (F0/0)	Fast Ethernet 0/1 (F0/1)	Serial 0/0/0 (S0/0/0)	Serial 0/0/1 (S0/0/1)
1900	Gigabit Ethernet 0/0 (G0/0)	Gigabit Ethernet 0/1 (G0/1)	Serial 0/0/0 (S0/0/0)	Serial 0/0/1 (S0/0/1)
2801	Fast Ethernet 0/0 (F0/0)	Fast Ethernet 0/1 (F0/1)	Serial 0/1/0 (S0/1/0)	Serial 0/1/1 (S0/1/1)
2811	Fast Ethernet 0/0 (F0/0)	Fast Ethernet 0/1 (F0/1)	Serial 0/0/0 (S0/0/0)	Serial 0/0/1 (S0/0/1)
2900	Gigabit Ethernet 0/0 (G0/0)	Gigabit Ethernet 0/1 (G0/1)	Serial 0/0/0 (S0/0/0)	Serial 0/0/1 (S0/0/1)
4221	Gigabit Ethernet 0/0/0 (G0/0/0)	Gigabit Ethernet 0/0/1 (G0/0/1)	Serial 0/1/0 (S0/1/0)	Serial 0/1/1 (S0/1/1)
4300	Gigabit Ethernet 0/0/0 (G0/0/0)	Gigabit Ethernet 0/0/1 (G0/0/1)	Serial 0/1/0 (S0/1/0)	Serial 0/1/1 (S0/1/1)

Note: To find out how the router is configured, look at the interfaces to identify the type of router and how many interfaces the router has. There is no way to effectively list all the combinations of configurations for each router class. This table includes identifiers for the possible combinations of Ethernet and Serial interfaces in the device. The table does not include any other type of interface, even though a specific router may contain one. An example of this might be an ISDN BRI interface. The string in parenthesis is the legal abbreviation that can be used in Cisco IOS commands to represent the interface.

Device Config Final – Notes

22.1.4 Lab - Troubleshoot Control Plane Policing (CoPP)

Topology

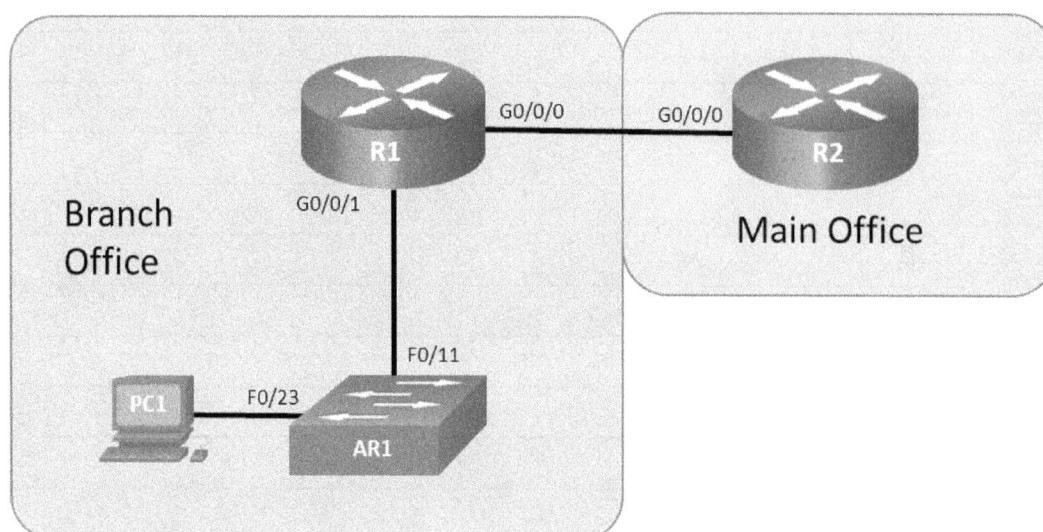

Addressing Table

Device	Interface	IP Address	Subnet Mask
R1	G0/0/0	172.16.12.1	255.255.255.252
	G0/0/1	10.10.1.1	255.255.255.0
R2	G0/0/0	172.16.12.2	255.255.255.252
A1	VLAN 1	10.10.1.4	255.255.255.0
PC1	NIC	10.10.1.5	255.255.255.0

Objectives

Troubleshoot network issues related to the configuration and operation of Control Plane Policing (CoPP).

Background/Scenario

Control Plane Policing (CoPP) is a protection feature for the router's control plane CPU. CoPP can granularly permit, drop, or rate-limit traffic to or from the CPU using a Modular QoS CLI (MQC) policy. The CoPP policy is applied to a dedicated control-plane "interface" which protects the CPU from unexpected extreme rates of traffic that could impact the stability of the router.

Note: The routers used with CCNP hands-on labs are Cisco 4221 with Cisco IOS XE Release 16.9.4 (universalk9 image). The switch used in the lab is a Cisco Catalyst 2960 with Cisco IOS Release 15.2(2) (lanbasek9 image). Other routers, switches, and Cisco IOS versions can be used. Depending on the model and Cisco IOS version, the commands available and the output produced might vary from what is shown in the labs. Refer to the Router Interface Summary Table at the end of the lab for the correct interface identifiers.

Note: Make sure that the routers and switches have been erased and have no startup configurations. If you are unsure, contact your instructor.

Required Resources

- 2 Routers (Cisco 4221 with Cisco IOS XE Release 16.9.4 universal image or comparable)

- 1 Switch (Cisco 2960 with Cisco IOS Release 15.2(2) lanbasek9 image or comparable)

- 1 PC (Choice of operating system with a terminal emulation program and a packet capture utility installed)

- Console cables to configure the Cisco IOS devices via the console ports

- Ethernet cables as shown in the topology

Instructions

Part 1: Trouble Ticket 22.1.4.1

Scenario:

At the main office, a decision was made to eliminate the use of Telnet for network device management. Rather than place ACLs on each interface, the main office network technician edited the existing CoPP configurations on the branch router R1, adding the restriction on Telnet by creating an ACL, class-map, and policy-map to drop all Telnet traffic to the router. The tech also added a traffic class for SSH access. While testing the new changes at the branch office, the branch network technician finds that Telnet is still possible.

Your tasks are to FIND the error(s), document your findings and the command(s) or method(s) used to fix them, FIX the issue(s) presented here and then test the network to ensure the following conditions are met:

1. the complaint received in the ticket is resolved

2. the control-plane policy-map keeps Telnet from succeeding either from the main office or from the branch management network.

Use the commands listed below to load the configuration files for this trouble ticket:

Device	Command
R1	`copy flash:/enarsi/22.1.4.1-r1-config.txt run`
R2	`copy flash:/enarsi/22.1.4.1-r2-config.txt run`
A1	`copy flash:/enarsi/22.1.4.1-a1-config.txt run`

- PC1 is on the management network and is configured with a static IP address from the addressing table.

- **aaa new-model** is enabled on router R1.

- Privileged EXEC password is **cisco12345cisco**.

- Passwords on all devices are **cisco1234**. If a username is required, use **admin**.

- After you have fixed the ticket, change the MOTD on EACH DEVICE using the following command:

    banner motd # This is $(hostname) FIXED from ticket <ticket number> #

- Save the configuration by issuing the **wri** command (on each device).

- Inform your instructor that you are ready for the next ticket.

- After the instructor approves your solution for this ticket, issue the privileged EXEC command **reset.now**. This script will clear your configurations and reload the devices.

Part 2: Trouble Ticket 22.1.4.2

Scenario:

While the main office network tech was editing the CoPP configuration on the branch R1 router, the tech noticed that there was not a separate class for SSH, that it was part of the MGMT class. The tech decided to add a traffic class for SSH access, so it would be easier to troubleshoot remote access issues. The branch technician reports that after the traffic class change was added, SSH seems much slower and less responsive than before.

Your tasks are to FIND the error(s), document your findings and the command(s) or method(s) used to fix them, FIX the issue(s) presented here and then test the network to ensure the following conditions are met:

1. the complaint received in the ticket is resolved

2. SSH traffic response issues are solved

Use the commands listed below to load the configuration files for this trouble ticket:

Device	Command
R1	`copy flash:/enarsi/22.1.4.2-r1-config.txt run`
R2	`copy flash:/enarsi/22.1.4.2-r2-config.txt run`
A1	`copy flash:/enarsi/22.1.4.2-a1-config.txt run`

- PC1 is on the management network and is configured with a static IP address from the addressing table.

- **aaa new-model** is enabled on router R1.

- Privileged EXEC password is **cisco12345cisco**.

- Passwords on all devices are **cisco1234**. If a username is required, use **admin**.

- After you have fixed the ticket, change the MOTD on EACH DEVICE using the following command:

    banner motd # This is $(hostname) FIXED from ticket <ticket number> #

- Save the configuration by issuing the **wri** command (on each device).

- Inform your instructor that you are finished.

- After the instructor approves your solution for this ticket, issue the privileged EXEC command **reset.now**. This script will clear your configurations and reload the devices.

Router Interface Summary Table

Router Model	Ethernet Interface #1	Ethernet Interface #2	Serial Interface #1	Serial Interface #2
1800	Fast Ethernet 0/0 (F0/0)	Fast Ethernet 0/1 (F0/1)	Serial 0/0/0 (S0/0/0)	Serial 0/0/1 (S0/0/1)
1900	Gigabit Ethernet 0/0 (G0/0)	Gigabit Ethernet 0/1 (G0/1)	Serial 0/0/0 (S0/0/0)	Serial 0/0/1 (S0/0/1)
2801	Fast Ethernet 0/0 (F0/0)	Fast Ethernet 0/1 (F0/1)	Serial 0/1/0 (S0/1/0)	Serial 0/1/1 (S0/1/1)
2811	Fast Ethernet 0/0 (F0/0)	Fast Ethernet 0/1 (F0/1)	Serial 0/0/0 (S0/0/0)	Serial 0/0/1 (S0/0/1)
2900	Gigabit Ethernet 0/0 (G0/0)	Gigabit Ethernet 0/1 (G0/1)	Serial 0/0/0 (S0/0/0)	Serial 0/0/1 (S0/0/1)
4221	Gigabit Ethernet 0/0/0 (G0/0/0)	Gigabit Ethernet 0/0/1 (G0/0/1)	Serial 0/1/0 (S0/1/0)	Serial 0/1/1 (S0/1/1)
4300	Gigabit Ethernet 0/0/0 (G0/0/0)	Gigabit Ethernet 0/0/1 (G0/0/1)	Serial 0/1/0 (S0/1/0)	Serial 0/1/1 (S0/1/1)

Note: To find out how the router is configured, look at the interfaces to identify the type of router and how many interfaces the router has. There is no way to effectively list all the combinations of configurations for each router class. This table includes identifiers for the possible combinations of Ethernet and Serial interfaces in the device. The table does not include any other type of interface, even though a specific router may contain one. An example of this might be an ISDN BRI interface. The string in parenthesis is the legal abbreviation that can be used in Cisco IOS commands to represent the interface.

Device Config Final – Notes

CHAPTER 23

Device Management and Management Tools Troubleshooting

23.1.2 Lab - Troubleshoot Device Access and File Transfer

Topology

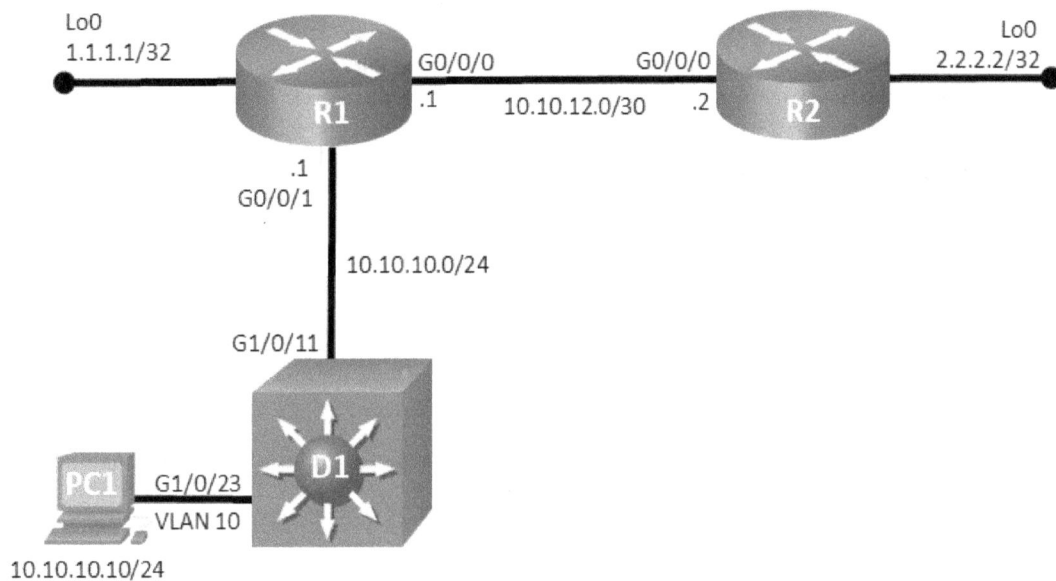

Addressing Table

Device	Interface	IP Address	Subnet Mask
R1	G0/0/0	10.10.12.1	255.255.255.0
	G0/0/1	10.10.10.1	255.255.255.0
R2	G0/0/0	10.10.12.2	255.255.255.0
D1	VLAN 10	10.10.10.2	255.255.255.0
PC1	NIC	10.10.10.10	255.255.255.0

Objectives

Troubleshoot device access and file transfer issues in the configurations.

Background/Scenario

In this topology, routers R1, R2, and switch D1 are configured with access and file transfer capabilities. You will be loading configurations with intentional errors onto the network. Your tasks are to FIND the error(s), document your findings and the command(s) or method(s) used to fix them, FIX the issue(s) presented here, and then test the network to ensure both of the following conditions are met:

1. the complaint received in the ticket is resolved

2. full functionality is restored

Note: The routers used with CCNP hands-on labs are Cisco 4221 with Cisco IOS XE Release 16.9.4 (universalk9 image). The switches used in the labs are Cisco Catalyst 3650 with Cisco IOS XE Release 16.9.4 (universalk9 image). Other routers, switches, and Cisco IOS versions can be used. Depending on the model and Cisco IOS version, the commands available and output produced might vary from what is shown in the labs. Refer to the Router Interface Summary Table at the end of the lab for the correct interface identifiers.

Note: Make sure that the switches have been erased and have no startup configurations. If you are unsure, contact your instructor.

Required Resources

- 2 Routers (Cisco 4221 with Cisco IOS XE Release 16.9.4 universal image or comparable)
- 1 Switch (Cisco 3560 with Cisco IOS XE Release 16.9.4 universal image or comparable)
- 1 PC (Choice of operating system with terminal emulation program installed)
- Console cables to configure the Cisco IOS devices via the console ports
- Ethernet cables as shown in the topology

Instructions

Part 1: Trouble Ticket 23.1.2.1

Scenario:

There have been reports received regarding the ability to copy configuration files from some of the devices to the TFTP server running on PC1.

Use the commands listed below to load the configuration files for this trouble ticket:

Device	Command
R1	`copy flash:/enarsi/23.1.2.1-r1-config.txt run`
R2	`copy flash:/enarsi/23.1.2.1-r2-config.txt run`
D1	`copy flash:/enarsi/23.1.2.1-d1-config.txt run`

- PC1 should be manually configured and able to ping its default gateway, as shown in the Addressing Table.

- PC1 needs to have the TFTP Server software running, make certain TFTP is configured to send and receive files.

- Passwords on all devices are **cisco12345**. If a username is required, use **admin**.

  Configuration files are to be copied from each of the devices to the TFTP server using the following commands:
  ```
  R1# copy running-config tftp://10.10.10.10/r1-config.txt
  R2# copy running-config tftp://10.10.10.10/r2-config.txt
  D1# copy running-config tftp://10.10.10.10/d1-config.txt
  ```
- After you have fixed the ticket, change the MOTD on EACH DEVICE using the following command:

  banner motd # This is $(hostname) FIXED from ticket <ticket number> #
- Save the configuration by issuing the **wri** command (on each device).
- Inform your instructor that you are ready for the next ticket.
- After the instructor approves your solution for this ticket, issue the **reset.now** privileged EXEC command on each device. This script will clear your configurations and reload the devices.

Part 2: Trouble Ticket 23.1.2.2

Scenario:

A network technician is attempting to copy the current configuration from router R1 to R2 flash using the command:
```
R1# copy running-config scp://admin@10.10.12.2/enarsi/r1-config.txt
```
The command fails, however, SSH appears to be working correctly.

Use the commands listed below to load the configuration files for this trouble ticket:

Device	Command
R1	copy flash:/enarsi/23.1.2.2-r1-config.txt run
R2	copy flash:/enarsi/23.1.2.2-r2-config.txt run
D1	copy flash:/enarsi/23.1.2.2-d1-config.txt run

- PC1 should be manually configured and able to ping its default gateway, as shown in the Addressing Table.
- Passwords on all devices are **cisco12345**. If a username is required, use **admin**.
- After you have fixed the ticket, change the MOTD on EACH DEVICE using the following command:

  banner motd # This is $(hostname) FIXED from ticket <ticket number> #
- Then save the configuration by issuing the **wri** command (on each device).
- Inform your instructor that you are ready for the next ticket.
- After the instructor approves your solution for this ticket, issue the **reset.now** privileged EXEC command. This script will clear your configurations and reload the devices.

Part 3: Trouble Ticket 23.1.2.3

Scenario:

All devices are to be accessible from the management PC using SSH. It has been reported that some devices are not permitting access.

Use the commands listed below to load the configuration files for this trouble ticket:

Device	Command
R1	`copy flash:/enarsi/23.1.2.3-r1-config.txt run`
R2	`copy flash:/enarsi/23.1.2.3-r2-config.txt run`
D1	`copy flash:/enarsi/23.1.2.3-d1-config.txt run`

- PC1 should be manually configured and able to ping its default gateway, as shown in the Addressing Table.

- Passwords on all devices are **cisco12345**. If a username is required, use **admin**.

- After you have fixed the ticket, change the MOTD on EACH DEVICE using the following command:

    banner motd # This is $(hostname) FIXED from ticket <ticket number> #

- Then save the configuration by issuing the **wri** command (on each device).

- Inform your instructor that you are ready for the next ticket.

- After the instructor approves your solution for this ticket, issue the **reset.now** privileged EXEC command. This script will clear your configurations and reload the devices.

Router Interface Summary Table

Router Model	Ethernet Interface #1	Ethernet Interface #2	Serial Interface #1	Serial Interface #2
1800	Fast Ethernet 0/0 (F0/0)	Fast Ethernet 0/1 (F0/1)	Serial 0/0/0 (S0/0/0)	Serial 0/0/1 (S0/0/1)
1900	Gigabit Ethernet 0/0 (G0/0)	Gigabit Ethernet 0/1 (G0/1)	Serial 0/0/0 (S0/0/0)	Serial 0/0/1 (S0/0/1)
2801	Fast Ethernet 0/0 (F0/0)	Fast Ethernet 0/1 (F0/1)	Serial 0/1/0 (S0/1/0)	Serial 0/1/1 (S0/1/1)
2811	Fast Ethernet 0/0 (F0/0)	Fast Ethernet 0/1 (F0/1)	Serial 0/0/0 (S0/0/0)	Serial 0/0/1 (S0/0/1)
2900	Gigabit Ethernet 0/0 (G0/0)	Gigabit Ethernet 0/1 (G0/1)	Serial 0/0/0 (S0/0/0)	Serial 0/0/1 (S0/0/1)
4221	Gigabit Ethernet 0/0/0 (G0/0/0)	Gigabit Ethernet 0/0/1 (G0/0/1)	Serial 0/1/0 (S0/1/0)	Serial 0/1/1 (S0/1/1)
4300	Gigabit Ethernet 0/0/0 (G0/0/0)	Gigabit Ethernet 0/0/1 (G0/0/1)	Serial 0/1/0 (S0/1/0)	Serial 0/1/1 (S0/1/1)

Note: To find out how the router is configured, look at the interfaces to identify the type of router and how many interfaces the router has. There is no way to effectively list all the combinations of configurations for each router class. This table includes identifiers for the possible combinations of Ethernet and Serial interfaces in the device. The table does not include any other type of interface, even though a specific router may contain one. An example of this might be an ISDN BRI interface. The string in parenthesis is the legal abbreviation that can be used in Cisco IOS commands to represent the interface.

Device Config Final – Notes

23.1.3 Lab - Troubleshoot SNMP and Logging Issues

Topology

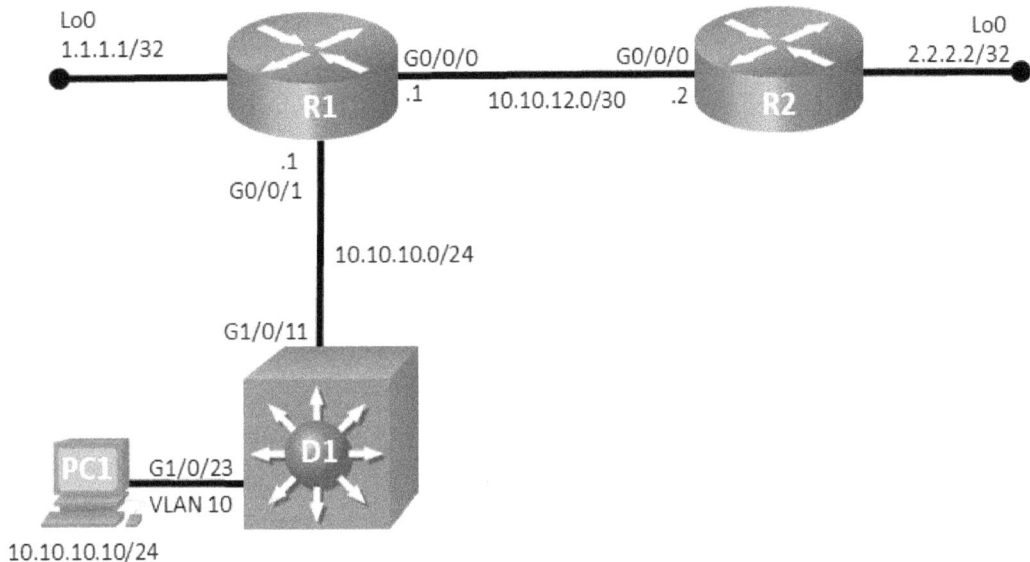

Addressing Table

Device	Interface	IP Address	Subnet Mask
R1	G0/0/0	10.10.12.1	255.255.255.0
	G0/0/1	10.10.10.1	255.255.255.0
R2	G0/0/0	10.10.12.2	255.255.255.0
D1	VLAN 10	10.10.10.2	255.255.255.0
PC1	NIC	10.10.10.10	255.255.255.0

Objectives

Troubleshoot the logging issues for the devices in the topology and make the necessary corrections.

Background/Scenario

In this topology, routers R1, R2, and switch D1 are configured with logging and SNMP. You will be loading configurations with intentional errors onto the network. Your tasks are to FIND the error(s), document your findings and the command(s) or method(s) used to fix them, FIX the issue(s) presented here, and then test the network to ensure both of the following conditions are met:

1. the complaint received in the ticket is resolved

2. full functionality is restored

Note: The routers used with CCNP hands-on labs are Cisco 4221 with Cisco IOS XE Release 16.9.4 (universalk9 image). The switches used in the labs are Cisco Catalyst 3650 with Cisco IOS XE Release 16.9.4 (universalk9 image). Other routers, switches, and Cisco IOS versions can be used. Depending on the model and Cisco IOS version, the commands available and the output produced might vary from what is shown in the labs. Refer to the Router Interface Summary Table at the end of the lab for the correct interface identifiers.

Note: Make sure that the devices have been erased and have no startup configurations. If you are unsure, contact your instructor.

Required Resources

- 2 Routers (Cisco 4221 with Cisco IOS XE Release 16.9.4 universal image or comparable)
- 1 Switch (Cisco 3560 with Cisco IOS XE Release 16.9.4 universal image or comparable)
- 1 PC (Choice of operating system with terminal emulation and syslog programs installed)
- Console cables to configure the Cisco IOS devices via the console ports
- Ethernet cables as shown in the topology

Part 1: Trouble Ticket 23.1.3.1

Scenario:

SNMP messages should be coming from router R1 and switch D1. To avoid service interruption, loopback interfaces have been configured on those devices to test SNMP operation. Disabling and enabling the loopback interface should generate an SNMP trap that is displayed on the Syslog server log screen. Correct any necessary configuration issues and verify traps are being logged from both devices.

Use the commands listed below to load the configuration files for this trouble ticket:

Device	Command
R1	`copy flash:/enarsi/23.1.3.1-r1-config.txt run`
R2	`copy flash:/enarsi/23.1.3.1-r2-config.txt run`
D1	`copy flash:/enarsi/23.1.3.1-d1-config.txt run`

- PC1 should be manually configured and able to ping its default gateway, as shown in the Addressing Table.
- PC1 needs to have syslog software running. In this example, the Kiwi Syslog Server software is used and the following settings are used for all the trouble tickets in this lab.

  Kiwi Syslog Server Settings:

  File->Setup->Inputs – Add addresses 10.10.10.1, 10.10.10.2, and 10.10.12.2

  File->Setup->Inputs->UDP – Checkbox for Listen for UDP messages checked

  File->Setup->Inputs->SNMP – Checkbox for Listen for SNMP Traps checked

  File->Setup->Inputs->SNMP – Add/Remove SNMP v3 Credentials dialog

  User Name – USER1

  Authentication Password – cisco12345

Algorithm - SHA

Private Password – cisco54321

Algorithm - AES

Security Level - Authentication & Privacy dropdown selected

Click **Add User**

Close the **Setup** dialog box.

- Passwords on all devices are **cisco12345**. If a username is required, use **admin**.

- After you have fixed the ticket, change the MOTD on EACH DEVICE using the following command:

  banner motd # This is $(hostname) FIXED from ticket <ticket number> #

- Then save the configuration by issuing the **wri** command (on each device).

- Inform your instructor that you are ready for the next ticket.

- After the instructor approves your solution for this ticket, issue the **reset.now** privileged EXEC command on each device. This script will clear your configurations and reload the devices.

- Clear the log messages on PC1 to prepare for the next ticket.

Part 2: Trouble Ticket 23.1.3.2

Scenario:

A network technician notices that logging messages from the routers are not consistent. All messages should record the time the event occurs. Both routers should record when changes are made to the devices. Use the loopback interfaces on the routers to determine if messages are being recorded correctly.

Use the commands listed below to load the configuration files for this trouble ticket:

Device	Command
R1	`copy flash:/enarsi/23.1.3.2-r1-config.txt run`
R2	`copy flash:/enarsi/23.1.3.2-r2-config.txt run`
D1	`copy flash:/enarsi/23.1.3.2-d1-config.txt run`

- PC1 should be manually configured and able to ping its default gateway, as shown in the Addressing Table.

- PC1 needs to have the Kiwi Syslog Server software running with the same setting used in Trouble Ticket 23.1.3.1.

- Passwords on all devices are **cisco12345**. If a username is required, use **admin**.

- After you have fixed the ticket, change the MOTD on EACH DEVICE using the following command:

  banner motd # This is $(hostname) FIXED from ticket <ticket number> #

- Then save the configuration by issuing the **wri** command (on each device).

- Inform your instructor that you are ready for the next ticket.

■ After the instructor approves your solution for this ticket, issue the **reset.now** privileged EXEC command. This script will clear your configurations and reload the devices.

Router Interface Summary Table

Router Model	Ethernet Interface #1	Ethernet Interface #2	Serial Interface #1	Serial Interface #2
1800	Fast Ethernet 0/0 (F0/0)	Fast Ethernet 0/1 (F0/1)	Serial 0/0/0 (S0/0/0)	Serial 0/0/1 (S0/0/1)
1900	Gigabit Ethernet 0/0 (G0/0)	Gigabit Ethernet 0/1 (G0/1)	Serial 0/0/0 (S0/0/0)	Serial 0/0/1 (S0/0/1)
2801	Fast Ethernet 0/0 (F0/0)	Fast Ethernet 0/1 (F0/1)	Serial 0/1/0 (S0/1/0)	Serial 0/1/1 (S0/1/1)
2811	Fast Ethernet 0/0 (F0/0)	Fast Ethernet 0/1 (F0/1)	Serial 0/0/0 (S0/0/0)	Serial 0/0/1 (S0/0/1)
2900	Gigabit Ethernet 0/0 (G0/0)	Gigabit Ethernet 0/1 (G0/1)	Serial 0/0/0 (S0/0/0)	Serial 0/0/1 (S0/0/1)
4221	Gigabit Ethernet 0/0/0 (G0/0/0)	Gigabit Ethernet 0/0/1 (G0/0/1)	Serial 0/1/0 (S0/1/0)	Serial 0/1/1 (S0/1/1)
4300	Gigabit Ethernet 0/0/0 (G0/0/0)	Gigabit Ethernet 0/0/1 (G0/0/1)	Serial 0/1/0 (S0/1/0)	Serial 0/1/1 (S0/1/1)

Note: To find out how the router is configured, look at the interfaces to identify the type of router and how many interfaces the router has. There is no way to effectively list all the combinations of configurations for each router class. This table includes identifiers for the possible combinations of Ethernet and Serial interfaces in the device. The table does not include any other type of interface, even though a specific router may contain one. An example of this might be an ISDN BRI interface. The string in parenthesis is the legal abbreviation that can be used in Cisco IOS commands to represent the interface.

Device Config Final – Notes

_____

_____

_____

_____

_____

_____

_____

_____

_____

_____

_____

_____

_____

23.1.4 Lab - Troubleshoot IP SLA and Netflow

Topology

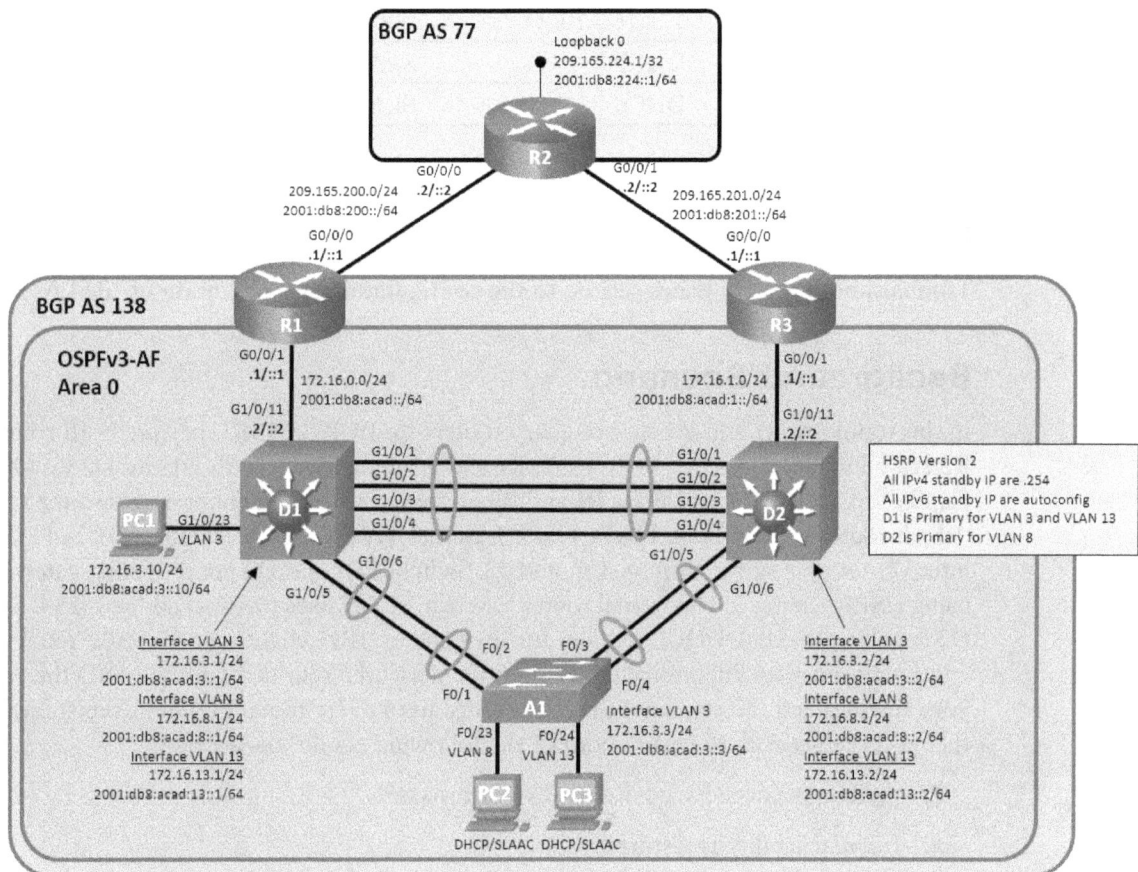

Addressing Table

Device	Interface	IPv4 Address/Mask	IPv6 Address/Prefix Length	Link-Local Address
R1	G0/0/0	209.165.200.1/24	2001:db8:200::1/64	fe80::1:1
	G0/0/1	172.16.0.1/24	2001:db8:acad::1/64	fe80::1:2
R2	G0/0/0	209.165.200.2/24	2001:db8:200::2/64	fe80::2:1
	G0/0/1	209.165.201.2/24	2001:db8:201::2/64	fe80::2:3
	Loopback 0	209.165.224.1/32	2001:db8:224::1/64	fe80::2:4
R3	G0/0/0	209.165.201.1/24	2001:db8:201::1/64	fe80::3:1
	G0/0/1	172.16.1.1/24	2001:db8:acad:1::1/64	fe80::3:2
D1	G1/0/11	172.16.0.2/24	2001:db8:acad::2/64	fe80::d1:1
	VLAN 3	172.16.3.1/24	2001:db8:acad:3::1/64	fe80::d1:2
	VLAN 8	172.16.8.1/24	2001:db8:acad:8::1/64	fe80::d1:3
	VLAN 13	172.16.13.1/24	2001:db8:acad:13::1/64	fe80::d1:4
D2	G1/0/11	172.16.1.2/24	2001:db8:acad:1::2/64	fe80::d2:1
	VLAN 3	172.16.3.2/24	2001:db8:acad:3::2/64	fe80::d2:2

Device	Interface	IPv4 Address/Mask	IPv6 Address/Prefix Length	Link-Local Address
	VLAN 8	172.16.8.2/24	2001:db8:acad:8::2/64	fe80::d2:3
	VLAN 13	172.16.13.2/24	2001:db8:acad:13::2/64	fe80::d2:4
A1	VLAN 3	172.16.3.3/24	2001:db8:acad:3::3/64	fe80::a1:1
PC1	NIC	172.16.3.10/24	2001:db8:acad:3::10/64	N/A
PC2	NIC	DHCP	SLAAC	
PC3	NIC	DHCP	SLAAC	

Objectives

Troubleshoot network issues related to the configuration and operation of IP SLAs and Netflow.

Background/Scenario

In this topology, R1 and R3 are boundary routers for BGP AS 138. They are both connected to R2. R2 is a boundary router for BGP AS 77. R1 and R3 are adjacent with D1 and D2 via OSPFv3 Address Families for both IPv4 and IPv6. R1 and R3 are both providing default routes to the OSPF network. The default routes are configured to be OSPF External Type 1 routes. Switches D1 and D2 are performing inter-VLAN routing for VLANs 3, 8, and 13. Switches D1 and D2 are providing gateway redundancy using HSRP version 2. The virtual router for each VLAN uses the host address .254. Switches D1 and D2 are also providing DHCP services for IPv4 clients. IPv6 clients use SLAAC. You will be loading configurations with intentional errors onto the network. Your tasks are to FIND the error(s), document your findings and the command(s) or method(s) used to fix them, FIX the issue(s) presented here and then test the network to ensure both of the following conditions are met:

1. the complaint received in the ticket is resolved

2. full reachability is restored

Note: The routers used with CCNP hands-on labs are Cisco 4221 with Cisco IOS XE Release 16.9.4 (universalk9 image). The switches used in the labs are Cisco Catalyst 3650 with Cisco IOS XE Release 16.9.4 (universalk9 image) and Cisco Catalyst 2960 with Cisco IOS Release 15.2(2) (lanbasek9 image). Other routers, switches, and Cisco IOS versions can be used. Depending on the model and Cisco IOS version, the commands available and the output produced might vary from what is shown in the labs. Refer to the Router Interface Summary Table at the end of the lab for the correct interface identifiers.

Note: Make sure that the devices have been erased and have no startup configurations. If you are unsure, contact your instructor.

Note: The default Switch Database Manager (SDM) template on a Catalyst 2960 does not support IPv6. You must change the default SDM template to the dual-ipv4-and-ipv6 default template using the **sdm prefer dual-ipv4-and-ipv6 default** global configuration command. Changing the template will require a reboot.

Required Resources

- 3 Routers (Cisco 4221 with Cisco IOS XE Release 16.9.4 universal image or comparable)

- 2 Switches (Cisco 3560 with Cisco IOS XE Release 16.9.4 universal image or comparable)

- 1 Switch (Cisco 2960 with Cisco IOS Release 15.2(2) lanbasek9 image or comparable)

- 3 PCs (Choice of operating system with terminal emulation program and a packet capturing utility installed)

- Console cables to configure the Cisco IOS devices via the console ports

- Ethernet cables as shown in the topology

Instructions

Part 1: Trouble Ticket 23.1.4.1

Scenario:

You tasked the junior network administrators working over the weekend to deploy and test IP SLAs on switches D1 and D2 so that they would relinquish the HSRP Active Role if an upstream interface were to go down. The reports you receive on Monday morning state that the SLAs are in place, but HSRP is not behaving as expected. They need your expertise to figure out what is wrong.

Use the commands listed below to load the configuration files for this trouble ticket:

Device	Command
R1	copy flash:/enarsi/23.1.4.1-r1-config.txt run
R2	copy flash:/enarsi/23.1.4.1-r2-config.txt run
R3	copy flash:/enarsi/23.1.4.1-r3-config.txt run
D1	copy flash:/enarsi/23.1.4.1-d1-config.txt run
D2	copy flash:/enarsi/23.1.3.1-d2-config.txt run
A1	copy flash:/enarsi/23.1.4.1-a1-config.txt run

- PC1 must have the addresses shown in the topology diagram statically assigned. PC2 and PC3 will receive their addresses dynamically.

- Passwords on all devices are **cisco12345**. If a username is required, use **admin**.

- When you have fixed the ticket, change the MOTD on EACH DEVICE using the following command:

  banner motd # This is $(hostname) FIXED from ticket <ticket number> #

- Save the configuration by issuing the **wri** command (on each device).

- Inform your instructor that you are ready for the next ticket.

- After the instructor approves your solution for this ticket, issue the **reset.now** privileged EXEC command. This script will clear your configurations and reload the devices.

Part 2: Trouble Ticket 23.1.4.2

Note: This ticket only works on 4000-series routers. If the routers in use are ISR G2 series (29/39xx series), use trouble ticket 23.1.4.3 instead.

Scenario:

Management is asking for detailed information on traffic flowing in and out of the network. They want this information to help shape updates to the organizational security policy, as well as get an idea about bandwidth utilization. Your intention is to configure Flexible Netflow to gather information on traffic entering and exiting the OSPF interfaces on R1 and R3. After a lot of work sorting out how to configure the technology, you thought you had it configured, but the collector at PC1 is still not receiving any data.

Use the commands listed below to load the configuration files for this trouble ticket:

Device	Command
R1	`copy flash:/enarsi/23.1.4.2-r1-config.txt run`
R2	`copy flash:/enarsi/23.1.4.2-r2-config.txt run`
R3	`copy flash:/enarsi/23.1.4.2-r3-config.txt run`
D1	`copy flash:/enarsi/23.1.4.2-d1-config.txt run`
D2	`copy flash:/enarsi/23.1.4.2-d2-config.txt run`
A1	`copy flash:/enarsi/23.1.4.2-a1-config.txt run`

- PC1 must have the addresses shown in the topology diagram statically assigned. PC2 and PC3 will receive their addresses dynamically.

- Passwords on all devices are **cisco12345**. If a username is required, use **admin**.

- When you have fixed the ticket, change the MOTD on EACH DEVICE using the following command:

  banner motd # This is $(hostname) FIXED from ticket <ticket number> #

- Save the configuration by issuing the **wri** command (on each device).

- Inform your instructor that you are ready for the next ticket.

- After the instructor approves your solution for this ticket, issue the **reset.now** privileged EXEC command. This script will clear your configurations and reload the devices.

Part 3: Trouble Ticket 23.1.4.3

Note: This ticket only works on ISR G2 series (29/39xx series) routers. If the routers in use are from the 4000-series, use trouble ticket 23.1.4.2 instead.

Scenario:

Management is asking for detailed information on traffic flowing out of the network. They want this information to help shape updates to the organizational security policy, as well as get an idea about bandwidth utilization. Your job is to configure Netflow to gather information on traffic entering and exiting the OSPF interfaces on R1 and R3. This is a new technology for you, but you think you have worked out how to configure it, unfortunately the collector at PC1 is still not receiving any data.

Use the commands listed below to load the configuration files for this trouble ticket:

Device	Command
R1	`copy flash:/enarsi/23.1.5.3-r1-config.txt run`
R2	`copy flash:/enarsi/23.1.5.3-r2-config.txt run`
R3	`copy flash:/enarsi/23.1.4.3-r3-config.txt run`
D1	`copy flash:/enarsi/23.1.4.3-d1-config.txt run`
D2	`copy flash:/enarsi/23.1.4.3-d2-config.txt run`
A1	`copy flash:/enarsi/23.1.3.3-a1-config.txt run`

- PC1 must have the addresses shown in the topology diagram statically assigned. PC2 and PC3 will receive their addresses dynamically.

- Passwords on all devices are **cisco12345**. If a username is required, use **admin**.

- When you have fixed the ticket, change the MOTD on EACH DEVICE using the following command:

   banner motd # This is $(hostname) FIXED from ticket <ticket number> #

- Then save the configuration by issuing the **wri** command (on each device).

- Inform your instructor that you are ready for the next ticket.

- After the instructor approves your solution for this ticket, issue the **reset.now** privileged EXEC command. This script will clear your configurations and reload the devices.

Router Interface Summary Table

Router Model	Ethernet Interface #1	Ethernet Interface #2	Serial Interface #1	Serial Interface #2
1800	Fast Ethernet 0/0 (F0/0)	Fast Ethernet 0/1 (F0/1)	Serial 0/0/0 (S0/0/0)	Serial 0/0/1 (S0/0/1)
1900	Gigabit Ethernet 0/0 (G0/0)	Gigabit Ethernet 0/1 (G0/1)	Serial 0/0/0 (S0/0/0)	Serial 0/0/1 (S0/0/1)
2801	Fast Ethernet 0/0 (F0/0)	Fast Ethernet 0/1 (F0/1)	Serial 0/1/0 (S0/1/0)	Serial 0/1/1 (S0/1/1)
2811	Fast Ethernet 0/0 (F0/0)	Fast Ethernet 0/1 (F0/1)	Serial 0/0/0 (S0/0/0)	Serial 0/0/1 (S0/0/1)
2900	Gigabit Ethernet 0/0 (G0/0)	Gigabit Ethernet 0/1 (G0/1)	Serial 0/0/0 (S0/0/0)	Serial 0/0/1 (S0/0/1)
4221	Gigabit Ethernet 0/0/0 (G0/0/0)	Gigabit Ethernet 0/0/1 (G0/0/1)	Serial 0/1/0 (S0/1/0)	Serial 0/1/1 (S0/1/1)
4300	Gigabit Ethernet 0/0/0 (G0/0/0)	Gigabit Ethernet 0/0/1 (G0/0/1)	Serial 0/1/0 (S0/1/0)	Serial 0/1/1 (S0/1/1)

Note: To find out how the router is configured, look at the interfaces to identify the type of router and how many interfaces the router has. There is no way to effectively list all the combinations of configurations for each router class. This table includes identifiers for the possible combinations of Ethernet and Serial interfaces in the device. The table does not include any other type of interface, even though a specific router may contain one. An example of this might be an ISDN BRI interface. The string in parenthesis is the legal abbreviation that can be used in Cisco IOS commands to represent the interface.

Device Config Final – Notes

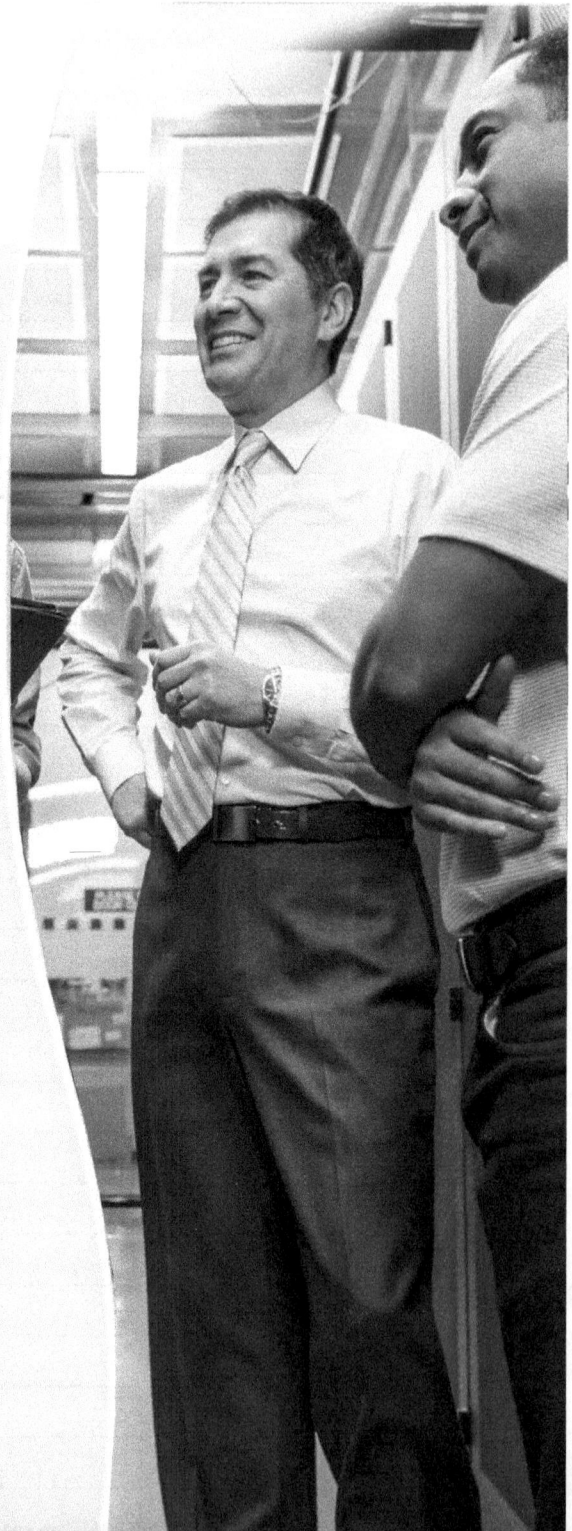

www.ingramcontent.com/pod-product-compliance
Lightning Source LLC
Chambersburg PA
CBHW060956210326
11590OD00031B/4040